Frontiers in Time Scales and Inequalities

SERIES ON CONCRETE AND APPLICABLE MATHEMATICS

ISSN: 1793-1142

Series Editor: Professor George A. Anastassiou
Department of Mathematical Sciences
University of Memphis
Memphis, TN 38152, USA

Published*

*To view the complete list of the published volumes in the series, please visit:
http://www.worldscientific/series/scaam

Series on Concrete and Applicable Mathematics – Vol. 17

Frontiers in Time Scales and Inequalities

George A. Anastassiou

University of Memphis, USA

World Scientific

NEW JERSEY · LONDON · SINGAPORE · BEIJING · SHANGHAI · HONG KONG · TAIPEI · CHENNAI · TOKYO

Published by

World Scientific Publishing Co. Pte. Ltd.
5 Toh Tuck Link, Singapore 596224
USA office: 27 Warren Street, Suite 401-402, Hackensack, NJ 07601
UK office: 57 Shelton Street, Covent Garden, London WC2H 9HE

Library of Congress Cataloging-in-Publication Data
Anastassiou, George A., 1952–
   Frontiers in time scales and inequalities / by George Anastassiou (University of Memphis, USA).
     pages cm. -- (Series on concrete and applicable mathematics (SCAM) ; vol. 17)
   Includes bibliographical references and index.
   ISBN 978-9814704434 (alk. paper)
   1. Approximation theory. I. Title.
   QA221.A63635 2015
   511'.4--dc23

                          2015025796

British Library Cataloguing-in-Publication Data
A catalogue record for this book is available from the British Library.

In-house Editors: V. Vishnu Mohan/Kwong Lai Fun

Typeset by Stallion Press
Email: enquiries@stallionpress.com

Printed in Singapore

Preface

In this monograph we present recent work of last four years of the author in discrete and fractional analysis. It is the natural outgrowth of his related publications. Chapters are self-contained and can be read independently and advanced courses can be taught out of this book. An extensive list of references is given per chapter.

The topics covered are diverse. We introduce the right delta and right nabla fractional calculus on time scales. We continue with right delta and right nabla discrete fractional calculus in the Caputo sense. Then we give representations formulae of functions on time scales and we present Ostrowski type inequalities, Landau type inequalities, Grüss type and comparison of means inequalities, all these over the time scales. We continue with integral operator inequalities and their multivariate vectorial versions using convexity of functions again all these over time scales. It follows Grüss and Ostrowski type inequalities involving s-convexity of functions, we examine also the general case when we involve several functions. Then we present general fractional Hermite–Hadamard type inequalities using m-convexity and (s, m)-convexity. We finish monograph by introducing the reduction method in fractional calculus and study its connection to fractional Ostrowski type inequalities.

This book's results are expected to find applications in many areas of pure and applied mathematics, especially in difference equations and fractional differential equations. As such this monograph is suitable for researchers, graduate students, and seminars of the above subjects, also to be in all science libraries.

The preparation of book took place during 2014–2015 in Memphis, Tennessee, USA.

I would like to thank Professor Razvan Mezei, of Lenoir-Rhyne University, for checking and reading the manuscript.

George A. Anastassiou
Department of Mathematical Sciences
University of Memphis
Memphis, TN 38152
USA

April 1, 2015

Contents

Chapter 1

Foundations of Right Delta Fractional Calculus on Time Scales

Here we develop the right delta fractional calculus on time scales. This chapter follows [5].

1.1 Background

For the basics of times scales please read [1], [2], [3], [6], [8], [9], [10], [11], [12], [13], [14].

Let $\mathbb{T}$ be a time scale, and ([10], p. 38) $g_k, h_k : \mathbb{T}^2 \to \mathbb{R}$, $k \in \mathbb{N}_0 = \mathbb{N} \cup \{0\}$, $s, t \in \mathbb{T} : g_0(t,s) = h_0(t,s) = 1$,

$$g_{k+1}(t,s) = \int_s^t g_k(\sigma(\tau), s)\,\Delta\tau, \tag{1.1}$$

$$h_{k+1}(t,s) = \int_s^t h_k(\tau, s)\,\Delta\tau, \quad \forall\, s, t \in \mathbb{T}.$$

We have

$$h_k^\Delta(t,s) = h_{k-1}(t,s), \tag{1.2}$$

$$g_k^\Delta(t,s) = g_{k-1}(\sigma(t), s), \quad k \in \mathbb{N}, t \in \mathbb{T}^k.$$

Also

$$g_1(t,s) = h_1(t,s) = t - s, \quad \forall\, s, t \in \mathbb{T}. \tag{1.3}$$

Here g_k, h_k are continuous in t.

Example 1.1. (see [10], p. 39-40)

i) When $\mathbb{T} = \mathbb{R}$,

$$g_k(t,s) = h_k(t,s) = \frac{(t-s)^k}{k!}, \quad \forall\, s, t \in \mathbb{R}. \tag{1.4}$$

ii) When $\mathbb{T} = \mathbb{Z}$, we get

$$h_k(t,s) = \frac{(t-s)^{(k)}}{k!},$$

and

$$g_k(t,s) = \frac{(t-s+k-1)^{(k)}}{k!}, \tag{1.5}$$

furthermore it holds

$$h_k(t,s) = (-1)^k g_k(s,t), \quad k \in \mathbb{N}_0, \ \forall \ s,t \in \mathbb{Z}. \tag{1.6}$$

We need

Theorem 1.1. *([10], p. 45) We have that*

$$h_n(t,s) = (-1)^n g_n(s,t), \tag{1.7}$$

for every $t \in \mathbb{T}$ and every $s \in \mathbb{T}^{k^n}$.
If $\mathbb{T} = \mathbb{T}^k$, then (1.7) is true for every $s,t \in \mathbb{T}$.

We need delta Taylor formula on time scales.

Theorem 1.2. *([7], [12]) Assume $\mathbb{T} = \mathbb{T}^k$, $f \in C^m_{rd}(\mathbb{T})$, $m \in \mathbb{N}$, $s,t \in \mathbb{T}$. Then*

$$f(t) = \sum_{k=0}^{m-1} f^{\Delta^k}(s) h_k(t,s) + R^s_m(f)(t), \tag{1.8}$$

where

$$R^s_m(f)(t) = \int_s^t h_{m-1}(t,\sigma(\tau)) f^{\Delta^m}(\tau) \Delta\tau. \tag{1.9}$$

We notice that

$$R^s_m(f)(t) = -\int_t^s h_{m-1}(t,\sigma(\tau)) f^{\Delta^m}(\tau) \Delta\tau$$

$$\overset{(1.7)}{=} (-1)^m \int_t^s g_{m-1}(\sigma(\tau),t) f^{\Delta^m}(\tau) \Delta\tau. \tag{1.10}$$

In this chapter we assume $\mathbb{T} = \mathbb{T}^k$.
We make

Definition 1.1. Let $\alpha \geq 0$. We consider the continuous functions

$$g_\alpha : \mathbb{T}^2 \to \mathbb{R} :$$

$$g_0(t,s) = 1,$$

$$g_{\alpha+1}(t,s) = \int_s^t g_\alpha(\sigma(\tau),s) \Delta\tau, \quad \forall \ s,t \in \mathbb{T}. \tag{1.11}$$

We are motivated a lot by the formula

$$\int_t^x \frac{(x-s)^{\mu-1}}{\Gamma(\mu)} \frac{(s-t)^{\nu-1}}{\Gamma(\nu)} ds = \frac{(x-t)^{\mu+\nu-1}}{\Gamma(\mu+\nu)}, \tag{1.12}$$

where $\mu, \nu > 0$ and Γ is the gamma function

Assumption 1.1. Let $\alpha, \beta > 1$ and $x \le t \le \tau$, $x, t, \tau \in \mathbb{T}$. We assume that

$$\int_x^{\sigma(\tau)} g_{\alpha-1}(\sigma(t), x) g_{\beta-1}(\sigma(\tau), t) \Delta t = g_{\alpha+\beta-1}(\sigma(\tau), x). \tag{1.13}$$

We call for $\alpha, \beta > 1$ and $x \le t \le \tau$,

$$\theta(x, \tau) := \int_\tau^{\sigma(\tau)} g_{\alpha-1}(\sigma(t), x) g_{\beta-1}(\sigma(\tau), t) \Delta t. \tag{1.14}$$

By Theorem 1.75, p. 28, [10] for $f \in C_{rd}(\mathbb{T})$ and $t \in \mathbb{T}^k$, we have

$$\int_t^{\sigma(t)} f(\tau) \Delta \tau = \mu(t) f(t), \tag{1.15}$$

where $\mu(t) := \sigma(t) - t$.

So by (1.15) we get that

$$\theta(x, \tau) = g_{\alpha-1}(\sigma(\tau), x) g_{\beta-1}(\sigma(\tau), \tau) \mu(\tau). \tag{1.16}$$

Definition 1.2. Let $f \in L_1([a, b) \cap \mathbb{T})$. We define the right forward graininess deviation functional of f as follows:

$$E(f, \alpha, \beta, b, \mathbb{T}, x) = \int_x^b f(\tau) \theta(x, \tau) \Delta \tau \tag{1.17}$$

$$= \int_x^b f(\tau) g_{\alpha-1}(\sigma(\tau), x) g_{\beta-1}(\sigma(\tau), \tau) \mu(\tau) \Delta \tau.$$

If $\mathbb{T} = \mathbb{R}$, then $\mu(\tau) = 0$ and hence $E(f, \alpha, \beta, b, \mathbb{R}, x) = 0$.

1.2 Results

We give

Definition 1.3. Let $a, b \in \mathbb{T}$, $\alpha \ge 1$ and $f : [a, b] \cap \mathbb{T} \to \mathbb{R}$. Here $f \in L_1([a, b) \cap \mathbb{T})$ (Lebesgue Δ-integrable function on $[a, b) \cap \mathbb{T}$). We define the right Δ-Riemann-Liouville type fractional integral

$$I_{b-}^\alpha f(t) := \int_t^b g_{\alpha-1}(\sigma(\tau), t) f(\tau) \Delta \tau, \tag{1.18}$$

for $t \in [a, b] \cap \mathbb{T}$. Here $\int_t^b \cdot \Delta \tau = \int_{[t,b)} \cdot \Delta \tau$.

By [9] we get that $I_{b-}^1 f(t) = \int_t^b f(\tau)\,\Delta\tau$ is absolutely continuous in $t \in [a, b] \cap \mathbb{T}$.

Lemma 1.1. *Let $\alpha > 1$, $f \in L_1([a, b) \cap \mathbb{T})$, $f : [a, b] \cap \mathbb{T} \to \mathbb{R}$. Assume that $g_{\alpha-1}(\sigma(\tau), t)$ is Lebesgue Δ-measurable on $([a, b] \cap \mathbb{T})^2$; $a, b \in \mathbb{T}$. Then $I_{b-}^\alpha f \in L_1([a, b] \cap \mathbb{T})$, that is $I_{b-}^\alpha f$ is finite a.e.*

Proof. By Tietze's extension theorem of General Topology we easily derive that the continuous function $g_{\alpha-1}$ on $([a, b] \cap \mathbb{T})^2$ is bounded, since its continuous extension $F_{\alpha-1}$ on $[a, b]^2$ is bounded. Notice here $([a, b] \cap \mathbb{T})^2$ is a closed subset of $[a, b]^2$.

So there exists $M > 0$ such that $|g_{\alpha-1}(s, t)| \leq M$, $\forall\,(s, t) \in ([a, b] \cap \mathbb{T})^2$.

Let here id denote the identity map. We see that
$(\sigma, id)(([a, b) \cap \mathbb{T}) \times ([a, b] \cap \mathbb{T})) \subseteq ([a, b] \cap \mathbb{T})^2$.

Therefore $|g_{\alpha-1}(\sigma(\tau), t)| \leq M$, $\forall\,(\tau, t) \in ([a, b) \cap \mathbb{T}) \times ([a, b] \cap \mathbb{T})$, since $(\sigma(\tau), t) \in ([a, b] \cap \mathbb{T})^2$.

Define $K : \Omega := ([a, b] \cap \mathbb{T})^2 \to \mathbb{R}$, by

$$K(\tau, t) := \begin{cases} g_{\alpha-1}(\sigma(\tau), t), & \text{if } a \leq t \leq \tau < b, \\ 0, & \text{if } a \leq \tau < t \leq b, \end{cases}$$

where $t, \tau \in \mathbb{T}$.

Clearly here K is Lebesgue Δ-measurable on Ω, since the restriction of a measurable function to a measurable subset of its domain is a measurable function, and the union of two measurable functions over disjoint domains is measurable. Notice here that $|K(\tau, t)| \leq M$, $\forall\,(\tau, t) \in ([a, b] \cap \mathbb{T})^2$.

Next we consider the repeated double Lebesgue Δ-integral

$$\int_a^b \left(\int_a^b |K(\tau, t)|\,|f(\tau)|\,\Delta t \right) \Delta\tau = \int_a^b |f(\tau)| \left(\int_a^b |K(\tau, t)|\,\Delta t \right) \Delta\tau$$

$$\leq M(b - a) \int_a^b |f(\tau)|\,\Delta\tau = M(b - a) \|f\|_{L_1([a,b)\cap\mathbb{T})} < \infty.$$

By Tonelli's theorem we derive that $(\tau, t) \to K(\tau, t) f(\tau)$ is Lebesgue Δ-integrable over Ω.

Let now the characteristic function

$$\chi_{[t,b)}(\tau) = \begin{cases} 1, & \text{if } \tau \in [t, b) \\ 0, & \text{else}, \end{cases}$$

where $\tau \in [a, b] \cap \mathbb{T}$.

Then the function $(\tau, t) \to \chi_{[t,b)}(\tau) K(\tau, t) f(\tau)$ is Lebesgue Δ-integrable over Ω.

Hence by Fubini's theorem we get that

$$\int_a^b \chi_{[t,b)}(\tau) K(\tau, t) f(\tau)\,\Delta\tau = \int_t^b g_{\alpha-1}(\sigma(\tau), t) f(\tau)\,\Delta\tau = I_{b-}^\alpha f(t)$$

is Lebesgue Δ-integrable on $[a, b] \cap \mathbb{T}$, proving the claim. $\square$

From now on we make

Assumption 1.2. We suppose that $g_{\alpha-1}\left(\sigma\left(\cdot\right),\cdot\right)$ is continuous on $\left(\left[a,b\right]\cap\mathbb{T}\right)^{2}$, for any $\alpha>1$.

We give the following semigroup property of right Δ-Riemann-Liouville type fractional integrals.

Theorem 1.3. *Let* $a,b\in\mathbb{T}$, $f\in L_{1}\left(\left[a,b\right)\cap\mathbb{T}\right)$; $\alpha,\beta>1$. *Then*

$$I_{b-}^{\alpha}I_{b-}^{\beta}f\left(x\right)=I_{b-}^{\alpha+\beta}f\left(x\right)-E\left(f,\alpha,\beta,b,\mathbb{T},x\right),\quad\forall\,x\in\left[a,b\right]\cap\mathbb{T}. \qquad (1.19)$$

Proof. Here we have

$$I_{b-}^{\beta}f\left(t\right)=\int_{t}^{b}g_{\beta-1}\left(\sigma\left(\tau\right),t\right)f\left(\tau\right)\Delta\tau.$$

We observe that

$$I_{b-}^{\alpha}I_{b-}^{\beta}f\left(x\right)=\int_{x}^{b}g_{\alpha-1}\left(\sigma\left(t\right),x\right)I_{b-}^{\beta}f\left(t\right)\Delta t$$

$$=\int_{x}^{b}g_{\alpha-1}\left(\sigma\left(t\right),x\right)\left(\int_{t}^{b}g_{\beta-1}\left(\sigma\left(\tau\right),t\right)f\left(\tau\right)\Delta\tau\right)\Delta t$$

$$=\int_{x}^{b}\left(\int_{t}^{b}g_{\alpha-1}\left(\sigma\left(t\right),x\right)g_{\beta-1}\left(\sigma\left(\tau\right),t\right)f\left(\tau\right)\Delta\tau\right)\Delta t=:\left(*\right).$$

Clearly here it holds

$$\left|g_{\alpha-1}\left(\sigma\left(t\right),x\right)\right|\leq M_{1},\quad\forall\,t,x\in\left[a,b\right]\cap\mathbb{T},$$

and

$$\left|g_{\beta-1}\left(\sigma\left(\tau\right),t\right)\right|\leq M_{2},\quad\forall\,\tau,t\in\left[a,b\right]\cap\mathbb{T},$$

where $M_{1},M_{2}>0$.

Hence

$$\left|I_{b-}^{\alpha}I_{b-}^{\beta}f\left(x\right)\right|\leq\int_{x}^{b}\left(\int_{t}^{b}\left|g_{\alpha-1}\left(\sigma\left(t\right),x\right)\right|\left|g_{\beta-1}\left(\sigma\left(\tau\right),t\right)\right|\left|f\left(\tau\right)\right|\Delta\tau\right)\Delta t$$

$$\leq M_{1}M_{2}\left(\int_{x}^{b}\left(\int_{t}^{b}\left|f\left(\tau\right)\right|\Delta\tau\right)\Delta t\right)\leq M_{1}M_{2}\left(\int_{x}^{b}\left(\int_{a}^{b}\left|f\left(\tau\right)\right|\Delta\tau\right)\Delta t\right)$$

$$\leq M_{1}M_{2}\left(b-a\right)\left\|f\right\|_{L_{1}\left(\left[a,b\right)\cap\mathbb{T}\right)}<\infty.$$

So that $I_{b-}^{\alpha}I_{b-}^{\beta}f\left(x\right)$ exists, $\forall\,x\in\left[a,b\right]\cap\mathbb{T}$. Consequently by Fubini's theorem we have

$$\left(*\right)=\int_{x}^{b}\left(\int_{x}^{\tau}g_{\alpha-1}\left(\sigma\left(t\right),x\right)g_{\beta-1}\left(\sigma\left(\tau\right),t\right)f\left(\tau\right)\Delta t\right)\Delta\tau$$

$$= \int_x^b f(\tau) \left(\int_x^\tau g_{\alpha-1}(\sigma(t), x) g_{\beta-1}(\sigma(\tau), t) \Delta t \right) \Delta \tau$$

(here $x \le t < \tau$)

$$= \int_x^b f(\tau) \left(\int_x^{\sigma(\tau)} g_{\alpha-1}(\sigma(t), x) g_{\beta-1}(\sigma(\tau), t) \Delta t \right)$$

$$- \int_\tau^{\sigma(\tau)} g_{\alpha-1}(\sigma(t), x) g_{\beta-1}(\sigma(\tau), t) \Delta t \right) \Delta \tau$$

$$\overset{\text{by ((1.13), (1.14))}}{=} \int_x^b g_{\alpha+\beta-1}(\sigma(\tau), x) f(\tau) \Delta \tau - \int_x^b f(\tau) \theta(x, \tau) \Delta \tau$$

$$= I_{b-}^{\alpha+\beta} f(x) - \int_x^b f(\tau) \theta(x, \tau) \Delta \tau$$

$$\overset{(1.17)}{=} I_{b-}^{\alpha+\beta} f(x) - E(f, \alpha, \beta, b, \mathbb{T}, x),$$

proving the claim (1.19).   □

We make

Remark 1.1. Let $\mu > 2 : m - 1 < \mu \le m$, $m \in \mathbb{N}$, i.e. $m = \lceil \mu \rceil$ (ceiling of the number), $\widetilde{\nu} = m - \mu$ ($0 \le \widetilde{\nu} < 1$). Let $f \in C_{rd}^m([a, b] \cap \mathbb{T})$. Clearly here ([11]) f^{Δ^m} is Lebesgue Δ-integrable function. We define the right delta fractional derivative on $\mathbb{T}$ of order $\mu - 1$ as follows

$$\Delta_{b-}^{\mu-1} f(t) = (-1)^m \left(I_{b-}^{\widetilde{\nu}+1} f^{\Delta^m} \right)(t) = (-1)^m \int_t^b g_{\widetilde{\nu}}(\sigma(\tau), t) f^{\Delta^m}(\tau) \Delta \tau, \quad (1.20)$$

$\forall t \in [a, b] \cap \mathbb{T}$.

Notice $\Delta_{b-}^{\mu-1} f \in C([a, b] \cap \mathbb{T})$, by a simple argument using the dominated convergence theorem in Lebesgue Δ-sense.

If $\mu = m$, then $\widetilde{\nu} = 0$, then

$$\Delta_{b-}^{m-1} f(t) = (-1)^m \int_t^b f^{\Delta^m}(\tau) \Delta \tau = (-1)^m \left(f^{\Delta^{m-1}}(b) - f^{\Delta^{m-1}}(t) \right). \quad (1.21)$$

More generally, by [9], given that $f^{\Delta^{m-1}}$ is everywhere finite and absolutely continuous on $[a, b] \cap \mathbb{T}$, then f^{Δ^m} exists Δ-a.e. and is Lebesgue Δ-integrable on $[t, b) \cap \mathbb{T}$, $\forall t \in [a, b] \cap \mathbb{T}$, and one can plug it into (1.20).

Remark 1.2. We observe that

$$\left(I_{b-}^{\mu-1} \Delta_{b-}^{\mu-1} f \right)(t) = (-1)^m \left(I_{b-}^{\mu-1} I_{b-}^{\widetilde{\nu}+1} f^{\Delta^m} \right)(t)$$

$$\overset{(1.19)}{=} (-1)^m \left(\left(I_{b-}^{\mu-1+\widetilde{\nu}+1} f^{\Delta^m} \right)(t) - E\left(f^{\Delta^m}, \mu - 1, \widetilde{\nu} + 1, b, \mathbb{T}, t \right) \right)$$

$$= (-1)^m \left(\left(I_{b-}^m f^{\Delta^m} \right)(t) - E\left(f^{\Delta^m}, \mu - 1, \widetilde{\nu} + 1, b, \mathbb{T}, t \right) \right). \tag{1.22}$$

Therefore

$$\left(I_{b-}^{\mu-1} \Delta_{b-}^{\mu-1} f \right)(t) + (-1)^m E\left(f^{\Delta^m}, \mu - 1, \widetilde{\nu} + 1, b, \mathbb{T}, t \right)$$

$$= (-1)^m \left(I_{b-}^m f^{\Delta^m}(t) \right) = (-1)^m \int_t^b g_{m-1}\left(\sigma(\tau), t \right) f^{\Delta^m}(\tau)\, \Delta\tau \tag{1.23}$$

$$\overset{(1.10)}{=} R_m^b(f)(t).$$

Now we can use (1.8) with $s = b$.

We have established the following delta time scales right fractional Taylor formula.

Theorem 1.4. *Assume* $\mathbb{T} = \mathbb{T}^k$, $f \in C_{rd}^m(\mathbb{T})$, $m \in \mathbb{N}$, $a, b \in \mathbb{T}$, *and* $\mu > 2 : m - 1 < \mu \le m$, $\widetilde{\nu} = m - \mu$; *also suppose Assumption 1.1, Assumption 1.2. Then*

$$f(t) = \sum_{k=0}^{m-1} f^{\Delta^k}(b) h_k(t, b) + \left(I_{b-}^{\mu-1} \Delta_{b-}^{\mu-1} f \right)(t)$$

$$+ (-1)^m E\left(f^{\Delta^m}, \mu - 1, \widetilde{\nu} + 1, b, \mathbb{T}, t \right), \tag{1.24}$$

$\forall\, t \in [a, b] \cap \mathbb{T}$.

Remark 1.3. One can rewrite (1.24) as follows

$$f(t) = \sum_{k=0}^{m-1} f^{\Delta^k}(b) h_k(t, b) + \int_t^b g_{\mu-2}\left(\sigma(\tau), t \right) \left(\Delta_{b-}^{\mu-1} f \right)(\tau)\, \Delta\tau$$

$$+ (-1)^m \int_t^b f^{\Delta^m}(\tau)\, g_{\mu-2}\left(\sigma(\tau), t \right) g_{\widetilde{\nu}}\left(\sigma(\tau), \tau \right) \mu(\tau)\, \Delta\tau, \tag{1.25}$$

$\forall\, t \in [a, b] \cap \mathbb{T}$.

Corollary 1.1. *In the assumptions of Theorem 1.4, additionally assume that* $f^{\Delta^k}(b) = 0$, $k = 0, 1, ..., m - 1$. *Then*

$$B(t) := f(t) + (-1)^{m+1} E\left(f^{\Delta^m}, \mu - 1, \widetilde{\nu} + 1, b, \mathbb{T}, t \right)$$

$$= \int_t^b g_{\mu-2}\left(\sigma(\tau), t \right) \left(\Delta_{b-}^{\mu-1} f \right)(\tau)\, \Delta\tau, \tag{1.26}$$

$\forall\, t \in [a, b] \cap \mathbb{T}$.

Remark 1.4. Notice (by [9]) that $\left(I_{b-}^{\mu-1} \Delta_{b-}^{\mu-1} f \right)(t)$ and $E(f^{\Delta^m}, \mu - 1, \widetilde{\nu} + 1, b, \mathbb{T}, t)$ are absolutely continuous functions on $[a, b] \cap \mathbb{T}$.

One can use (1.25) and (1.26) to establish right fractional delta inequalities on time scales of Poincaré type, Sobolev type, Opial type, Ostrowski type, Hilbert-Pachpatte type, etc, analogous to [4].

Our theory is not void because it is fulfilled when $\mathbb{T} = \mathbb{R}$, etc, see also [4].

Bibliography

1. R. Agarwal, M. Bohner, *Basic calculus on time scales and some of its applications*, Results Math. 35 (1-2) (1999), 3-22.
2. R. Agarwal, M. Bohner, A. Peterson, *Inequalities on time scales: a survey*, Math. Inequalities Appl. 4 (4) (2001), 535-557.
3. G. Anastassiou, *Time scales inequalities*, Inter. J. Difference Equations 5 (1) (2010), 1-23.
4. G. Anastassiou, *Principles of delta fractional calculus on time scales and inequalities*, Math. Comput. Modelling 52 (3-4) (2010), 556-566.
5. G. Anastassiou, *Elements of right delta fractional calculus on time scales*, J. Concrete Applicable Math. 10 (3-4) (2012), 159-167.
6. M. Bohner, G.S. Guseinov, *Multiple Lebesgue integration on time scales*, Adv. Difference Equations 2006 (2006), Article ID 26391, 1-12.
7. M. Bohner, G. Guseinov, *The convolution on time scales*, Abstr. Appl. Anal. 2007 (2007), Article ID 58373, 24 pages.
8. M. Bohner, G. Guseinov, *Double integral calculus of variations on time scales*, Comput. Math. Appl. 54 (2007), 45-57.
9. M. Bohner, H. Luo, *Singular second-order multipoint dynamic boundary value problems with mixed derivatives*, Adv. Difference Equations 2006 (2006), Article ID 54989, 1-15.
10. M. Bohner, A. Peterson, *Dynamic Equations on Time Scales: An Introduction with Applications*, Birkhaüser, Boston (2001).
11. G. Guseinov, *Integration on time scales*, J. Math. Anal. Appl. 285 (2003), 107-127.
12. R. Higgins, A. Peterson, *Cauchy functions and Taylor's formula for time scales* $\mathbb{T}$, in B. Aulbach, S. Elaydi, G. Ladas (eds.), Proc. Sixth Internat. Conf. Difference Equations, New Progress in Difference Equations, Augsburg, Germany, 2001, pp. 299-308, Chapman & Hall/CRC (2004).
13. S. Hilger, *Ein Maßketten kalkül mit Anwendung auf Zentrumsmannig-faltigkeiten*, Ph.D. thesis, Universität Würzburg, Germany (1988).
14. W. Liu, Q. A. Ngô, W. Chen, *Ostrowski type inequalities on time scales for double integrals*, Acta Appl. Math. 106 (2009), 229-239.

Bibliography

Chapter 2

Principles of Right Nabla Fractional Calculus on Time Scales

Here we develop the right nabla fractional calculus on time scales. We introduce the related Riemann-Liouville type fractional integral and Caputo like fractional derivative and prove a fractional Taylor formula with integral remainder. It follows [3].

2.1   Background

For the basics of times scales the reader is referred to [2], [4], [5], [6], [7], [8], [9], [10], [11], [12], [13].

Let $\mathbb{T}$ be a time scale, and $\widehat{h}_k : \mathbb{T}^2 \to \mathbb{R}$, $k \in \mathbb{N}_0 = \mathbb{N} \cup \{0\}$, such that $\forall \ s, t \in \mathbb{T}$, $\widehat{h}_0 (t, s) = 1$,

$$\widehat{h}_{k+1} (t, s) = \int_s^t \widehat{h}_k (\tau, s) \nabla \tau. \tag{2.1}$$

Here $\widehat{h}_k$ are ld-continuous in t, and

$$\widehat{h}_k^\nabla (t, s) = \widehat{h}_{k-1} (t, s), \quad k \in \mathbb{N}, \ t \in \mathbb{T}_k,$$

with $\widehat{h}_1 (t, s) = t - s$, $\forall \ s, t \in \mathbb{T}$.

From [4], we write down Taylor's formula in terms of nabla polynomials

Theorem 2.1. *Assume that* $\mathbb{T} = \mathbb{T}_k$. *Let* $f \in C_{ld}^m (\mathbb{T}, \mathbb{R})$, $m \in \mathbb{N}$, $b, t \in \mathbb{T}$. *Then*

$$f (t) = \sum_{k=0}^{m-1} \widehat{h}_k (t, b) f^{\nabla^k} (b) + \int_b^t \widehat{h}_{m-1} (t, \rho (\tau)) f^{\nabla^m} (\tau) \nabla \tau. \tag{2.2}$$

Call

$$R_m^b (f) (t) := \int_b^t \widehat{h}_{m-1} (t, \rho (\tau)) f^{\nabla^m} (\tau) \nabla \tau = - \int_t^b \widehat{h}_{m-1} (t, \rho (\tau)) f^{\nabla^m} (\tau) \nabla \tau. \tag{2.3}$$

Following [4], we define

$$\widehat{g}_0 (t, s) = 1,$$

$$\widehat{g}_{n+1}(t,s) = \int_s^t \widehat{g}_n(\rho(\tau),s)\,\nabla\tau, \quad n \in \mathbb{N}, \, s,t \in \mathbb{T}. \tag{2.4}$$

Notice here

$$\widehat{g}_{n+1}^{\nabla}(t,s) = \widehat{g}_n(\rho(t),s), \quad t \in \mathbb{T}_k,$$

$$\widehat{g}_1(t,s) = t - s, \quad \forall\, s,t \in \mathbb{T}.$$

From [4] we need

Theorem 2.2. *If* $\mathbb{T} = \mathbb{T}_k = \mathbb{T}^k$, *and* $n \in \mathbb{N}_0$, *then*

$$\widehat{h}_n(t,s) = (-1)^n \, \widehat{g}_n(s,t), \quad \forall\, s,t \in \mathbb{T}. \tag{2.5}$$

By Theorem 2.2 we get that

$$R_m^b(f)(t) = (-1)^m \int_t^b \widehat{g}_{m-1}(\rho(\tau),t)\, f^{\nabla^m}(\tau)\,\nabla\tau. \tag{2.6}$$

We make

Definition 2.1. Let $\alpha \geq 0$ real number. We consider the continuous functions

$$\widehat{g}_\alpha : \mathbb{T}^2 \to \mathbb{R},$$

such that

$$\widehat{g}_0(t,s) = 1,$$

$$\widehat{g}_{\alpha+1}(t,s) = \int_s^t \widehat{g}_\alpha(\rho(\tau),s)\,\nabla\tau, \quad \forall\, s,t \in \mathbb{T}. \tag{2.7}$$

We motivated a lot by the formula

$$\int_t^x \frac{(x-s)^{\mu-1}}{\Gamma(\mu)} \frac{(s-t)^{\nu-1}}{\Gamma(\nu)}\,ds = \frac{(x-t)^{\mu+\nu-1}}{\Gamma(\mu+\nu)}, \tag{2.8}$$

where $\mu,\nu > 0$ and Γ the gamma function.

We make

Assumption 2.1. Let $\alpha,\beta > 1$ and $x < t \leq \tau$, $x,t,\tau \in \mathbb{T}$. We assume that

$$\int_x^{\rho(\tau)} \widehat{g}_{\alpha-1}(\rho(t),x)\,\widehat{g}_{\beta-1}(\rho(\tau),t)\,\nabla t = \widehat{g}_{\alpha+\beta-1}(\rho(\tau),x). \tag{2.9}$$

We call for $\alpha,\beta > 1$ and $x < t \leq \tau$,

$$\gamma(x,\tau) := \int_{\rho(\tau)}^{\tau} \widehat{g}_{\alpha-1}(\rho(t),x)\,\widehat{g}_{\beta-1}(\rho(\tau),t)\,\nabla t.$$

It holds

$$\gamma(x,\tau) = \nu(\tau)\,\widehat{g}_{\alpha-1}(\rho(\tau),x)\,\widehat{g}_{\beta-1}(\rho(\tau),\tau), \tag{2.10}$$

where $\nu(\tau) := \tau - \rho(\tau)$, the backward graininess, see [10], p. 332, under the assumption $\mathbb{T} = \mathbb{T}_k$.

2.2 Results

We need

Definition 2.2. Let $a, b \in \mathbb{T}$, $\alpha \geq 1$ and $f : [a, b] \cap \mathbb{T} \to \mathbb{R}$. Here $f \in L_1 ((a, b] \cap \mathbb{T})$ (Lebesgue ∇-integrable function on $(a, b] \cap \mathbb{T}$). We define the right ∇-Riemann-Liouville type fractional integral

$$J_{b-}^{\alpha} f(t) := \int_t^b \widehat{g}_{\alpha-1}(\rho(\tau), t) f(\tau) \nabla \tau, \qquad (2.11)$$

for $t \in [a, b] \cap \mathbb{T}$. Here $\int_t^b \cdot \nabla \tau = \int_{(t, b]} \cdot \nabla \tau$.

By [9] we get that $J_{b-}^1 f(t) = \int_t^b f(\tau) \nabla \tau$ is absolutely continuous in $t \in [a, b] \cap \mathbb{T}$.

Lemma 2.1. *Let* $\alpha > 1$, $f \in L_1 ((a, b] \cap \mathbb{T})$, $f : [a, b] \cap \mathbb{T} \to \mathbb{R}$. *Assume that* $\widehat{g}_{\alpha-1}(\rho(\tau), t)$ *is Lebesgue ∇-measurable on* $([a, b] \cap \mathbb{T})^2$; $a, b \in \mathbb{T}$. *Then* $J_{b-}^{\alpha} f \in L_1 ([a, b] \cap \mathbb{T})$, *that is* $J_{b-}^{\alpha} f$ *is finite a.e.*

Proof. By Tietze's extension theorem of General Topology we easily derive that the continuous function $\widehat{g}_{\alpha-1}$ on $([a, b] \cap \mathbb{T})^2$ is bounded, since its continuous extension $G_{\alpha-1}$ on $[a, b]^2$ is bounded. Notice that $([a, b] \cap \mathbb{T})^2$ is a closed subset of $[a, b]^2$.

So there exists $M > 0$ such that $|\widehat{g}_{\alpha-1}(s, t)| \leq M$, $\forall (s, t) \in ([a, b] \cap \mathbb{T})^2$.

Let id denote the identity map. We see that $(\rho, id) (((a, b] \cap \mathbb{T}) \times ([a, b] \cap \mathbb{T})) \subseteq ([a, b] \cap \mathbb{T})^2$.

Therefore $|\widehat{g}_{\alpha-1}(\rho(\tau), t)| \leq M$, $\forall (\tau, t) \in ((a, b] \cap \mathbb{T}) \times ([a, b] \cap \mathbb{T})$, since $(\rho(\tau), t) \in ([a, b] \cap \mathbb{T})^2$.

Define $K : \Omega := ([a, b] \cap \mathbb{T})^2 \to \mathbb{R}$, by

$$K(\tau, t) := \begin{cases} \widehat{g}_{\alpha-1}(\rho(\tau), t), & \text{if } a \leq t < \tau \leq b, \\ 0, & \text{if } a \leq \tau \leq t \leq b, \end{cases}$$

where $t, \tau \in \mathbb{T}$.

Clearly K is Lebesgue ∇-measurable on Ω, since the restriction of a measurable function to a measurable subset of its domain is measurable function and the union of two measurable functions over disjoint domains is measurable. Notice that $|K(\tau, t)| \leq M$, $\forall (\tau, t) \in ([a, b] \cap \mathbb{T})^2$.

Next we consider the repeated double Lebesgue ∇-integral

$$\int_a^b \left(\int_a^b |K(\tau, t)| |f(\tau)| \nabla t \right) \nabla \tau = \int_a^b |f(\tau)| \left(\int_a^b |K(\tau, t)| \nabla t \right) \nabla \tau$$

$$\leq M(b - a) \int_a^b |f(\tau)| \nabla \tau = M(b - a) \|f\|_{L_1((a, b] \cap \mathbb{T})} < \infty.$$

By Tonelli's theorem we derive that $(\tau, t) \to K(\tau, t) f(\tau)$ is Lebesgue ∇-integrable over Ω.

Let now the characteristic function

$$\chi_{(t,b]}(\tau) = \begin{cases} 1, & \text{if } \tau \in (t,b] \\ 0, & \text{else}, \end{cases}$$

where $\tau \in [a,b] \cap \mathbb{T}$.

Then the function $(\tau,t) \to \chi_{(t,b]}(\tau) K(\tau,t) f(\tau)$ is Lebesgue ∇-integrable over Ω.

Hence by Fubini's theorem we get that

$$\int_a^b \chi_{(t,b]}(\tau) K(\tau,t) f(\tau) \nabla\tau = \int_t^b \widehat{g}_{\alpha-1}(\rho(\tau),t) f(\tau) \nabla\tau = J_{b-}^\alpha f(t)$$

is Lebesgue ∇-integrable on $[a,b] \cap \mathbb{T}$, proving the claim. $\qquad\qquad\square$

We make

Assumption 2.2. From now on we assume that $\widehat{g}_{\alpha-1}(\rho(\cdot),\cdot)$ is continuous on $([a,b] \cap \mathbb{T})^2$, for any $\alpha > 1$.

We give

Definition 2.3. Let $f \in L_1((a,b] \cap \mathbb{T})$. We define the right backward graininess deviation functional of f as follows

$$\theta(f,\alpha,\beta,b,\mathbb{T},x) := \int_x^b f(\tau)\gamma(x,\tau)\nabla\tau. \qquad (2.12)$$

It holds

$$\theta(f,\alpha,\beta,b,\mathbb{T},x) = \int_x^b f(\tau)\nu(\tau)\widehat{g}_{\alpha-1}(\rho(\tau),x)\widehat{g}_{\beta-1}(\rho(\tau),\tau)\nabla\tau, \qquad (2.13)$$

under the assumption $\mathbb{T} = \mathbb{T}_k$.

If $\mathbb{T} = \mathbb{R}$, then $\theta(f,\alpha,\beta,b,\mathbb{T},x) = 0$.

We give the following semigroup property of right ∇-Riemann-Liouville type fractional integrals.

Theorem 2.3. *Let the time scale* $\mathbb{T}$ *such that* $a,b \in \mathbb{T}$, $f \in L_1((a,b] \cap \mathbb{T})$; $\alpha,\beta > 1$. *Then*

$$J_{b-}^\alpha J_{b-}^\beta f(x) = J_{b-}^{\alpha+\beta} f(x) + \theta(f,\alpha,\beta,b,\mathbb{T},x), \quad \forall\, x \in [a,b] \cap \mathbb{T}. \qquad (2.14)$$

Proof. For $\beta > 1$ we have

$$J_{b-}^\beta f(t) = \int_t^b \widehat{g}_{\beta-1}(\rho(\tau),t) f(\tau)\nabla\tau.$$

We observe that

$$J_{b-}^\alpha J_{b-}^\beta f(x) = \int_x^b \widehat{g}_{\alpha-1}(\rho(t),x) J_{b-}^\beta f(t)\nabla t$$

$$= \int_x^b \widehat{g}_{\alpha-1}\left(\rho\left(t\right),x\right)\left(\int_t^b \widehat{g}_{\beta-1}\left(\rho\left(\tau\right),t\right)f\left(\tau\right)\nabla\tau\right)\nabla t$$

$$= \int_x^b \left(\int_t^b \widehat{g}_{\alpha-1}\left(\rho\left(t\right),x\right)\widehat{g}_{\beta-1}\left(\rho\left(\tau\right),t\right)f\left(\tau\right)\nabla\tau\right)\nabla t =: (*).$$

Clearly here it holds

$$\left|\widehat{g}_{\alpha-1}\left(\rho\left(t\right),x\right)\right| \le M_1, \quad \forall\, t, x \in [a,b]\cap\mathbb{T},$$

and

$$\left|\widehat{g}_{\beta-1}\left(\rho\left(\tau\right),t\right)\right| \le M_2, \quad \forall\, \tau, t \in [a,b]\cap\mathbb{T},$$

where $M_1, M_2 > 0$.

Hence

$$\left|J_{b-}^{\alpha} J_{b-}^{\beta} f\left(x\right)\right| \le \int_x^b \left(\int_t^b \left|\widehat{g}_{\alpha-1}\left(\rho\left(t\right),x\right)\right|\left|\widehat{g}_{\beta-1}\left(\rho\left(\tau\right),t\right)\right|\left|f\left(\tau\right)\right|\nabla\tau\right)\nabla t$$

$$\le M_1 M_2 \left(\int_x^b \left(\int_t^b \left|f\left(\tau\right)\right|\nabla\tau\right)\nabla t\right) \le M_1 M_2 \left(\int_x^b \left(\int_a^b \left|f\left(\tau\right)\right|\nabla\tau\right)\nabla t\right)$$

$$\le M_1 M_2 \left(b - a\right)\left\|f\right\|_{L_1((a,b]\cap\mathbb{T})} < \infty.$$

Therefore $J_{b-}^{\alpha} J_{b-}^{\beta} f\left(x\right)$ exists, $\forall\, x \in [a,b]\cap\mathbb{T}$. Consequently by Fubini's theorem we have

$$(*) = \int_x^b \left(\int_x^\tau \widehat{g}_{\alpha-1}\left(\rho\left(t\right),x\right)\widehat{g}_{\beta-1}\left(\rho\left(\tau\right),t\right)f\left(\tau\right)\nabla t\right)\nabla\tau$$

$$= \int_x^b f\left(\tau\right)\left(\int_x^\tau \widehat{g}_{\alpha-1}\left(\rho\left(t\right),x\right)\widehat{g}_{\beta-1}\left(\rho\left(\tau\right),t\right)\nabla t\right)\nabla\tau$$

$(x < t \le \tau)$

$$\overset{(2.9)}{=} \int_x^b f\left(\tau\right)\left(\widehat{g}_{\alpha+\beta-1}\left(\rho\left(\tau\right),x\right) + \int_{\rho(\tau)}^\tau \widehat{g}_{\alpha-1}\left(\rho\left(t\right),x\right)\widehat{g}_{\beta-1}\left(\rho\left(\tau\right),t\right)\nabla t\right)\nabla\tau$$

$$= \int_x^b \widehat{g}_{\alpha+\beta-1}\left(\rho\left(\tau\right),x\right)f\left(\tau\right)\nabla\tau + \int_x^b f\left(\tau\right)\gamma\left(x,\tau\right)\nabla\tau$$

$$= J_{b-}^{\alpha+\beta} f\left(x\right) + \int_x^b f\left(\tau\right)\gamma\left(x,\tau\right)\nabla\tau.$$

So we have that

$$J_{b-}^{\alpha} J_{b-}^{\beta} f\left(x\right) = J_{b-}^{\alpha+\beta} f\left(x\right) + \int_x^b f\left(\tau\right)\gamma\left(x,\tau\right)\nabla\tau$$

$$= J_{b-}^{\alpha+\beta} f\left(x\right) + \theta\left(f,\alpha,\beta,b,\mathbb{T},x\right),$$

proving the claim. $\qquad\qquad\square$

We make

Remark 2.1. Let $\mu > 2 : m - 1 < \mu \leq m \in \mathbb{N}$, i.e. $m = \lceil \mu \rceil$ (ceiling of number), $\widetilde{\nu} = m - \mu$ $(0 \leq \widetilde{\nu} < 1)$. Let $f \in C^m_{ld}\left([a,b] \cap \mathbb{T}\right)$. Clearly here ([11]) f^{∇^m} is a Lebesgue ∇-integrable function. We define the right nabla fractional derivative on $\mathbb{T}$ of order $\mu - 1$ as follows:

$$\nabla^{\mu-1}_{b-} f\left(t\right) = \left(-1\right)^m \left(J^{\widetilde{\nu}+1}_{b-} f^{\nabla^m}\right)\left(t\right) = \left(-1\right)^m \int_t^b \widehat{g}_{\widetilde{\nu}}\left(\rho\left(\tau\right), t\right) f^{\nabla^m}\left(\tau\right) \nabla\tau, \quad (2.15)$$

$\forall\, t \in [a,b] \cap \mathbb{T}$.

Notice $\nabla^{\mu-1}_{b-} f \in C\left([a,b] \cap \mathbb{T}\right)$, by a simple argument using the dominated convergence theorem in Lebesgue ∇-sense.

If $\mu = m$, then $\widetilde{\nu} = 0$, then

$$\nabla^{m-1}_{b-} f\left(t\right) = \left(-1\right)^m \int_t^b f^{\nabla^m}\left(\tau\right) \nabla\tau = \left(-1\right)^m \left(f^{\nabla^{m-1}}\left(b\right) - f^{\nabla^{m-1}}\left(t\right)\right). \quad (2.16)$$

More generally, by [9], given that $f^{\nabla^{m-1}}$ is everywhere finite and absolutely continuous on $[a,b] \cap \mathbb{T}$, then f^{∇^m} exists ∇-a.e. and is Lebesgue ∇-integrable on $(t,b] \cap \mathbb{T}$, $\forall\, t \in [a,b] \cap \mathbb{T}$, and one can plug it into (2.15).

Remark 2.2. We observe that

$$J^{\mu-1}_{b-} \nabla^{\mu-1}_{b-} f\left(t\right) = \left(-1\right)^m \left(J^{\mu-1}_{b-} J^{\widetilde{\nu}+1}_{b-} f^{\nabla^m}\left(t\right)\right)$$

$$\overset{(2.14)}{=} \left(-1\right)^m \left(J^{\mu-1+\widetilde{\nu}+1}_{b-} f^{\nabla^m}\left(t\right) + \theta\left(f^{\nabla^m}, \mu - 1, \widetilde{\nu} + 1, b, \mathbb{T}, t\right)\right)$$

$$= \left(-1\right)^m \left(J^m_{b-} f^{\nabla^m}\left(t\right) + \theta\left(f^{\nabla^m}, \mu - 1, \widetilde{\nu} + 1, b, \mathbb{T}, t\right)\right). \quad (2.17)$$

Hence we proved that

$$J^{\mu-1}_{b-} \nabla^{\mu-1}_{b-} f\left(t\right) + \left(-1\right)^{m+1} \theta\left(f^{\nabla^m}, \mu - 1, \widetilde{\nu} + 1, b, \mathbb{T}, t\right)$$

$$= \left(-1\right)^m \left(J^m_{b-} f^{\nabla^m}\left(t\right)\right) = \left(-1\right)^m \left(\int_t^b \widehat{g}_{m-1}\left(\rho\left(\tau\right), t\right) f^{\nabla^m}\left(\tau\right) \nabla\tau\right)$$

$$\overset{(2.6)}{=} \left(R^b_m\left(f\right)\right)\left(t\right), \quad (2.18)$$

under the assumption $\mathbb{T} = \mathbb{T}_k = \mathbb{T}^k$.

We have established the following right nabla time scales Taylor formula.

Theorem 2.4. *Assume* $\mathbb{T} = \mathbb{T}_k = \mathbb{T}^k$. *Let* $f \in C^m_{ld}\left(\mathbb{T}\right)$, $m \in \mathbb{N}$, $a, b \in \mathbb{T}$, *with* $\mu > 2 : m - 1 < \mu \leq m$, $\widetilde{\nu} = m - \mu$. *Then*

$$f\left(t\right) = \sum_{k=0}^{m-1} \widehat{h}_k\left(t, b\right) f^{\nabla^k}\left(b\right) + J^{\mu-1}_{b-} \nabla^{\mu-1}_{b-} f\left(t\right)$$

$$+ \left(-1\right)^{m+1} \theta\left(f^{\nabla^m}, \mu - 1, \widetilde{\nu} + 1, b, \mathbb{T}, t\right), \quad (2.19)$$

$\forall\, t \in [a,b] \cap \mathbb{T}$.

Remark 2.3. One can rewrite (2.19) as follows

$$f(t) = \sum_{k=0}^{m-1} \widehat{h}_k(t, b) f^{\nabla^k}(b)$$

$$+ (-1)^{m+1} \int_t^b f^{\nabla^m}(\tau) \nu(\tau) \widehat{g}_{\mu-2}(\rho(\tau), t) \widehat{g}_{\widetilde{\nu}}(\rho(\tau), \tau) \nabla\tau$$

$$+ \int_t^b \widehat{g}_{\mu-2}(\rho(\tau), t) \left(\nabla_{b-}^{\mu-1} f\right)(\tau) \nabla\tau, \tag{2.20}$$

$\forall\, t \in [a, b] \cap \mathbb{T}$.

Corollary 2.1. *In the assumptions of Theorem 2.4, additionally assume that* $f^{\nabla^k}(b) = 0$, $k = 0, 1, ..., m-1$. *Then*

$$A(t) := f(t) + (-1)^m \theta\left(f^{\nabla^m}, \mu - 1, \widetilde{\nu} + 1, b, \mathbb{T}, t\right)$$

$$= \int_t^b \widehat{g}_{\mu-2}(\rho(\tau), t) \left(\nabla_{b-}^{\mu-1} f\right)(\tau) \nabla\tau, \tag{2.21}$$

$\forall\, t \in [a, b] \cap \mathbb{T}$.

Remark 2.4. Notice (by [9]) that $\left(J_{b-}^{\mu-1} \nabla_{b-}^{\mu-1} f\right)(t)$ and $\theta(f^{\nabla^m}, \mu-1, \widetilde{\nu}+1, b, \mathbb{T}, t)$ are absolutely continuous functions on $[a, b] \cap \mathbb{T}$.

One can use (2.20) and (2.21) to establish right fractional nabla inequalities on time scales of Poincaré type, Sobolev type, Opial type, Ostrowski type and Hilbert-Pachpatte type, etc, analogous to [1]

Our theory is not void because it is valid when $\mathbb{T} = \mathbb{R}$, see also [1].

Bibliography

1. G.A. Anastassiou, *Foundations of nabla fractional calculus on time scales and inequalities*, Comput. Math. Appl. 59 (12) (2010), 3750-3762.
2. G.A. Anastassiou, *Nabla time scales inequalities*, Editor Al. Paterson, special issue on Time Scales, Internat. J. Dynam. Syst. Difference Equations 3 (1/2) 2011 (2011), 59-83.
3. G.A. Anastassiou, *Basics of right nabla fractional calculus on time scales*, J. Nonlinear Evolution Equations Appl. 2011 (8) (2011), pp. 111-118 (electronic).
4. D.R. Anderson, *Taylor polynomials for nabla dynamic equations on times scales*, Panamer. Math. J. 12 (4) (2002), 17-27.
5. D. Anderson, J. Bullok, L. Erbe, A. Peterson, H. Tran, *Nabla dynamic equations on time scales*, Panamer. Math. J. 13 (1) (2003), 1-47.
6. F. Atici, D. Biles, A. Lebedinsky, *An application of time scales to economics*, Math. Comput. Modelling 43 (2006), 718-726.
7. M. Bohner, G.S. Guseinov, *Multiple Lebesgue integration on time scales*, Adv. Difference Equations 2006 (2006), Article ID 26391, 1-12.
8. M. Bohner, G. Guseinov, *Double integral calculus of variations on time scales*, Comput. Math. Appl. 54 (2007), 45-57.
9. M. Bohner, H. Luo, *Singular second-order multipoint dynamic boundary value problems with mixed derivatives*, Adv. Difference Equations 2006 (2006), Article ID 54989, 1-15.
10. M. Bohner, A. Peterson, *Dynamic Equations on Time Scales: An Introduction with Applications*, Birkhaüser, Boston (2001).
11. G. Guseinov, *Integration on time scales*, J. Math. Anal. Appl. 285 (2003), 107-127.
12. S. Hilger, *Ein Maßketten kalkül mit Anwendung auf Zentrumsmannig-faltigkeiten*, Ph.D. thesis, Universität Würzburg, Germany (1998).
13. N. Martins, D. Torres, *Calculus of variations on time scales with nabla derivatives*, Nonlinear Anal. 71 (12) (2009), 763-773.

Chapter 3

About Right Delta Discrete Fractionality

Here we define a Caputo like right discrete delta fractional difference and we produce a right discrete delta fractional Taylor formula for the first time. We estimate the remainder. Then we produce related right discrete delta fractional Ostrowski, Poincaré and Sobolev type inequalities. It follows [3].

3.1 Introduction and Background

Here we work on the time scale $\mathbb{T} = a + \mathbb{Z}$, where $a \in \mathbb{R}$. We consider functions $f : (a + \mathbb{Z}) \to \mathbb{R}$. If a function f is defined on a subset of $a + \mathbb{Z}$, then one can extend it to all of $a + \mathbb{Z}$, by assigning zero values to f on the complement with respect to $a + \mathbb{Z}$ of that subset. Let $t \in \mathbb{R}$, $n \in \mathbb{N}$, the falling factorial is defined by $t^{(n)} = t(t-1)\cdots(t-n+1) = \prod_{i=0}^{n-1}(t-i)$, and in general $t^{(\alpha)} = \frac{\Gamma(t+1)}{\Gamma(t+1-\alpha)}$, where $\alpha \in \mathbb{R}$, with Γ the gamma function $\Gamma(\nu) = \int_0^\infty e^{-t} t^{\nu-1} dt$, $\nu > 0$; $t^{(0)} = 1$.

From the time scales theory [6], p. 29, [7], we know that the delta integral on $(a + \mathbb{Z})$

$$\int_{a^*}^{b^*} f(t) \Delta t = \sum_{t=a^*}^{b^*-1} f(t), \quad a^* < b^*, \tag{3.1}$$

$a^*, b^* \in (a + \mathbb{Z})$, $f : (a + \mathbb{Z}) \to \mathbb{R}$.

Let $t \in (a + \mathbb{Z})$, then the forward difference

$$\Delta f(t) := f(t+1) - f(t) = f^\Delta(t),$$

the delta time scale derivative, see [6], p. 5, and

$$\Delta^k f(t) = \sum_{l=0}^{k} \binom{k}{l} (-1)^{k-l} f(t+l) = f^{\Delta^k}(t),$$

the kth order delta time scale derivative, see [6], p. 14.

Notice here that if f is restricted on $[a^*, b^*] \cap (a + \mathbb{Z})$, then $\Delta^k f$ runs on $[a^*, b^* - k] \cap (a + \mathbb{Z})$.

For a general time scale $\mathbb{T}$, see [6], p. 38, we define

$$h_k \,:\, \mathbb{T}^2 \to \mathbb{R}, \; k \in \mathbb{N}_0 = \mathbb{N} \cup \{0\}, \; h_0(t,s) = 1, \; \forall \, s, t \in \mathbb{T}, \qquad (3.2)$$

$$h_{k+1}(t,s) = \int_s^t h_k(\tau,s)\,\Delta\tau, \; \forall \, s, t \in \mathbb{T}.$$

We have that the delta derivative $h_k^{\Delta}(t,s) = h_{k-1}(t,s)$, $k \in \mathbb{N}$, $t \in \mathbb{T}^k$, (for the definitions of $\mathbb{T}_k$, $\mathbb{T}^k$, see [6], p. 331, p. 2, respectively) and $h_1(t,s) = t - s$, $\forall$ $s, t \in \mathbb{T}$.

Notice here that

$$(a + \mathbb{Z}) = (a + \mathbb{Z})_k = (a + \mathbb{Z})^k. \qquad (3.3)$$

We need

Lemma 3.1. *On* $(a + \mathbb{Z})$ *we have*

$$h_k(t,s) = \frac{(t-s)^{(k)}}{k!}, \; \forall \, k \in \mathbb{N}_0. \qquad (3.4)$$

Lemma 3.2. *It holds on* $(a + \mathbb{Z})$ *that*

$$\left\{ \frac{(t-s)^{(k+1)}}{(k+1)!} \right\}^{\Delta_t} = \frac{(t-s)^{(k)}}{k!}, \quad k \in \mathbb{N}_0. \qquad (3.5)$$

Proof. We have that

$$\left\{ \frac{(t-s)^{(k+1)}}{(k+1)!} \right\}^{\Delta_t} = \frac{(t+1-s)^{(k+1)}}{(k+1)!} - \frac{(t-s)^{(k+1)}}{(k+1)!}$$

$$= \frac{1}{(k+1)!} \left\{ ((t-s)+1)^{(k+1)} - (t-s)^{(k+1)} \right\}$$

$$= \frac{1}{(k+1)!} \left\{ \prod_{i=0}^{k} [((t-s)+1) - i] - \prod_{i=0}^{k} ((t-s) - i) \right\}$$

$$= \frac{1}{(k+1)!} \left\{ \prod_{i=0}^{k} [((t-s) - (i-1))] - \prod_{i=0}^{k} ((t-s) - i) \right\}$$

$$= \frac{1}{(k+1)!} \left\{ ((t-s)+1)\,[(t-s)\,((t-s)-1)\,((t-s)-2)\,((t-s)-3) \right.$$

$$\cdots ((t-s) - (k-1))] - [(t-s)\,((t-s)-1)\,((t-s)-2)\,((t-s)-3)$$

$$\cdots ((t-s) - (k-1))]\,((t-s) - k) \}$$

$$= \frac{(t-s)\,((t-s)-1)\,((t-s)-2)\cdots((t-s)-(k-1))}{(k+1)!}$$

$$\cdot (((t-s)+1) - ((t-s) - k))$$

$$= \frac{(t-s)\,((t-s)-1)\,((t-s)-2)\cdots((t-s)-k+1)}{k!} = \frac{(t-s)^{(k)}}{k!}.$$

$$\square$$

Proof. of Lemma 3.1.

Notice that

$$h_0(t,s) = \frac{(t-s)^{(0)}}{0!} = (t-s)^{(0)} = 1.$$

Assume (3.4) correct for k. Then

$$h_{k+1}(t,s) = \int_s^t h_k(\tau,s)\,\Delta\tau = \int_s^t \frac{(\tau-s)^{(k)}}{k!}\,\Delta\tau = \frac{(t-s)^{(k+1)}}{(k+1)!},$$

proving the claim. $\qquad\qquad\qquad\qquad\qquad\qquad\qquad\qquad\qquad\square$

Let $f:(a+\mathbb{Z})\to\mathbb{R}$, $m\in\mathbb{N}$. Then by delta Taylor formula on time scales (see [5], [7]), applied on $(a+\mathbb{Z})$, see also (3.4), we get

$$f(t) = \sum_{k=0}^{m-1}\Delta^k f(s)\frac{(t-s)^{(k)}}{k!} + \int_s^t \frac{(t-\tau-1)^{(m-1)}}{(m-1)!}\Delta^m f(\tau)\,\Delta\tau, \qquad (3.6)$$

$\forall\, t,s\in(a+\mathbb{Z})$.

For $s=b^*\in(a+\mathbb{Z})$, $t\in[a^*,b^*]\cap(a+\mathbb{Z})$, where $a^*<b^*$, we get that

$$f(t) = \sum_{k=0}^{m-1}\Delta^k f(b^*)\frac{(t-b^*)^{(k)}}{k!} - \int_t^{b^*} \frac{(t-s-1)^{(m-1)}}{(m-1)!}\Delta^m f(s)\,\Delta s$$

$$= \sum_{k=0}^{m-1}\Delta^k f(b^*)\frac{(t-b^*)^{(k)}}{k!} - \frac{1}{(m-1)!}\sum_{s=t}^{b^*-1}(t-s-1)^{(m-1)}\Delta^m f(s). \qquad (3.7)$$

We call the remainder

$$R^*(t) = -\frac{1}{(m-1)!}\sum_{s=t}^{b^*-1}(t-s-1)^{(m-1)}\Delta^m f(s). \qquad (3.8)$$

We need

Proposition 3.1. *For $s,t\in(a+\mathbb{Z})$, $m\in\mathbb{N}$, it holds*

$$(t-s-1)^{(m-1)} = (-1)^{m-1}(s+m-t-1)^{(m-1)}. \qquad (3.9)$$

Proof. We notice that

$$(t-s-1)^{(m-1)} = \prod_{i=0}^{m-2}(t-s-1-i)$$

$$= (t-s-1)(t-s-2)(t-s-3)\cdots(t-s-(m-1))$$

$$= (-(s-t)-1)(-(s-t)-2)(-(s-t)-3)\cdots(-(s-t)-(m-1))$$

$$= (-1)^{m-1}(((s-t)+1)((s-t)+2)((s-t)+3)\cdots((s-t)+(m-1)))$$

$$= (-1)^{m-1}(s+m-t-1)^{(m-1)}.$$

Indeed it is

$$(s+m-t-1)^{(m-1)} = \prod_{i=0}^{m-2}(s+m-t-1-i)$$

$$= (s+m-t-1)(s+m-t-2)(s+m-t-3)\cdots(s+m-t-1-m+2)$$

$$= (s+m-t-1)(s+m-t-2)(s+m-t-3)\cdots(s-t+1).$$

$\qquad\qquad\qquad\qquad\qquad\qquad\qquad\qquad\qquad\qquad\qquad\qquad\qquad\square$

We need

Definition 3.1. (see [4])

Let $\nu > 0$, the right fractional sum here is given by

$$\left(\Delta_{b^*-1}^{-\nu} f\right)(t) := \frac{1}{\Gamma(\nu)} \sum_{s=t+\nu}^{b^*-1} (s-t-1)^{(\nu-1)} f(s), \qquad (3.10)$$

$$\left(\Delta_{b^*-1}^{0} f\right)(t) := f(t),$$

where f is restricted on $[a^*, b^*] \cap (a + \mathbb{Z})$.

Notice $\left(\Delta_{b^*-1}^{-\nu} f\right)$ is defined on $\{a^* - \nu,\ a^* - \nu + 1,\ a^* - \nu + 2, \ldots, b^* - 1 - \nu\}$. Here one can take $a^* = -\infty$.

We also need

Theorem 3.1. *(see [8]) Let $\mu, \nu \geq 0$. Then*

$$\left(\Delta_{b^*-1-\nu}^{-\mu} \Delta_{b^*-1}^{-\nu} f\right)(t) = \left(\Delta_{b^*-1}^{-(\mu+\nu)} f\right)(t), \qquad (3.11)$$

where $t \in \{a^* - (\mu + \nu),\ a^* - (\mu + \nu) + 1, \ldots, b^* - 1 - (\mu + \nu)\}$.

Remark 3.1. So far we have based on (3.9) that

$$R^*(t) = -\frac{1}{(m-1)!} \sum_{s=t}^{b^*-1} (t-s-1)^{(m-1)} \Delta^m f(s)$$

$$= \frac{(-1)^m}{(m-1)!} \sum_{s=t}^{b^*-1} \left((s+m-t-1)^{(m-1)}\right) \Delta^m f(s)$$

$$= \frac{(-1)^m}{(m-1)!} \sum_{s=(t-m)+m}^{b^*-1} \left((s-(t-m)-1)^{(m-1)}\right) \Delta^m f(s)$$

(call $t' = t - m$)

$$= \frac{(-1)^m}{(m-1)!} \sum_{s=t'+m}^{b^*-1} \left((s-t'-1)^{(m-1)}\right) \Delta^m f(s)$$

(notice here $t' \in \{a^* - m, \ldots, b^* - 1 - m\}$)

$$\overset{(3.10)}{=} (-1)^m \left(\Delta_{b^*-1}^{-m} (\Delta^m f)\right)(t') = (-1)^m \left(\Delta_{b^*-1}^{-m} (\Delta^m f)\right)(t-m).$$

So we have proved

Theorem 3.2. *It holds*

$$R^*(t) = (-1)^m \left(\Delta_{b^*-1}^{-m} (\Delta^m f)\right)(t-m), \qquad (3.12)$$

where $t \in [a^*, b^* - 1] \cap (a + \mathbb{Z})$, *with* $a^* \leq b^* - 1$; $a^*, b^* \in (a + \mathbb{Z})$, $f : (a + \mathbb{Z}) \to \mathbb{R}$, a^* *could be* $-\infty$.

3.2 Main Results

We give

Definition 3.2. Let $\mu > 0$, $m - 1 < \mu \leq m$, $m \in \mathbb{N}$, $m = \lceil \mu \rceil$ (ceiling of number), $\nu := m - \mu$, that is $\mu + \nu = m$.

The μ-th order delta right fractional difference (Caputo way) is given by

$$\left(\Delta^{\mu}_{(b^*-1)-} f \right)(t) := (-1)^m \left(\Delta^{-\nu}_{(b^*-1)} \left(\Delta^m f \right) \right)(t)$$

$$= \frac{(-1)^m}{\Gamma(\nu)} \sum_{s=t+\nu}^{b^*-1} (s - t - 1)^{(\nu-1)} \left(\Delta^m f \right)(s), \tag{3.13}$$

where $t \leq b^* - 1 - \nu$, $b^* \in (a + \mathbb{Z})$, $t \in (a - \nu + \mathbb{Z})$, $a \in \mathbb{R}$.

If $\mu = m \in \mathbb{N}$, then $\left(\Delta^{\mu}_{(b^*-1)-} f \right)(t) = (-1)^m \left(\Delta^m f \right)(t)$.

Theorem 3.3. *It holds*

$$R^*(t) = \left(\Delta^{-\mu}_{(b^*-1)-\nu} \left(\Delta^{\mu}_{(b^*-1)-} f \right) \right)(t - m), \tag{3.14}$$

for $\mu > 0$, $m - 1 < \mu \leq m$, $m \in \mathbb{N}$, $t \in [a^*, b^* - 1] \cap (a + \mathbb{Z})$, $a^* \leq b^* - 1$; $a^*, b^* \in (a + \mathbb{Z})$, $f : (a + \mathbb{Z}) \to \mathbb{R}$; a^* could be $-\infty$.

Proof. Let $t \in [a^*, b^* - 1] \cap (a + \mathbb{Z})$, $a^* \leq b^* - 1$; $a^*, b^* \in (a + \mathbb{Z})$, $f : (a + \mathbb{Z}) \to \mathbb{R}$. Then $t - m \in [a^* - m, b^* - 1 - m] \cap (a + \mathbb{Z})$. We observe that

$$\left(\Delta^{-\mu}_{(b^*-1)-\nu} \left(\Delta^{\mu}_{(b^*-1)-} f \right) \right)(t - m)$$

$$= (-1)^m \left(\Delta^{-\mu}_{(b^*-1)-\nu} \left(\Delta^{-\nu}_{(b^*-1)} \left(\Delta^m f \right) \right) \right)(t - m)$$

$$= (-1)^m \left(\Delta^{-(\mu+\nu)}_{(b^*-1)} \left(\Delta^m f \right) \right)(t - m)$$

$$= (-1)^m \left(\Delta^{-m}_{b^*-1} \left(\Delta^m f \right) \right)(t - m) \overset{(3.12)}{=} R^*(t).$$

$\square$

We have proved the following delta right discrete fractional Taylor formula.

Theorem 3.4. *Let $a \in \mathbb{R}$, $f : (a + \mathbb{Z}) \to \mathbb{R}$. Let $a^*, b^* \in (a + \mathbb{Z})$, $a^* < b^*$. Let $\mu > 0 : m - 1 < \mu \leq m$, $m \in \mathbb{N}$, $(m = \lceil \mu \rceil)$, $\nu = m - \mu$, $t \in [a^*, b^* - 1] \cap (a + \mathbb{Z})$. Then*

$$f(t) = \sum_{k=0}^{m-1} \Delta^k f(b^*) \frac{(t - b^*)^{(k)}}{k!} + R^*(t), \tag{3.15}$$

where

$$R^*(t) = \left(\Delta^{-\mu}_{(b^*-1)-\nu} \left(\Delta^{\mu}_{(b^*-1)-} f \right) \right)(t - m)$$

$$= \frac{1}{\Gamma(\mu)} \sum_{s=t-m+\mu}^{(b^*-1-\nu)} (s - t + m - 1)^{(\mu-1)} \left(\Delta^{\mu}_{(b^*-1)-} f \right)(s). \tag{3.16}$$

Above a^* could be $-\infty$.

Corollary 3.1. *In the assumptions of Theorem 3.4, assume that* $\Delta^k f\left(b^*\right) = 0$, $k = 0, 1, \ldots, m-1$. *Then*

$$f(t) = \frac{1}{\Gamma(\mu)} \sum_{s=t-m+\mu}^{(b^*-1-\nu)} (s-t+m-1)^{(\mu-1)} \left(\Delta^\mu_{(b^*-1)-} f\right)(s). \qquad (3.17)$$

We need

Proposition 3.2. *([2]) It holds*

$$\sum_{s=t+\mu}^{b^*-\nu} (s-t-1)^{(\mu-1)} = \frac{(b^*-\nu-t)^{(\mu)}}{\mu} > 0. \qquad (3.18)$$

So by (3.18) we get

Proposition 3.3. *We have*

$$\sum_{s=(t-m)+\mu}^{(b^*-1)-\nu} (s-(t-m)-1)^{(\mu-1)} = \frac{(b^*-1-\nu-t+m)^{(\mu)}}{\mu} > 0. \qquad (3.19)$$

We give the estimate

Theorem 3.5. *All as in Theorem 3.4. Then*

$$\left| f(t) - \sum_{k=0}^{m-1} \Delta^k f(b^*) \frac{(t-b^*)^{(k)}}{k!} \right| = |R^*(t)|$$

$$\leq \frac{(b^*-1-\nu-t+m)^{(\mu)}}{\Gamma(\mu+1)} \max_{s \in \{t-m+\mu,\ldots,b^*-1-\nu\}} \left| \left(\Delta^\mu_{(b^*-1)-} f\right)(s) \right|. \qquad (3.20)$$

Proof. By (3.16) we get

$$|R^*(t)| \leq \frac{1}{\Gamma(\mu)} \left(\sum_{s=t-m+\mu}^{b^*-1-\nu} (s-(t-m)-1)^{(\mu-1)} \right)$$

$$\cdot \max_{s \in \{t-m+\mu,\ldots,b^*-1-\nu\}} \left| \left(\Delta^\mu_{(b^*-1)-} f\right)(s) \right|$$

$$\overset{(3.19)}{=} \frac{(b^*-1-\nu-t+m)^{(\mu)}}{\Gamma(\mu+1)} \max_{s \in \{t-m+\mu,\ldots,b^*-1-\nu\}} \left| \left(\Delta^\mu_{(b^*-1)-} f\right)(s) \right|.$$

$$\square$$

We need

Lemma 3.3. *([1], p. 580) Let* $a > \nu$, $a, \nu > -1$, $a, \nu \in \mathbb{R}$, $a \leq b$. *Then*

$$\sum_{r=a}^{b} r^{(\nu)} = \left(\frac{(b+1)^{(\nu+1)} - a^{(\nu+1)}}{\nu+1} \right). \qquad (3.21)$$

We give a related Ostrowski inequality.

Theorem 3.6. *Let $a \in \mathbb{R}$, $f : (a + \mathbb{Z}) \to \mathbb{R}$. Let $a^*, b^* \in (a + \mathbb{Z})$, $b^* - a^* \geq 2$. Let $\mu > 0 : m - 1 < \mu \leq m$, $m \in \mathbb{N}$, $\nu = m - \mu$, $t \in [a^*, b^* - 1] \cap (a + \mathbb{Z})$. Assume $\Delta^k f(b^*) = 0$, $k = 1, \ldots, m - 1$. Then*

$$\left| \frac{1}{b^* - a^*} \sum_{t=a^*}^{b^*-1} f(t) - f(b^*) \right| \qquad (3.22)$$

$$\leq \frac{(b^* - a^* + \mu)^{(\mu+1)}}{(b^* - a^*)\,\Gamma(\mu + 2)} \max_{s \in \{a^* - m + \mu, \ldots, b^* - 1 - \nu\}} \left| \left(\Delta^\mu_{(b^* - 1)_-} f \right)(s) \right|.$$

Proof. Using (3.15) and (3.16), since $\Delta^k f(b^*) = 0$, $k = 1, \ldots, m - 1$, we get $f(t) - f(b^*) = R^*(t)$, $t \in [a^*, b^* - 1] \cap (a + \mathbb{Z})$. Then we observe

$$E_1 := \left| \frac{1}{b^* - a^*} \sum_{t=a^*}^{b^*-1} f(t) - f(b^*) \right| = \frac{1}{b^* - a^*} \left| \sum_{t=a^*}^{b^*-1} (f(t) - f(b^*)) \right|$$

$$\leq \frac{1}{b^* - a^*} \sum_{t=a^*}^{b^*-1} |f(t) - f(b^*)| = \frac{1}{b^* - a^*} \sum_{t=a^*}^{b^*-1} |R^*(t)|$$

$$\overset{(3.20)}{\leq} \frac{1}{(b^* - a^*)\,\Gamma(\mu + 1)} \left(\sum_{t=a^*}^{b^*-1} (b^* - 1 - \nu - t + m)^{(\mu)} \right)$$

$$\cdot \max_{s \in \{a^* - m + \mu, \ldots, b^* - 1 - \nu\}} \left| \left(\Delta^\mu_{(b^* - 1)_-} f \right)(s) \right| =: (*).$$

Call $r = b^* - 1 - \nu - t + m$. Since $a^* \leq t \leq b^* - 1$, then $-a^* \geq -t \geq 1 - b^*$ and $b^* - 1 - \nu - a^* + m \geq b^* - 1 - \nu - t + m \geq b^* - 1 - \nu + 1 - b^* + m = m - \nu = \mu$.
Therefore

$$\mu \leq r \leq b^* - a^* - 1 + \mu.$$

We would like to calculate

$$\sum_{t=a^*}^{b^*-1} (b^* - 1 - \nu - t + m)^{(\mu)} = \sum_{r=\mu}^{b^* - a^* - 1 + \mu} r^{(\mu)}$$

$$= \mu^{(\mu)} + \sum_{r=\mu+1}^{b^* - a^* - 1 + \mu} r^{(\mu)} = \Gamma(\mu + 1) + \sum_{r=\mu+1}^{b^* - a^* - 1 + \mu} r^{(\mu)}$$

(since $b^* - a^* \geq 2$ we get that $\mu + 1 \leq b^* - a^* - 1 + \mu$, so that we can apply (3.21))

$$= \Gamma(\mu + 1) + \frac{(b^* - a^* + \mu)^{(\mu+1)} - (\mu + 1)^{(\mu+1)}}{\mu + 1}$$

$$= \Gamma\left(\mu+1\right) + \frac{\left(b^*-a^*+\mu\right)^{(\mu+1)}}{\mu+1} - \frac{\Gamma\left(\mu+2\right)}{\mu+1}$$

$$= \Gamma\left(\mu+1\right) + \frac{\left(b^*-a^*+\mu\right)^{(\mu+1)}}{\mu+1} - \Gamma\left(\mu+1\right) = \frac{\left(b^*-a^*+\mu\right)^{(\mu+1)}}{\mu+1}.$$

Consequently it holds

$$\sum_{t=a^*}^{b^*-1} \left(b^*-1-\nu-t+m\right)^{(\mu)} = \frac{\left(b^*-a^*+\mu\right)^{(\mu+1)}}{\mu+1}.$$

So we have

$$(*) = \frac{\left(b^*-a^*+\mu\right)^{(\mu+1)}}{\left(b^*-a^*\right)\Gamma\left(\mu+2\right)} \max_{s\in\{a^*-m+\mu,\ldots,b^*-1-\nu\}} \left|\left(\Delta_{(b^*-1)-}^\mu f\right)(s)\right| =: E_2.$$

Hence $E_1 \leq E_2$, proving the claim.   □

A related Poincaré type inequality follows

Theorem 3.7. *All as in Corollary 3.1. Let* $p,q > 1 : \frac{1}{p} + \frac{1}{q} = 1$. *Then*

$$\sum_{t=a^*}^{b^*-1} |f(t)|^q \leq \frac{1}{\left(\Gamma\left(\mu\right)\right)^q} \left(\sum_{t=a^*}^{b^*-1} \left(\sum_{s=t-m+\mu}^{b^*-1-\nu} (s-t+m-1)^{p(\mu-1)}\right)^{\frac{q}{p}}\right)$$

$$\cdot \left(\sum_{s=a^*-m+\mu}^{b^*-1-\nu} \left|\left(\Delta_{(b^*-1)-}^\mu f\right)(s)\right|^q\right). \tag{3.23}$$

Proof.   Notice here $s-t+m \geq \mu > 0$ and $s-t+m-\mu \geq 0$ and $s-t+m-\mu+1 > 0$, so that

$$(s-t+m-1)^{(\mu-1)} = \frac{\Gamma\left(s-t+m\right)}{\Gamma\left(s-t+m-\mu+1\right)} > 0.$$

By (3.17) we get

$$|f(t)| \leq \frac{1}{\Gamma\left(\mu\right)} \sum_{s=t-m+\mu}^{b^*-1-\nu} (s-t+m-1)^{(\mu-1)} \left|\left(\Delta_{(b^*-1)-}^\mu f\right)(s)\right|$$

$$\leq \frac{1}{\Gamma\left(\mu\right)} \left(\sum_{s=t-m+\mu}^{b^*-1-\nu} (s-t+m-1)^{p(\mu-1)}\right)^{\frac{1}{p}} \left(\sum_{s=t-m+\mu}^{b^*-1-\nu} \left|\left(\Delta_{(b^*-1)-}^\mu f\right)(s)\right|^q\right)^{\frac{1}{q}}$$

$$\leq \frac{1}{\Gamma\left(\mu\right)} \left(\sum_{s=t-m+\mu}^{b^*-1-\nu} (s-t+m-1)^{p(\mu-1)}\right)^{\frac{1}{p}} \left(\sum_{s=a^*-m+\mu}^{b^*-1-\nu} \left|\left(\Delta_{(b^*-1)-}^\mu f\right)(s)\right|^q\right)^{\frac{1}{q}}.$$

Therefore we found

$$|f(t)| \leq \frac{1}{\Gamma\left(\mu\right)} \left(\sum_{s=t-m+\mu}^{b^*-1-\nu} (s-t+m-1)^{p(\mu-1)}\right)^{\frac{1}{p}} \tag{3.24}$$

$$\cdot \left(\sum_{s=a^*-m+\mu}^{b^*-1-\nu} \left| \left(\Delta_{(b^*-1)-}^{\mu} f \right)(s) \right|^q \right)^{\frac{1}{q}}.$$

Hence

$$|f(t)|^q \le \frac{1}{(\Gamma(\mu))^q} \left(\sum_{s=t-m+\mu}^{b^*-1-\nu} (s-t+m-1)^{p(\mu-1)} \right)^{\frac{q}{p}} \tag{3.25}$$

$$\cdot \left(\sum_{s=a^*-m+\mu}^{b^*-1-\nu} \left| \left(\Delta_{(b^*-1)-}^{\mu} f \right)(s) \right|^q \right).$$

Applying $\sum_{t=a^*}^{b^*-1}$ to both sides of (3.25) we derive (3.23). $\qquad\square$

We finish with a related Sobolev type inequality

Theorem 3.8. *All as in Theorem 3.7. Let* $r \ge 1$. *Then*

$$\left(\sum_{t=a^*}^{b^*-1} |f(t)|^r \right)^{\frac{1}{r}} \le \frac{1}{\Gamma(\mu)} \left(\sum_{t=a^*}^{b^*-1} \left(\sum_{s=t-m+\mu}^{b^*-1-\nu} (s-t+m-1)^{p(\mu-1)} \right)^{\frac{r}{p}} \right)^{\frac{1}{r}}$$

$$\cdot \left(\sum_{s=a^*-m+\mu}^{b^*-1-\nu} \left| \left(\Delta_{(b^*-1)-}^{\mu} f \right)(s) \right|^q \right)^{\frac{1}{q}}. \tag{3.26}$$

Proof. Raising (3.24) to the power r we obtain

$$|f(t)|^r \le \frac{1}{(\Gamma(\mu))^r} \left(\sum_{s=t-m+\mu}^{b^*-1-\nu} (s-t+m-1)^{p(\mu-1)} \right)^{\frac{r}{p}}$$

$$\cdot \left(\sum_{s=a^*-m+\mu}^{b^*-1-\nu} \left| \left(\Delta_{(b^*-1)-}^{\mu} f \right)(s) \right|^q \right)^{\frac{r}{q}}.$$

Hence it holds

$$\sum_{t=a^*}^{b^*-1} |f(t)|^r \le \frac{1}{(\Gamma(\mu))^r} \left(\sum_{t=a^*}^{b^*-1} \left(\sum_{s=t-m+\mu}^{b^*-1-\nu} (s-t+m-1)^{p(\mu-1)} \right)^{\frac{r}{p}} \right)$$

$$\cdot \left(\sum_{s=a^*-m+\mu}^{b^*-1-\nu} \left| \left(\Delta_{(b^*-1)-}^{\mu} f \right)(s) \right|^q \right)^{\frac{r}{q}}. \tag{3.27}$$

Raising (3.27) to the power $\frac{1}{r}$ we derive (3.26). $\qquad\square$

Bibliography

1. G.A. Anastassiou, *Intelligent Mathematics: Computational Analysis*, Springer, Heidelberg, 2011.
2. G.A. Anastassiou, *Right nabla discrete fractional calculus*, Intern. J. Difference Equations 6 (2) (2011), 91-104.
3. G. A. Anastassiou, *Right delta discrete fractionality*, Dynamic Systems Appl. 20 (2011), 531-540.
4. N. Bastos, R. Ferreira, D. Torres, *Discrete-time fractional variational problems*, Signal Processing 91 (2011), 513-524.
5. M. Bohner, G. Guseinov, *The convolution on time scales*, Abstr. Appl. Anal. 2007 (2007), Article ID 58373, 24 pages.
6. M. Bohner, A. Peterson, *Dynamic equations on Time Scales: An Introduction with Applications*, Birkhaüser, Boston, 2001.
7. R. Higgins, A. Peterson, *Cauchy functions and Taylor's formula for time scales* $\mathbb{T}$, in B. Aulbach, S. Elaydi, G. Ladas (eds.), Proc. Sixth Internat. Conf. Difference Equations, New Progress in Difference Equations, Augsburg, Germany, 2001, pp. 299-308, Chapman & Hall/CRC (2004).
8. R. Ferreira, D. Tores, *Fractional h-difference equations arising from the calculus of variations*, Applicable Anal. Discrete Math. 5 (2011), 110-121.

Chapter 4

About Right Nabla Discrete Fractional Calculus

Here we define a Caputo like right discrete delta fractional difference and we produce a right discrete delta fractional Taylor formula. We estimate the remainder. Then we produce related right discrete delta fractional Ostrowski, Poincaré and Sobolev type inequalities. It follows [2].

4.1 Introduction and Background

Here we work on the time scale $\mathbb{T} = a + \mathbb{Z}$, where $a \in \mathbb{R}$. We consider functions $f : (a + \mathbb{Z}) \to \mathbb{R}$. If a function f is defined on a subset of $a + \mathbb{Z}$, then one can extend it to all of $a + \mathbb{Z}$, by assigning zero values to f on the complement, with respect to $a + \mathbb{Z}$, of that subset. Let $t \in \mathbb{R}$, the rising factorial is defined as

$$t^{\bar{n}} = t(t+1)(t+2)...(t+n-1), \ n \in \mathbb{N}$$

and $t^{\bar{0}} = 1$. Let $\alpha \in \mathbb{R}$, then in general define

$$t^{\bar{\alpha}} = \frac{\Gamma(t+\alpha)}{\Gamma(t)}$$

where $t \in \mathbb{R} - \{\dots, -2, -1, 0\}$, $0^{\bar{\alpha}} = 0$, where Γ is the gamma function;

$$\Gamma(\nu) = \int_0^\infty e^{-t} t^{\nu-1} dt, \ \nu > 0.$$

Note that

$$\nabla(t^{\bar{\alpha}}) = \alpha t^{\overline{\alpha-1}}$$

where $\nabla y(t) = y(t) - y(t-1) \ (= y^\nabla(t)$ the time scale nabla discrete derivative if $t \in (a + \mathbb{Z}))$. Also $\nabla y(t)$ is called the backward difference. We further define the falling factorial

$$t^{(n)} = t(t-1)...(t-n+1), \ n \in \mathbb{N},$$

and in general

$$t^{\bar{\alpha}} = \frac{\Gamma(t+1)}{\Gamma(t+1-\alpha)}.$$

Notice that
$$t^{\bar{\alpha}} = (t + \alpha - 1)^{(\alpha)}. \tag{4.1}$$
From time scales theory [5], p. 333, we know that
$$\int_{a^*}^{b^*} f(t)\nabla t = \sum_{t=a^*+1}^{b^*} f(t), \quad a^* < b^* \tag{4.2}$$
where $a^*, b^* \in (a+\mathbb{Z})$. Also for $f : (a+\mathbb{Z}) \to \mathbb{R}$ we define the higher order backward difference
$$\nabla^k f(t) = \sum_{l=0}^{k} (-1)^l \binom{k}{l} f(t-l) = f^{\nabla^k}(t), \quad k \in \mathbb{N}$$
the kth order time scale nabla derivative of f. If f runs on $[a^*, b^*] \cap (a + \mathbb{Z})$, then f^{∇^k} runs on $[a^* + k, b^*] \cap (a + \mathbb{Z})$. For general time scale $\mathbb{T}$, see [3],[5] we define
$$\hat{h}_k : \mathbb{T}^2 \to \mathbb{R}, \quad k \in \mathbb{N}_0 = \mathbb{N} \cup \{0\},$$
$$\hat{h}_0(t,s) = 1, \quad \forall s, t \in \mathbb{T},$$
$$\hat{h}_{k+1}(t,s) = \int_s^t \hat{h}_k(\tau, s)\nabla\tau, \quad \forall s, t \in \mathbb{T}. \tag{4.3}$$
We have that the nabla derivative
$$\hat{h}_k^{\nabla}(t,s) = \hat{h}_{k-1}(t,s), \quad k \in \mathbb{N}, \ t \in \mathbb{T}_k,$$
(for $\mathbb{T}_k$, $\mathbb{T}^k$ see [3], [5]) and $\hat{h}_1(t,s) = t - s$, $\forall s, t \in \mathbb{T}$. Notice here that
$$(a + \mathbb{Z}) = (a + \mathbb{Z})_k = (a + \mathbb{Z})^k. \tag{4.4}$$

Lemma 4.1. *We have that*
$$\hat{h}_k(t,s) = \frac{(t-s)^{\overline{k}}}{k!}, \quad \forall s, t \in (a + \mathbb{Z}). \tag{4.5}$$

Lemma 4.2. *It holds on $a + \mathbb{Z}$ that*
$$\left\{ \frac{(t-s)^{\overline{k+1}}}{(k+1)!} \right\}^{\nabla_t} = \frac{(t-s)^{\overline{k}}}{k!}, \quad k \in \mathbb{N}_0 \tag{4.6}$$
(∇, means nabla derivative with respect to t).

Proof.
$$\left(\frac{(t-s)^{\overline{k+1}}}{(k+1)!} \right)^{\nabla_t} = \frac{1}{(k+1)!} \left\{ (t-s)^{\overline{k+1}} - (t-s-1)^{\overline{k+1}} \right\}$$
$$= \frac{1}{(k+1)!} \{ (t-s)(t-s+1)(t-s+2)...(t-s+k)$$
$$- (t-s-1)(t-s)(t-s+1)...(t-s+k-1) \}$$
$$= \frac{1}{(k+1)!} \{ (t-s)(t-s+1)...(t-s+k-1) \}$$
$$\{ (t-s+k) - (t-s-1) \}$$
$$= \frac{1}{k!} \{ (t-s)(t-s+1)...(t-s+k-1) \}$$
$$= \frac{(t-s)^{\overline{k}}}{k!}.$$
$$\square$$

Proof. [Proof of Lemma 4.1] We see that

$$\hat{h}_0(t,s) = \frac{(t-s)^{\overline{0}}}{0!} = 1, \quad \forall s, t \in (a + \mathbb{Z}).$$

Assume (4.5) is correct. Then

$$\hat{h}_{k+1}(t,s) = \int_s^t \hat{h}_k(\tau, s)\nabla\tau = \int_s^t \frac{(\tau-s)^{\overline{k}}}{k!}\nabla\tau$$

$$= \frac{(t-s)^{\overline{k+1}}}{(k+1)!}.$$

$\square$

Again here we consider $f : (a + \mathbb{Z}) \to \mathbb{R}$ and let $m \in \mathbb{N}$ By nabla time scales Taylor formula, see [3] and Lemma 4.1, applied to $(a + \mathbb{Z})$ we obtain the Taylor formula:

$$f(t) = \sum_{k=0}^{m-1} \nabla^k f(s)\frac{(t-s)^{\overline{k}}}{k!} + \int_s^t \frac{(t-\tau+1)^{\overline{m-1}}}{(m-1)!}\nabla^m f(\tau)\nabla\tau, \qquad (4.7)$$

$\forall t, s \in (a + \mathbb{Z})$. Here $\nabla^0 f = f$.

We specialize (4.7) for $t \in [a^*, b^*] \cap (a + \mathbb{Z})$ and $s = b^* \in (a + \mathbb{Z})$, $a^* \in (a + \mathbb{Z})$, where $a^* < b^*$.

Hence we have

$$f(t) = \sum_{k=0}^{m-1} \nabla^k f(b^*)\frac{(t-b^*)^{\overline{k}}}{k!} + \int_{b^*}^t \frac{(t-\tau+1)^{\overline{m-1}}}{(m-1)!}\nabla^m f(\tau)\nabla\tau. \qquad (4.8)$$

That is

$$f(t) = \sum_{k=0}^{m-1} \nabla^k f(b^*)\frac{(t-b^*)^{\overline{k}}}{k!} - \int_t^{b^*} \frac{(t-\tau+1)^{\overline{m-1}}}{(m-1)!}\nabla^m f(\tau)\nabla\tau. \qquad (4.9)$$

Using (4.2) we can write (4.9) as follows

$$f(t) = \sum_{k=0}^{m-1} \nabla^k f(b^*)\frac{(t-b^*)^{\overline{k}}}{k!} - \frac{1}{(m-1)!}\sum_{s=t+1}^{b^*}(t-s+1)^{\overline{m-1}}\nabla^m f(s). \qquad (4.10)$$

We call the remainder of (4.10) as

$$R^*(t) := -\frac{1}{(m-1)!}\sum_{s=t+1}^{b^*}(t-s+1)^{\overline{m-1}}\nabla^m f(s), \qquad (4.11)$$

$\forall t \in [a^*, b^*) \cap (a + \mathbb{Z})$.

We need

Lemma 4.3. *It holds*

$$(t-s+1)^{\overline{m-1}} = (-1)^{m-1}(s-t-m+1)^{\overline{m-1}}. \qquad (4.12)$$

Proof. Notice that

$$
\begin{aligned}
(t-s+1)^{\overline{m-1}} &= (t-s+1)(t-s+1+1)(t-s+1+2) \\
&\quad \cdots (t-s+1+m-1-1) \\
&= (-(s-t)+1)(-(s-t)+2)(-(s-t)+3) \\
&\quad \cdots (-(s-t)+m-1) \\
&= (-1)^{m-1}(s-t-1)(s-t-2)(s-t-3) \\
&\quad \cdots (s-t-m+1) \\
&= (-1)^{m-1}(s-t-m+1) \\
&\quad \cdots (s-t-3)(s-t-2)(s-t-1) \\
&= (-1)^{m-1}(s-t-(m-1))^{\overline{m-1}} \\
&= (-1)^{m-1}(s-t-m+1)^{\overline{m-1}}.
\end{aligned}
$$

Indeed we have

$$
\begin{aligned}
(t-s-(m-1))^{\overline{m-1}} &= (s-t-m+1)(s-t-m+1+1)(s-t-m+1+2) \\
&\quad \cdots (s-t-m+1+m-2) \\
&= (s-t-m+1)(s-t-m+2)(s-t-m+3) \\
&\quad \cdots (s-t-3)(s-t-2)(s-t-1).
\end{aligned}
$$

$\square$

Because on (4.12) we obtain

$$
R^*(t) = \frac{(-1)^m}{(m-1)!} \sum_{s=t+1}^{b^*} (s-t-m+1)^{\overline{m-1}} \nabla^m f(s) \tag{4.13}
$$

$$
= \frac{(-1)^m}{(m-1)!} \sum_{s=t+1}^{b^*} (s-m-t+1)^{\overline{m-1}} \nabla^m f(s),
$$

$\forall t \in [a^*, b^*] \cap (a + \mathbb{Z})$. Here $s \geqslant t+1$, that is $s-t \geqslant 1$, and $b^* = (t+1)\bmod(1)$.

Remark 4.1. We notice that

$$
\begin{aligned}
(\lambda-m+1)^{\overline{m-1}} &= (\lambda+1-m)^{\overline{m-1}} \\
&= (\lambda+1-m)(\lambda+1-m+1)(\lambda+1-m+2)(\lambda+1-m+3) \\
&\quad \cdots (\lambda+1-m+m-3)(\lambda+1-m+m-2) \\
&= (\lambda-(m-1))(\lambda-(m-2))(\lambda-(m-3))(\lambda-(m-4)) \\
&\quad \cdots (\lambda-2)(\lambda-1),
\end{aligned}
$$

for any $\lambda \in \mathbb{N}$
So, when $\lambda = 1, 2, \ldots, m-1$ we get that

$$
(\lambda-m+1)^{\overline{m-1}} = 0.
$$

Therefore

$$\sum_{s=t+1}^{t+m-1} (s-m-t+1)^{\overline{m-1}} \nabla^m f(s) = 0, \qquad (4.14)$$

since $s - t = 1, 2, \ldots, m - 1$.
Conclusion, we get that

$$R^*(t) = \frac{(-1)^m}{(m-1)!} \sum_{s=t+m}^{b^*} (s-(m+t)+1)^{\overline{m-1}} \nabla^m f(s), \qquad (4.15)$$

which requires $t \leqslant b^* - m$. Since

$$(s-m-t+1)^{\overline{m-1}} = (s-m-t+1+m-2)^{(m-1)} \qquad (4.16)$$
$$= (s-t-1)^{(m-1)},$$

we can rewrite (4.15) as,

$$R^*(t) = \frac{(-1)^m}{(m-1)!} \sum_{s=t+m}^{b^*} (s-t-1)^{(m-1)} \nabla^m f(s), \qquad (4.17)$$

where $t \in [a^*, b^* - m] \cap (a + \mathbb{Z})$, $m \in \mathbb{N}$. Here we restricted f on $[a^*, b^*] \cap (a + \mathbb{Z})$, which implies that $\nabla^m f$ is restricted on $[a^* + m, b^*] \cap (a + \mathbb{Z})$.

We need

Definition 4.1. (see [4]) Let $\nu > 0$, the right fractional sum is given by

$$(\Delta_{b^*}^{-\nu} f)(t) := \frac{1}{\Gamma(\nu)} \sum_{s=t+\nu}^{b^*} (s-t-1)^{(\nu-1)} f(s), \qquad (4.18)$$

$$(\Delta_{b^*}^0 f)(t) := f(t),$$

where f is restricted on $[a^*, b^*] \cap (a + \mathbb{Z})$.
Here $(\Delta_{b^*}^{-\nu} f)$ is defined on

$$\{a^* - \nu, a^* - \nu + 1, a^* - \nu + 2, \ldots, b^* - \nu\}.$$

Here one can take $a^* = -\infty$.

We also need

Theorem 4.1. (see [6]) Let $\mu, \nu \geqslant 0$. Then

$$(\Delta_{b^*-\nu}^{-\mu} \Delta_{b^*}^{-\nu} f)(t) = (\Delta_{b^*}^{-(\mu+\nu)} f)(t), \qquad (4.19)$$

where $t \in \{a^* - (\mu+\nu), a^* - (\mu+\nu) + 1, \ldots, b^* - (\mu+\nu)\}$.

We have established

Theorem 4.2. *It holds*

$$R^*(t) = (-1)^m (\Delta_{b^*}^{-m} (\nabla^m f))(t), \qquad (4.20)$$

where $t \in \{a^*, a^* + 1, \ldots, b^* - m\}$, $a^* \leqslant b^* - m$, when f is restricted on $[a^*, b^*] \cap (a + \mathbb{Z})$, $a^*, b^* \in (a + \mathbb{Z})$.

4.2   Main Results

We give

Definition 4.2. Let $\mu > 0$, $m - 1 < \mu \leq m$, $m \in \mathbb{N}$, $m = \lceil \mu \rceil$ (ceiling of number), $\nu := m - \mu$, that is $\mu + \nu = m$.
The $\mu - th$ order nabla right fractional difference (Caputo way) is given by

$$(\nabla_{b^*-}^{\mu} f)(t) : = (-1)^m (\Delta_{b^*}^{-\nu} (\nabla^m f))(t) \tag{4.21}$$

$$= \frac{(-1)^m}{\Gamma(\nu)} \sum_{s=t+\nu}^{b^*} (s - t - 1)^{(\nu-1)} (\nabla^m f)(s),$$

where $t \leq b^* - \nu$, $b^* \in (a + \mathbb{Z})$, $t \in (a - \nu + \mathbb{Z})$, $a \in \mathbb{R}$

If $\mu = m \in \mathbb{N}$, then

$$(\nabla_{b^*-}^{\mu} f)(t) = (-1)^m (\nabla^m f)(t).$$

Remark 4.2. If we restrict $f : (a + \mathbb{Z}) \to \mathbb{R}$ on $[a^* - m, b^*]$, where $a^*, b^* \in (a + \mathbb{Z})$, $a^* < b^*$, $m \in \mathbb{N}$, then $\nabla^m f$ is defined on $[a^*, b^*] \cap (a + \mathbb{Z})$, and then $\nabla_{b^*-}^{\mu} f$ is defined on $\{a^* - \nu, a^* - \nu + 1, \ldots, b^* - \nu\}$.

We give

Theorem 4.3. Let $a^*, b^* \in (a + \mathbb{Z})$, $a^* \leq b^* - m$, and $t \in \{a^*, a^* + 1, \ldots, b^* - m\}$. Then

$$R^*(t) = \left(\Delta_{b^*-\nu}^{-\mu} \left(\nabla_{b^*-}^{\mu} f \right) \right)(t), \tag{4.22}$$

where f is restricted on $[a^*, b^*] \cap (a + \mathbb{Z})$, and $\mu > 0$, $m - 1 < \mu \leq m$, $m \in \mathbb{N}$

Proof.   Let $m - 1 < \mu < m$. With the help of (4.19) and (4.20), we see that

$$\left(\Delta_{b^*-\nu}^{-\mu} \left(\nabla_{b^*-}^{\mu} f \right) \right)(t) = (-1)^m \left(\Delta_{b^*-\nu}^{-\mu} \Delta_{b^*}^{-\nu} \left(\nabla^m f \right) \right)(t)$$

$$= (-1)^m \left(\Delta_{b^*}^{-(\mu+\nu)} \left(\nabla^m f \right) \right)(t)$$

$$= (-1)^m \left(\Delta_{b^*}^{-m} \left(\nabla^m f \right) \right)(t)$$

$$= R^*(t).$$

If $\mu = m$, then (4.22) is trivial.   □

We have proved the following nabla right discrete fractional Taylor formula.

Theorem 4.4. Let $a \in \mathbb{R}$, $f : (a + \mathbb{Z}) \to \mathbb{R}$. Let $\mu > 0$, $m - 1 < \mu \leq m$, $(m = \lceil \mu \rceil)$, $m \in \mathbb{N}$, $\nu = m - \mu$. Let $a^*, b^* \in (a + \mathbb{Z}) : a^* \leq b^* - m$. Here a^* could be $-\infty$. Let $t \in \{a^*, a^* + 1, \ldots, b^* - m\}$. Then

$$f(t) = \sum_{k=0}^{m-1} \nabla^k f(b^*) \frac{(t - b^*)^{\overline{k}}}{k!} + R^*(t), \tag{4.23}$$

where the remainder

$$R^*(t) = \left(\Delta_{b^*-\nu}^{-\mu} \left(\nabla_{b^*-}^{\mu} f \right) \right)(t) \tag{4.24}$$

$$= \frac{1}{\Gamma(\mu)} \sum_{s=t+\mu}^{b^*-\nu} (s-t-1)^{(\mu-1)} (\nabla_{b^*-}^{\mu} f)(s).$$

Remark 4.3. Using (4.1) we get that

$$(s-t-1)^{(\mu-1)} = (s-t-\mu+1)^{\overline{\mu-1}}. \tag{4.25}$$

Therefore by (4.24) we also derive

$$R^*(t) = \frac{1}{\Gamma(\mu)} \sum_{s=t+\mu}^{b^*-\nu} (s-t-\mu+1)^{\overline{\mu-1}} (\nabla_{b^*-}^{\mu} f)(s). \tag{4.26}$$

Corollary 4.1. *Same assumptions as in Theorem 4.11. Assume further that*

$$\nabla^k f(b^*) = 0, \ for \ k = 0,1,\dots,m-1.$$

Then

$$f(t) = \frac{1}{\Gamma(\mu)} \sum_{s=t+\mu}^{b^*-\nu} (s-t-1)^{(\mu-1)} (\nabla_{b^*-}^{\mu} f)(s). \tag{4.27}$$

We need

Proposition 4.1. *It holds*

$$\sum_{s=t+\mu}^{b^*-\nu} (s-t-1)^{(\mu-1)} = \frac{(b^*-\nu-t)^{(\mu)}}{\mu} > 0. \tag{4.28}$$

Proof. We proved in [1], p. 579 that,

$$\frac{\Gamma(x+1)}{\Gamma(x-k+1)} = \frac{1}{(k+1)} \left(\frac{\Gamma(x+2)}{\Gamma(x-k+1)} - \frac{\Gamma(x+1)}{\Gamma(x-k)} \right),$$

$x > k, \ x, k \in \mathbb{R}; \ k > -1, \ x > -1.$

We have that

$$\sum_{s=t+\mu}^{b^*-\nu} (s-t-1)^{(\mu-1)} = \sum_{s=t+\mu}^{b^*-\nu} \frac{\Gamma(s-t)}{\Gamma(s-t-\mu+1)}$$

$$= \sum_{s=t+\mu+1}^{b^*-\nu} \frac{\Gamma(s-t)}{\Gamma(s-t-\mu+1)} + \Gamma(\mu).$$

Here $\mu > 0$ and $t + \mu + 1 \leqslant s \leqslant b^* - \nu$. See that $x := s - t - 1 \geqslant \mu$, $k := \mu - 1 > -1$, $x > k$, and $x - k = s - t - \mu$. Therefore

$$\frac{\Gamma(s-t)}{\Gamma(s-t-\mu+1)} = \frac{\Gamma(x+1)}{\Gamma(x+1-k)} = \frac{\Gamma(x+1)}{\Gamma(x-k+1)}$$

$$= \frac{1}{\mu}\left(\frac{\Gamma(x+2)}{\Gamma(x-k+1)} - \frac{\Gamma(x+1)}{\Gamma(x-k)}\right)$$

$$= \frac{1}{\mu}\left(\frac{\Gamma(s-t+1)}{\Gamma(s-t-\mu+1)} - \frac{\Gamma(s-t)}{\Gamma(s-t-\mu)}\right).$$

Consequently we have

$$\sum_{s=t+\mu+1}^{b^*-\nu} \frac{\Gamma(s-t)}{\Gamma(s-t-\mu+1)} = \frac{1}{\mu}\sum_{s=t+\mu+1}^{b^*-\nu}\left(\frac{\Gamma(s-t+1)}{\Gamma(s-t+1-\mu)} - \frac{\Gamma(s-t)}{\Gamma(s-t-\mu)}\right)$$

$$(s-t \geqslant \mu+1) \quad \text{(telescoping sum)}$$

$$= \frac{1}{\mu}\big[(\Gamma(\mu+2) - \Gamma(\mu+1))$$

$$+ \left(\frac{\Gamma(\mu+3)}{2} - \Gamma(\mu+2)\right)$$

$$+ \left(\frac{\Gamma(\mu+4)}{3!} - \frac{\Gamma(\mu+3)}{2!}\right)$$

$$+ \left(\frac{\Gamma(\mu+5)}{4!} - \frac{\Gamma(\mu+4)}{3!}\right)$$

$$\text{(so for } s = b^* - \nu = t+\mu+1+\lambda, \; \lambda \in \mathbb{N}, \text{ that is } s-t = \mu+1+\lambda)$$

$$+ \cdots + \frac{\Gamma(\mu+1+\lambda+1)}{\Gamma(2+\lambda)} - \frac{\Gamma(\mu+1+\lambda)}{\Gamma(1+\lambda)}\Big]$$

$$= \frac{\Gamma(\mu+\lambda+2)}{\mu\Gamma(2+\lambda)} - \frac{\Gamma(\mu+1)}{\mu}$$

$$\text{(By } b^* - \nu = t+\mu+1+\lambda, \text{ then } b^* - \nu + 1 = t+\mu+2+\lambda \text{ and}$$

$$b^* - \nu + 1 - t = \mu+2+\lambda, \text{ that is } \lambda + 2 = b^* - \nu + 1 - t - \mu.$$

$$\text{Also } \frac{\Gamma(\mu+1)}{\mu} = \Gamma(\mu).)$$

$$= \frac{\Gamma(b^* - \nu - t + 1)}{\mu\Gamma(b^* - \nu - t + 1 - \mu)} - \Gamma(\mu)$$

$$= \frac{(b^* - \nu - t)^{(\mu)}}{\mu} - \Gamma(\mu).$$

So we have proved

$$\sum_{s=t+\mu}^{b^*-\nu} (s-t-1)^{(\mu-1)} = \frac{(b^* - \nu - t)^{(\mu)}}{\mu}.$$

The last is positive because

$$b^* - \nu - t - \mu - 1 \geqslant 0$$

and

$$b^* - \nu - t - \mu + 1 \geqslant 2,$$

also $b^* - \nu - t + 1 \geqslant \mu + 2 > 0$, and $b^* - \nu - t > 0$. $\qquad\qquad\qquad\square$

We give the estimate

Theorem 4.5. *In the assumptions of Theorem 4.11 we have that*

$$\left| f(t) - \sum_{k=0}^{m-1} \nabla^k f(b^*) \frac{(t-b^*)^{\overline{k}}}{k!} \right| \tag{4.29}$$

$$= |R^*(t)| \leqslant \frac{(b^* - \nu - t)^{(\mu)}}{\Gamma(\mu+1)} \max_{s \in \{t+\mu,\dots,b^*-\nu\}} \left| \nabla_{b^*-}^{\mu} f(s) \right|.$$

Proof. By (4.24) and (4.28) we get

$$|R^*(t)| \leqslant \frac{1}{\Gamma(\mu)} \left| \sum_{s=t+\mu}^{b^*-\nu} (s-t-1)^{(\mu-1)} \right| \max_{s \in \{t+\mu,\dots,b^*-\nu\}} \left| \nabla_{b^*-}^{\mu} f(s) \right|$$

$$= \frac{1}{\Gamma(\mu)} \frac{(b^*-\nu-t)^{(\mu)}}{\mu} \max_{s \in \{t+\mu,\dots,b^*-\nu\}} \left| \nabla_{b^*-}^{\mu} f(s) \right|$$

$$= \frac{(b^*-\nu-t)^{(\mu)}}{\Gamma(\mu+1)} \max_{s \in \{t+\mu,\dots,b^*-\nu\}} \left| \nabla_{b^*-}^{\mu} f(s) \right|.$$

$\qquad\qquad\qquad\square$

We need

Lemma 4.4. *([1], p. 580) Let $a > \nu$, $a, \nu > -1$, $a, \nu \in \mathbb{R}$, $a \leqslant b$. Then*

$$\sum_{r=a}^{b} r^{(\nu)} = \left(\frac{(b+1)^{(\nu+1)} - a^{(\nu+1)}}{\nu+1} \right). \tag{4.30}$$

We give a related Ostrowski inequality.

Theorem 4.6. *Let $a \in \mathbb{R}$, $f : (a + \mathbb{Z}) \to \mathbb{R}$. Let $\mu > 0, m - 1 < \mu \leqslant m, m \in \mathbb{N}$, $\nu = m - \mu$. Let $a^*, b^* \in (a + \mathbb{Z}) : b^* - a^* \geqslant m + 1$. Assume that $\nabla^k f(b^*) = 0$, $k = 1, \dots, m - 1$. Then*

$$\left| \frac{1}{(b^* - m - a^* + 1)} \sum_{t=a^*}^{b^*-m} f(t) - f(b^*) \right| \tag{4.31}$$

$$\leqslant \frac{(b^* - \nu - a^* + 1)^{(\mu+1)}}{\Gamma(\mu+2)(b^* - m - a^* + 1)} \left(\max_{s \in \{a^*+\mu,\dots,b^*-\nu\}} \left| \nabla_{b^*-}^{\mu} f(s) \right| \right).$$

Proof. Using (4.23) and (4.24), since $\nabla^k f(b^*) = 0$, $k = 1, \dots, m - 1$, we get

$$f(t) - f(b^*) = R^*(t).$$

Then we observe that

$$
E \; := \; \left| \frac{1}{(b^* - m - a^* + 1)} \sum_{t=a^*}^{b^*-m} f(t) - f(b^*) \right|
$$

$$
= \frac{1}{(b^* - m - a^* + 1)} \left| \sum_{t=a^*}^{b^*-m} (f(t) - f(b^*)) \right|
$$

$$
\leqslant \frac{1}{(b^* - m - a^* + 1)} \left(\sum_{t=a^*}^{b^*-m} |f(t) - f(b^*)| \right)
$$

$$
= \frac{1}{(b^* - m - a^* + 1)} \sum_{t=a^*}^{b^*-m} |R^*(t)|
$$

(by (4.29))

$$
\leqslant \frac{1}{(b^* - m - a^* + 1)} \left(\sum_{t=a^*}^{b^*-m} (b^* - \nu - t)^{(\mu)} \right) \frac{\max_{s \in \{a^* + \mu, \ldots, b^* - \nu\}} \left| \nabla_{b^*-}^{\mu} f(s) \right|}{\Gamma(\mu + 1)}
$$

$$
=: (*)
$$

Call $\tau := b^* - \nu - t$. Here $a^* \leqslant t \leqslant b^* - m$, and $a^* \geqslant -t \geqslant m - b^*$, which implies that $\mu \leqslant \tau \leqslant b^* - \nu - a^*$. Also $\mu^{(\mu)} = \Gamma(\mu + 1)$.

We calculate

$$
\sum_{t=a^*}^{b^*-m} (b^* - \nu - t)^{(\mu)} = \sum_{\tau=\mu}^{b^*-\nu-a^*} \tau^{(\mu)}
$$

$$
= \mu^{(\mu)} + \sum_{\tau=\mu+1}^{b^*-\nu-a^*} \tau^{(\mu)}
$$

$$
= \Gamma(\mu + 1) + \sum_{\tau=\mu+1}^{b^*-\nu-a^*} \tau^{(\mu)}
$$

(notice $b^* - a^* \geqslant m + 1$ by assumption, hence $\mu + 1 \leqslant b^* - \nu - a^*$)

(by (4.30))

$$
= \Gamma(\mu + 1) + \left(\frac{(b^* - \nu - a^* + 1)^{(\mu+1)} - (\mu + 1)^{(\mu+1)}}{\mu + 1} \right)
$$

$$
= \Gamma(\mu + 1) + \left(\frac{(b^* - \nu - a^* + 1)^{(\mu+1)}}{\mu + 1} - \frac{\Gamma(\mu + 2)}{\mu + 1} \right)
$$

$$
= \Gamma(\mu + 1) + \frac{(b^* - \nu - a^* + 1)^{(\mu+1)}}{\mu + 1} - \Gamma(\mu + 1)
$$

$$
= \frac{(b^* - \nu - a^* + 1)^{(\mu+1)}}{\mu + 1}.
$$

So we have proved that

$$\sum_{t=a^*}^{b^*-m} (b^* - \nu - t)^{(\mu)} = \frac{(b^* - \nu - a^* + 1)^{(\mu+1)}}{\mu + 1}.$$

Therefore

$$(*) = \frac{1}{(b^* - m - a^* + 1)} \left(\frac{(b^* - \nu - a^* + 1)^{(\mu+1)}}{\mu + 1} \right)$$

$$\frac{\max_{s \in \{a^*+\mu, \ldots, b^*-\nu\}} \left| \nabla_{b^*-}^{\mu} f(s) \right|}{\Gamma(\mu + 1)}.$$

That is proving

$$E \leqslant \frac{(b^* - \nu - a^* + 1)^{(\mu+1)} \max_{s \in \{a^*+\mu, \ldots, b^*-\nu\}} \left| \nabla_{b^*-}^{\mu} f(s) \right|}{\Gamma(\mu + 2)(b^* - m - a^* + 1)},$$

which establishes the theorem. □

A related Poincaré type inequality follows

Theorem 4.7. *Here all are as in Theorem 4.11. Assume*

$$\nabla^k f(b^*) = 0, k = 0, 1, \ldots, m - 1.$$

Let $p, q > 1 : \frac{1}{p} + \frac{1}{q} = 1$. Then

$$\sum_{t=a^*}^{b^*-m} |f(t)|^q \leqslant \frac{1}{(\Gamma(\mu))^q} \tag{4.32}$$

$$\cdot \left(\sum_{t=a^*}^{b^*-m} \left(\left(\sum_{s=t+\mu}^{b^*-\nu} (s - t - 1)^{p(\mu-1)} \right)^{\frac{q}{p}} \right) \right)$$

$$\cdot \left(\sum_{s=a^*+\mu}^{b^*-\nu} \left| \nabla_{b^*-}^{\mu} f(s) \right|^q \right).$$

Proof. Notice here $s - t \geqslant \mu > 0$ and $s - t - \mu \geqslant 0$ and $s - t - \mu + 1 > 0$, so that

$$(s - t - 1)^{(\mu-1)} = \frac{\Gamma(s - t)}{\Gamma(s - t - \mu + 1)} > 0.$$

By (4.27) we get

$$|f(t)| \leqslant \frac{1}{\Gamma(\mu)} \sum_{s=t+\mu}^{b^*-\nu} (s - t - 1)^{(\mu-1)} \left| (\nabla_{b^*-}^{\mu} f)(s) \right|$$

$$\leqslant \frac{1}{\Gamma(\mu)} \left(\sum_{s=t+\mu}^{b^*-\nu} (s - t - 1)^{p(\mu-1)} \right)^{\frac{1}{p}} \left(\sum_{s=t+\mu}^{b^*-\nu} \left| (\nabla_{b^*-}^{\mu} f)(s) \right|^q \right)^{\frac{1}{q}}$$

$$\leqslant \frac{1}{\Gamma(\mu)} \left(\sum_{s=t+\mu}^{b^*-\nu} (s - t - 1)^{p(\mu-1)} \right)^{\frac{1}{p}} \left(\sum_{s=a^*+\mu}^{b^*-\nu} \left| (\nabla_{b^*-}^{\mu} f)(s) \right|^q \right)^{\frac{1}{q}}.$$

Therefore we found

$$|f(t)| \leqslant \frac{1}{\Gamma(\mu)} \left(\sum_{s=t+\mu}^{b^*-\nu} (s-t-1)^{p(\mu-1)} \right)^{\frac{1}{p}} \left(\sum_{s=a^*+\mu}^{b^*-\nu} |(\nabla_{b^*-}^\mu f)(s)|^q \right)^{\frac{1}{q}} \quad (4.33)$$

Hence

$$|f(t)|^q \leqslant \frac{1}{(\Gamma(\mu))^q} \left(\sum_{s=t+\mu}^{b^*-\nu} (s-t-1)^{p(\mu-1)} \right)^{\frac{q}{p}} \left(\sum_{s=a^*+\mu}^{b^*-\nu} |(\nabla_{b^*-}^\mu f)(s)|^q \right). \quad (4.34)$$

Applying $\sum_{t=a^*}^{b^*-m}$ to both sides of (4.34) we derive (4.32). $\qquad\square$

We finish with a related Sobolev type inequality.

Theorem 4.8. *All as in Theorem 4.7, and $r \geqslant 1$. Then*

$$\left(\sum_{t=a^*}^{b^*-m} |f(t)|^r \right)^{\frac{1}{r}} \leqslant \frac{1}{\Gamma(\mu)} \quad (4.35)$$

$$\cdot \left(\sum_{t=a^*}^{b^*-m} \left(\left(\sum_{s=t+\mu}^{b^*-\nu} (s-t-1)^{p(\mu-1)} \right)^{\frac{r}{p}} \right) \right)^{\frac{1}{r}}$$

$$\cdot \left(\sum_{s=a^*+\mu}^{b^*-\nu} |\nabla_{b^*-}^\mu f(s)|^q \right)^{\frac{1}{q}}.$$

Proof. Raising (4.33) to the power r we obtain

$$|f(t)|^r \leqslant \frac{1}{(\Gamma(\mu))^r} \left(\sum_{s=t+\mu}^{b^*-\nu} (s-t-1)^{p(\mu-1)} \right)^{\frac{r}{p}} \left(\sum_{s=a^*+\mu}^{b^*-\nu} |(\nabla_{b^*-}^\mu f)(s)|^q \right)^{\frac{r}{q}}.$$

Hence it holds

$$\sum_{t=a^*}^{b^*-m} |f(t)|^r \leqslant \frac{1}{(\Gamma(\mu))^r} \quad (4.36)$$

$$\cdot \left(\sum_{t=a^*}^{b^*-m} \left(\sum_{s=t+\mu}^{b^*-\nu} (s-t-1)^{p(\mu-1)} \right)^{\frac{r}{p}} \right)$$

$$\cdot \left(\sum_{s=a^*+\mu}^{b^*-\nu} |(\nabla_{b^*-}^\mu f)(s)|^q \right)^{\frac{r}{q}}.$$

Raising (4.36) to the power $\frac{1}{r}$ we derive (4.35). $\qquad\square$

Bibliography

1. G.A. Anastassiou, *Intelligent Mathematics: Computational Analysis*, Springer Heidelberg, 2011.
2. G.A. Anastassiou, *Right nabla discrete fractional calculus*, Intern. J. Difference Equations 6 (2) (2011), 91-104.
3. D.R. Anderson, *Taylor polynomials for nabla dynamic equations on time scales*, Panamer. Math. J. 12 (4) (2002), 17-27.
4. N. Bastos, R. Ferreira, D. Torres, *Discrete-time fractional variational problems*, Signal Processing 91 (2011), 513-524.
5. M. Bohner, A. Peterson, *Dynamic Equations on Time Scales: An Introduction with Applications*, Birkhaüser, Boston, 2011.
6. R. Ferreira, D. Torres, *Fractional h-difference equations arising from the calculus of variations*, Applicable Anal. Discrete Math. 5 (2011), 110-121.

Chapter 5

Representations and Ostrowski Inequalities over Time Scales

Here we give univariate and multivariate representations of Montgomery type for hybrid functions on Time scales. Based on these we establish univariate and multivariate Ostrowski type inequalities on Time scales domains. These compare the average of a function to its values. The estimates involve the higher order delta and nabla derivatives and partial derivatives. We finish with applications on the time scales $\mathbb{R}$ and $\mathbb{Z}$. It follows [11].

5.1 Introduction

In 1938, A. Ostrowski [28] proved the following important inequality:

Theorem 5.1. *Let $f : [a, b] \to \mathbb{R}$ be continuous on $[a, b]$ and differentiable on (a, b) whose derivative $f' : (a, b) \to \mathbb{R}$ is bounded on (a, b), i.e. $\|f\|'_\infty := \sup\limits_{t \in (a,b)} |f'(t)| < \infty$. Then*

$$\left| \frac{1}{b-a} \int_a^b f(t)\, dt - f(x) \right| \leq \left[\frac{1}{4} + \frac{\left(x - \frac{a+b}{2}\right)^2}{(b-a)^2} \right] (b-a) \|f'\|_\infty,$$

for any $x \in [a, b]$. The constant $\frac{1}{4}$ is the best possible.

Since then there has been a lot of activity around these inequalities with important applications to Numerical Analysis and Probability.

This chapter is also greatly motivated by the following result:

Theorem 5.2. *(see [3]) Let $f \in C^1 \left(\prod_{i=1}^k [a_i, b_i] \right)$, where $a_i < b_i$; $a_i, b_i \in \mathbb{R}$, $i = 1, \ldots, k$, and let $\overrightarrow{x_0} := (x_{01}, \ldots, x_{0k}) \in \prod_{i=1}^k [a_i, b_i]$ be fixed. Then*

$$\left| \frac{1}{\prod_{i=1}^k (b_i - a_i)} \int_{a_1}^{b_1} \cdots \int_{a_i}^{b_i} \cdots \int_{a_k}^{b_k} f(z_1, \ldots, z_k)\, dz_1 ... dz_k - f(\overrightarrow{x_0}) \right|$$

$$\leq \sum_{i=1}^k \left(\frac{(x_{0i} - a_i)^2 + (b_i - x_{0i})^2}{2(b_i - a_i)} \right) \left\| \frac{\partial f}{\partial z_i} \right\|_\infty.$$

The last inequality is sharp, the optimal function is

$$f^*\left(z_1, \ldots, z_k\right) := \sum_{i=1}^{k} |z_i - x_{0i}|^{\alpha_i}, \quad \alpha_i > 1.$$

We are also motivated by [4], [5], [6].

In this chapter we establish univariate and multivariate Montgomery identities on Time Scales, then based on these we prove univariate and multivariate Ostrowski type inequalities on Time Scales boxes. These involve the high order time scales derivatives and partial derivatives of the engaged function. The estimates are with respect to the norms $\|\cdot\|_p$, $1 \leq p \leq \infty$. We also give applications. Our studied functions are hybrid, in the sense that each variable of the function can be continuous or discrete or both, and these variables may come from various general time scales. To keep the chapter of limited size, for basics on Time Scales we refer the reader to [1], [2], [7], [10], [13], [14], [15], [18], [19], [20], [22], [23], [24], [25], [26], [27].

5.2   Main Results

We need the following Montgomery identities.

Proposition 5.1. *([21]) Let $a, b, s, t \in \mathbb{T}$, $a < b$ and $f \in C^1_{rd}\left([a, b]\right)$. Set*

$$p\left(t, s\right) := \begin{cases} s - a, & a \leq s < t, \\ s - b, & t \leq s \leq b. \end{cases}$$

Then

$$f\left(t\right) = \frac{1}{b-a} \int_a^b f\left(\sigma\left(t\right)\right) \Delta s + \frac{1}{b-a} \int_a^b p\left(t, s\right) f^\Delta\left(s\right) \Delta s, \quad \forall\, t \in [a, b] \cap \mathbb{T}. \quad (5.1)$$

Proposition 5.2. *Let $a, b, x, t \in \mathbb{T}$, $a < b$ and $f \in C^1_{ld}\left([a, b]\right)$. Set*

$$p^*\left(x, t\right) := \begin{cases} t - a, & a \leq t \leq x, \\ t - b, & x < t \leq b. \end{cases}$$

Then

$$f\left(x\right) = \frac{1}{b-a} \int_a^b f\left(\rho\left(t\right)\right) \nabla t + \frac{1}{b-a} \int_a^b p^*\left(x, t\right) f^\nabla\left(t\right) \nabla t, \quad \forall\, x \in [a, b] \cap \mathbb{T}. \quad (5.2)$$

Proof.   We apply integration by parts twice. We have

$$\int_a^x \left(t - a\right) f^\nabla\left(t\right) \nabla t = \left(t - a\right) f\left(t\right) \big|_a^x - \int_a^x f\left(\rho\left(t\right)\right) \left(t - a\right)^\nabla \nabla t$$

$$= \left(x - a\right) f\left(x\right) - \int_a^x f\left(\rho\left(t\right)\right) \nabla t,$$

and

$$\int_x^b \left(t - b\right) f^\nabla\left(t\right) \nabla t = \left(t - b\right) f\left(t\right) \big|_x^b - \int_x^b f\left(\rho\left(t\right)\right) \left(t - b\right)^\nabla \nabla t$$

$$= (b - x) f(x) - \int_x^b f(\rho(t)) \nabla t.$$

Therefore by adding the last two identities we get

$$\int_a^x (t - a) f^{\nabla}(t) \nabla t + \int_x^b (t - b) f^{\nabla}(t) \nabla t$$

$$= (b - a) f(x) - \int_a^b f(\rho(t)) \nabla t,$$

and

$$f(x) - \frac{1}{b-a} \int_a^b f(\rho(t)) \nabla t$$

$$= \frac{1}{b-a} \left[\int_a^x (t - a) f^{\nabla}(t) \nabla t + \int_x^b (t - b) f^{\nabla}(t) \nabla t \right].$$

Consequently we obtain

$$f(x) - \frac{1}{b-a} \int_a^b f(\rho(t)) \nabla t = \frac{1}{b-a} \int_a^b p^*(x,t) f^{\nabla}(t) \nabla t,$$

proving the claim. $\qquad\qquad\square$

We present the representation result

Theorem 5.3. *Let* $a, b \in \mathbb{T}$, $a < b$ *and* $f \in C_{rd}^n([a,b])$, $n \in \mathbb{N}$. *Let* $x \in [a,b] \cap \mathbb{T}$. *Define the kernel*

$$P(r,s) := \begin{cases} \frac{s-a}{b-a}, & a \le s < r, \\ \frac{s-b}{b-a}, & r \le s \le b, \end{cases}$$

where $r, s \in [a,b] \cap \mathbb{T}$. *Then*

$$f(t) = \frac{1}{b-a} \int_a^b f(\sigma(s_1)) \Delta s_1 \tag{5.3}$$

$$+ \frac{1}{b-a} \left(\sum_{k=1}^{n-1} \int_{[a,b]^{k+1}} P(x,s_1) \prod_{i=1}^{k-1} P(s_i, s_{i+1}) f^{\Delta^k}(\sigma(s_{k+1})) \Delta s_{k+1} \Delta s_k ... \Delta s_1 \right)$$

$$+ \int_{[a,b]^n} P(x,s_1) \prod_{i=1}^{n-1} P(s_i, s_{i+1}) f^{\Delta^n}(s_n) \Delta s_n ... \Delta s_1.$$

We make the conventions $\sum_{k=1}^0 \cdot = 0$, $\prod_{i=1}^0 \cdot = 1$.

Proof. Here we use repeatedly Proposition 5.1. For basics see [22], p. 22, Theorem 1.60(iv).

We have by (5.1) that

$$f(x) = \frac{1}{b-a} \int_a^b f(\sigma(s_1)) \Delta s_1 + \int_a^b P(x, s_1) f^\Delta(s_1) \Delta s_1.$$

Doing the same for f^Δ we get

$$f^\Delta(s_1) = \frac{1}{b-a} \int_a^b f^\Delta(\sigma(s_2)) \Delta s_2 + \int_a^b P(s_1, s_2) f^{\Delta^2}(s_2) \Delta s_2.$$

Hence

$$f(x) = \frac{1}{b-a} \int_a^b f(\sigma(s_1)) \Delta s_1$$

$$+ \frac{1}{b-a} \int_a^b \left(\int_a^b P(x, s_1) f^\Delta(\sigma(s_2)) \Delta s_2 \right) \Delta s_1$$

$$+ \int_a^b \left(\int_a^b P(x, s_1) P(s_1, s_2) f^{\Delta^2}(s_2) \Delta s_2 \right) \Delta s_1.$$

We also have

$$f^{\Delta^2}(s_2) = \frac{1}{b-a} \int_a^b f^{\Delta^2}(\sigma(s_3)) \Delta s_3 + \int_a^b P(s_2, s_3) f^{\Delta^3}(s_3) \Delta s_3,$$

resulting to

$$f(x) = \frac{1}{b-a} \int_a^b f(\sigma(s_1)) \Delta s_1$$

$$+ \frac{1}{b-a} \int_a^b \left(\int_a^b P(x, s_1) f^\Delta(\sigma(s_2)) \Delta s_2 \right) \Delta s_1$$

$$+ \frac{1}{b-a} \int_{[a,b]^3} P(x, s_1) P(s_1, s_2) f^{\Delta^2}(\sigma(s_3)) \Delta s_3 \Delta s_2 \Delta s_1$$

$$+ \int_{[a,b]^3} P(x, s_1) P(s_1, s_2) P(s_2, s_3) f^{\Delta^3}(s_3) \Delta s_3 \Delta s_2 \Delta s_1.$$

Similarly it holds

$$f^{\Delta^3}(s_3) = \frac{1}{b-a} \int_a^b f^{\Delta^3}(\sigma(s_4)) \Delta s_4 + \int_a^b P(s_3, s_4) f^{\Delta^4}(s_4) \Delta s_4,$$

resulting to

$$f(x) = \frac{1}{b-a} \int_a^b f(\sigma(s_1)) \Delta s_1$$

$$+\frac{1}{b-a}\int_{[a,b]^2} P(x,s_1)\, f^\Delta(\sigma(s_2))\,\Delta s_2 \Delta s_1$$

$$+\frac{1}{b-a}\int_{[a,b]^3} P(x,s_1)\, P(s_1,s_2)\, f^{\Delta^2}(\sigma(s_3))\,\Delta s_3 \Delta s_2 \Delta s_1$$

$$+\frac{1}{b-a}\int_{[a,b]^4} P(x,s_1)\, P(s_1,s_2)\, P(s_2,s_3)\, f^{\Delta^3}(\sigma(s_4))\,\Delta s_4 \Delta s_3 \Delta s_2 \Delta s_1$$

$$+\int_{[a,b]^4} P(x,s_1)\, P(s_1,s_2)\, P(s_2,s_3)\, P(s_3,s_4)\, f^{\Delta^4}(s_4)\,\Delta s_4 \Delta s_3 \Delta s_2 \Delta s_1,$$

etc, proving the claim. $\qquad\square$

We also give

Theorem 5.4. *Let $a,b \in \mathbb{T}$, $a < b$ and $f \in C^n_{ld}([a,b])$, $n \in \mathbb{N}$. Let $x \in [a,b] \cap \mathbb{T}$. Define the kernel*

$$P^*(r,s) := \begin{cases} \frac{s-a}{b-a}, & a \le s \le r, \\ \frac{s-b}{b-a}, & r < s \le b, \end{cases}$$

where $r,s \in [a,b] \cap \mathbb{T}$. Then

$$f(x) = \frac{1}{b-a}\int_a^b f(\rho(s_1))\,\nabla s_1 \tag{5.4}$$

$$+\frac{1}{b-a}\left(\sum_{k=1}^{n-1}\int_{[a,b]^{k+1}} P^*(x,s_1)\prod_{i=1}^{k-1} P^*(s_i,s_{i+1})\, f^{\nabla^k}(\rho(s_{k+1}))\,\nabla s_{k+1}\nabla s_k...\nabla s_1\right)$$

$$+\int_{[a,b]^n} P^*(x,s_1)\prod_{i=1}^{n-1} P^*(s_i,s_{i+1})\, f^{\nabla^n}(s_n)\,\nabla s_n...\nabla s_1.$$

We make conventions $\sum_{k=1}^0 \cdot = 0$, $\prod_{i=1}^0 \cdot = 1$.

Proof. Similar to Theorem 5.3, using (5.2) repeatedly. $\qquad\square$

Next we give multivariate Δ-representations

Theorem 5.5. *Let the time scales $(\mathbb{T}_i, \sigma_i, \Delta_i)$, $i = 1,2,3$ and $a_i < b_i$; $a_i, b_i \in \mathbb{T}_i$, $i = 1,2,3$. Here σ_i, $i = 1,2,3$ are continuous.*
  We consider $f \in C^3\left(\prod_{i=1}^3 ([a_i,b_i] \cap \mathbb{T}_i)\right)$. Let $(x_1, x_2, x_3) \in \prod_{i=1}^3 ([a_i,b_i] \cap \mathbb{T}_i)$. Define the kernels $p_i : [a_i,b_i]^2 \to \mathbb{R}$:

$$p_i(x_i,s_i) := \begin{cases} s_i - a_i, & s_i \in [a_i,x_i), \\ s_i - b_i, & s_i \in [x_i,b_i], \end{cases}$$

for $i = 1, 2, 3.$

    Denote

$$\int_{\prod_{i=1}^{j} [a_i,b_i]} \cdot = \int_{\prod_{i=1}^{j} ([a_i,b_i] \cap \mathbb{T}_i)} \cdot ,$$

for $j = 1, 2, 3.$

    Then

$$f(x_1, x_2, x_3) = \frac{1}{\prod_{i=1}^{3} (b_i - a_i)}$$

$$\cdot \left\{ \int_{\prod_{i=1}^{3} [a_i,b_i]} f(\sigma_1(s_1), \sigma_2(s_2), \sigma_3(s_3)) \Delta_3 s_3 \Delta_2 s_2 \Delta_1 s_1 \right.$$

$$+ \int_{\prod_{i=1}^{3} [a_i,b_i]} p_1(x_1, s_1) \frac{\partial f(s_1, \sigma_2(s_2), \sigma_3(s_3))}{\Delta_1 s_1} \Delta_3 s_3 \Delta_2 s_2 \Delta_1 s_1$$

$$+ \int_{\prod_{i=1}^{3} [a_i,b_i]} p_2(x_2, s_2) \frac{\partial f(\sigma_1(s_1), s_2, \sigma_3(s_3))}{\Delta_2 s_2} \Delta_3 s_3 \Delta_2 s_2 \Delta_1 s_1$$

$$+ \int_{\prod_{i=1}^{3} [a_i,b_i]} p_3(x_3, s_3) \frac{\partial f(\sigma_1(s_1), \sigma_2(s_2), s_3)}{\Delta_3 s_3} \Delta_3 s_3 \Delta_2 s_2 \Delta_1 s_1 \qquad (5.5)$$

$$+ \int_{\prod_{i=1}^{3} [a_i,b_i]} p_1(x_1, s_1) p_2(x_2, s_2) \frac{\partial^2 f(s_1, s_2, \sigma_3(s_3))}{\Delta_2 s_2 \Delta_1 s_1} \Delta_3 s_3 \Delta_2 s_2 \Delta_1 s_1$$

$$+ \int_{\prod_{i=1}^{3} [a_i,b_i]} p_1(x_1, s_1) p_3(x_3, s_3) \frac{\partial^2 f(s_1, \sigma_2(s_2), s_3)}{\Delta_3 s_3 \Delta_1 s_1} \Delta_3 s_3 \Delta_2 s_2 \Delta_1 s_1$$

$$+ \int_{\prod_{i=1}^{3} [a_i,b_i]} p_2(x_2, s_2) p_3(x_3, s_3) \frac{\partial^2 f(\sigma_1(s_1), s_2, s_3)}{\Delta_3 s_3 \Delta_2 s_2} \Delta_3 s_3 \Delta_2 s_2 \Delta_1 s_1$$

$$+ \int_{\prod_{i=1}^{3} [a_i,b_i]} p_1(x_1, s_1) p_2(x_2, s_2) p_3(x_3, s_3) \frac{\partial^3 f(s_1, s_2, s_3)}{\Delta_3 s_3 \Delta_2 s_2 \Delta_1 s_1} \Delta_3 s_3 \Delta_2 s_2 \Delta_1 s_1 \left. \right\} .$$

Proof. Here we use repeatedly the following Delta Montgomery identity (5.1) on times scales

$$g(u) = \frac{1}{\beta - \alpha} \int_\alpha^\beta g(\sigma(z)) \Delta z + \frac{1}{\beta - \alpha} \int_\alpha^\beta k(u, z) g^\Delta(z) \Delta z,$$

where $k : [\alpha, \beta]^2 \to \mathbb{R}$ is defined by

$$k(u, z) := \begin{cases} z - \alpha, & \text{if } z \in [\alpha, u) \\ z - \beta, & \text{if } z \in [u, \beta], \end{cases}$$

and $g \in C_{rd}^1([\alpha, \beta])$.

For basics here see [16], [17], [18], [19], [26].

We observe that

$$f(x_1, x_2, x_3) = A_0 + B_0,$$

where

$$A_0 := \frac{1}{b_1 - a_1} \int_{a_1}^{b_1} f(\sigma_1(s_1), x_2, x_3) \Delta_1 s_1,$$

and

$$B_0 := \frac{1}{b_1 - a_1} \int_{a_1}^{b_1} p_1(x_1, s_1) \frac{\partial f(s_1, x_2, x_3)}{\Delta_1 s_1} \Delta_1 s_1.$$

Furthermore we have

$$f(\sigma_1(s_1), x_2, x_3) = A_1 + B_1,$$

where

$$A_1 := \frac{1}{b_2 - a_2} \int_{a_2}^{b_2} f(\sigma_1(s_1), \sigma_2(s_2), x_3) \Delta_2 s_2,$$

and

$$B_1 := \frac{1}{b_2 - a_2} \int_{a_2}^{b_2} p_2(x_2, s_2) \frac{\partial f(\sigma_1(s_1), s_2, x_3)}{\Delta_2 s_2} \Delta_2 s_2.$$

Also we get that

$$f(\sigma_1(s_1), \sigma_2(s_2), x_3) = \frac{1}{b_3 - a_3} \int_{a_3}^{b_3} f(\sigma_1(s_1), \sigma_2(s_2), \sigma_3(s_3)) \Delta_3 s_3$$

$$+ \frac{1}{b_3 - a_3} \int_{a_3}^{b_3} p_3(x_3, s_3) \frac{\partial f(\sigma_1(s_1), \sigma_2(s_2), s_3)}{\Delta_3 s_3} \Delta_3 s_3.$$

Next we put things together, and we have

$$A_1 = \frac{1}{(b_2 - a_2)(b_3 - a_3)} \int_{a_2}^{b_2} \int_{a_3}^{b_3} f(\sigma_1(s_1), \sigma_2(s_2), \sigma_3(s_3)) \Delta_3 s_3 \Delta_2 s_2$$

$$+ \frac{1}{(b_2 - a_2)(b_3 - a_3)} \int_{a_2}^{b_2} \int_{a_3}^{b_3} p_3(x_3, s_3) \frac{\partial f(\sigma_1(s_1), \sigma_2(s_2), s_3)}{\Delta_3 s_3} \Delta_3 s_3 \Delta_2 s_2.$$

And

$$A_0 = \frac{1}{b_1 - a_1} \int_{a_1}^{b_1} (A_1 + B_1) \Delta_1 s_1$$

$$= \frac{1}{\prod_{i=1}^{3} (b_i - a_i)} \int_{\prod_{i=1}^{3} [a_i, b_i]} f(\sigma_1(s_1), \sigma_2(s_2), \sigma_3(s_3)) \Delta_3 s_3 \Delta_2 s_2 \Delta_1 s_1$$

$$+ \frac{1}{\prod_{i=1}^{3} (b_i - a_i)} \int_{\prod_{i=1}^{3} [a_i, b_i]} p_3(x_3, s_3) \frac{\partial f(\sigma_1(s_1), \sigma_2(s_2), s_3)}{\Delta_3 s_3} \Delta_3 s_3 \Delta_2 s_2 \Delta_1 s_1$$

$$+ \frac{1}{\prod_{i=1}^{2} (b_i - a_i)} \int_{\prod_{i=1}^{2} [a_i, b_i]} p_2(x_2, s_2) \frac{\partial f(\sigma_1(s_1), s_2, x_3)}{\Delta_2 s_2} \Delta_2 s_2 \Delta_1 s_1.$$

Also we obtain

$$\frac{\partial f(\sigma_1(s_1), s_2, x_3)}{\Delta_2 s_2} = \frac{1}{b_3 - a_3} \int_{a_3}^{b_3} \frac{\partial f(\sigma_1(s_1), s_2, \sigma_3(s_3))}{\Delta_2 s_2} \Delta_3 s_3$$

$$+ \frac{1}{b_3 - a_3} \int_{a_3}^{b_3} p_3(x_3, s_3) \frac{\partial f(\sigma_1(s_1), s_2, s_3)}{\Delta_3 s_3 \Delta_2 s_2} \Delta_3 s_3.$$

Consequently we get

$$A_0 = \frac{1}{\prod_{i=1}^{3} (b_i - a_i)} \int_{\prod_{i=1}^{3} [a_i, b_i]} f(\sigma_1(s_1), \sigma_2(s_2), \sigma_3(s_3)) \Delta_3 s_3 \Delta_2 s_2 \Delta_1 s_1$$

$$+ \frac{1}{\prod_{i=1}^{3} (b_i - a_i)} \int_{\prod_{i=1}^{3} [a_i, b_i]} p_3(x_3, s_3) \frac{\partial f(\sigma_1(s_1), \sigma_2(s_2), s_3)}{\Delta_3 s_3} \Delta_3 s_3 \Delta_2 s_2 \Delta_1 s_1$$

$$+ \frac{1}{\prod_{i=1}^{3} (b_i - a_i)} \int_{\prod_{i=1}^{3} [a_i, b_i]} p_2(x_2, s_2) \frac{\partial f(\sigma_1(s_1), s_2, \sigma_3(s_3))}{\Delta_2 s_2} \Delta_3 s_3 \Delta_2 s_2 \Delta_1 s_1$$

$$+ \frac{1}{\prod_{i=1}^{3} (b_i - a_i)} \int_{\prod_{i=1}^{3} [a_i, b_i]} p_2(x_2, s_2) p_3(x_3, s_3) \frac{\partial^2 f(\sigma_1(s_1), s_2, s_3)}{\Delta_3 s_3 \Delta_2 s_2} \Delta_3 s_3 \Delta_2 s_2 \Delta_1 s_1.$$

Similarly we obtain that

$$\frac{\partial f(s_1, x_2, x_3)}{\Delta_1 s_1} = \frac{1}{b_2 - a_2} \int_{a_2}^{b_2} \frac{\partial f(s_1, \sigma_2(s_2), x_3)}{\Delta_1 s_1} \Delta_2 s_2$$

$$+\frac{1}{b_2 - a_2} \int_{a_2}^{b_2} p_2(x_2, s_2) \frac{\partial^2 f(s_1, s_2, x_3)}{\Delta_2 s_2 \Delta_1 s_1} \Delta_2 s_2,$$

$$\frac{\partial f(s_1, \sigma_2(s_2), x_3)}{\Delta_1 s_1} = \frac{1}{b_3 - a_3} \int_{a_3}^{b_3} \frac{\partial f(s_1, \sigma_2(s_2), \sigma_3(s_3))}{\Delta_1 s_1} \Delta_3 s_3$$

$$+\frac{1}{b_3 - a_3} \int_{a_3}^{b_3} p_3(x_3, s_3) \frac{\partial^2 f(s_1, \sigma_2(s_2), s_3)}{\Delta_3 s_3 \Delta_1 s_1} \Delta_3 s_3,$$

and

$$\frac{\partial^2 f(s_1, s_2, x_3)}{\Delta_2 s_2 \Delta_1 s_1} = \frac{1}{b_3 - a_3} \int_{a_3}^{b_3} \frac{\partial^2 f(s_1, s_2, \sigma_3(s_3))}{\Delta_2 s_2 \Delta_1 s_1} \Delta_3 s_3$$

$$+\frac{1}{b_3 - a_3} \int_{a_3}^{b_3} p_3(x_3, s_3) \frac{\partial^3 f(s_1, s_2, s_3)}{\Delta_3 s_3 \Delta_2 s_2 \Delta_1 s_1} \Delta_3 s_3.$$

Consequently we obtain

$$\frac{\partial f(s_1, x_2, x_3)}{\Delta_1 s_1} = \frac{1}{(b_2 - a_2)(b_3 - a_3)} \int_{a_2}^{b_2} \int_{a_3}^{b_3} \frac{\partial f(s_1, \sigma_2(s_2), \sigma_3(s_3))}{\Delta_1 s_1} \Delta_3 s_3 \Delta_2 s_2$$

$$+\frac{1}{(b_2 - a_2)(b_3 - a_3)} \int_{a_2}^{b_2} \int_{a_3}^{b_3} p_3(x_3, s_3) \frac{\partial^2 f(s_1, \sigma_2(s_2), s_3)}{\Delta_3 s_3 \Delta_1 s_1} \Delta_3 s_3 \Delta_2 s_2$$

$$+\frac{1}{(b_2 - a_2)(b_3 - a_3)} \int_{a_2}^{b_2} \int_{a_3}^{b_3} p_2(x_2, s_2) \frac{\partial^2 f(s_1, s_2, \sigma_3(s_3))}{\Delta_2 s_2 \Delta_1 s_1} \Delta_3 s_3 \Delta_2 s_2$$

$$+\frac{1}{(b_2 - a_2)(b_3 - a_3)} \int_{a_2}^{b_2} \int_{a_3}^{b_3} p_2(x_2, s_2) p_3(x_3, s_3) \frac{\partial^3 f(s_1, s_2, s_3)}{\Delta_3 s_3 \Delta_2 s_2 \Delta_1 s_1} \Delta_3 s_3 \Delta_2 s_2.$$

Therefore we derive

$$B_0 = \frac{1}{\prod_{i=1}^{3}(b_i - a_i)} \int_{\prod_{i=1}^{3}[a_i, b_i]} p_1(x_1, s_1) \frac{\partial f(s_1, \sigma_2(s_2), \sigma_3(s_3))}{\Delta_1 s_1} \Delta_3 s_3 \Delta_2 s_2 \Delta_1 s_1$$

$$+\frac{1}{\prod_{i=1}^{3}(b_i - a_i)} \int_{\prod_{i=1}^{3}[a_i, b_i]} p_1(x_1, s_1) p_3(x_3, s_3) \frac{\partial^2 f(s_1, \sigma_2(s_2), s_3)}{\Delta_3 s_3 \Delta_1 s_1} \Delta_3 s_3 \Delta_2 s_2 \Delta_1 s_1$$

$$+\frac{1}{\prod_{i=1}^{3}(b_i - a_i)} \int_{\prod_{i=1}^{3}[a_i, b_i]} p_1(x_1, s_1) p_2(x_2, s_2) \frac{\partial^2 f(s_1, s_2, \sigma_3(s_3))}{\Delta_2 s_2 \Delta_1 s_1} \Delta_3 s_3 \Delta_2 s_2 \Delta_1 s_1$$

$$+\frac{1}{\prod_{i=1}^{3}(b_i - a_i)} \int_{\prod_{i=1}^{3}[a_i, b_i]} p_1(x_1, s_1) p_2(x_2, s_2) p_3(x_3, s_3)$$

$$\cdot \frac{\partial^3 f(s_1, s_2, s_3)}{\Delta_3 s_3 \Delta_2 s_2 \Delta_1 s_1} \Delta_3 s_3 \Delta_2 s_2 \Delta_1 s_1,$$

proving the claim. □

We continue with the generalization

Theorem 5.6. *Let the time scales* $(\mathbb{T}_i, \sigma_i, \Delta_i)$, $i = 1, 2, \ldots, n$, $n \in \mathbb{N}$ *and* $a_i < b_i$; $a_i, b_i \in \mathbb{T}_i$, $i = 1, \ldots, n$. *Here* σ_i, $i = 1, \ldots, n$ *are continuous. We consider* $f \in C^n \left(\prod_{i=1}^n ([a_i, b_i] \cap \mathbb{T}_i) \right)$. *Let* $(x_1, \ldots, x_n) \in \prod_{i=1}^n ([a_i, b_i] \cap \mathbb{T}_i)$. *Define the kernels* $p_i : [a_i, b_i]^2 \to \mathbb{R}$:

$$p_i(x_i, s_i) := \begin{cases} s_i - a_i, & s_i \in [a_i, x_i), \\ s_i - b_i, & s_i \in [x_i, b_i], \end{cases}$$

for $i = 1, 2, \ldots, n$.
Denote

$$\int_{\prod_{i=1}^j [a_i, b_i]} \cdot = \int_{\prod_{i=1}^j ([a_i, b_i] \cap \mathbb{T}_i)} \cdot, \quad j = 1, \ldots, n.$$

Then

$$f(x_1, \ldots, x_n) = \frac{1}{\prod_{i=1}^n (b_i - a_i)}$$

$$\cdot \left\{ \int_{\prod_{i=1}^n [a_i, b_i]} f(\sigma_1(s_1), \sigma_2(s_2), \ldots, \sigma_n(s_n)) \, \Delta_n s_n \Delta_{n-1} s_{n-1} \ldots \Delta_1 s_1 \right.$$

$$+ \sum_{j=1}^n \int_{\prod_{i=1}^n [a_i, b_i]} p_j(x_j, s_j) \frac{\partial f(\sigma_1(s_1), \ldots, \sigma_{j-1}(s_{j-1}), s_j, \sigma_{j+1}(s_{j+1}), \ldots, \sigma_n(s_n))}{\Delta_j s_j}$$

$$\cdot \Delta_n s_n \Delta_{n-1} s_{n-1} \ldots \Delta_1 s_1$$

$$+ \sum_{\substack{l_1 = 1 \\ j < k}}^{\binom{n}{2}} \left(\int_{\prod_{i=1}^n [a_i, b_i]} p_j(x_j, s_j) p_k(x_k, s_k) \right. \tag{5.6}$$

$$\cdot \frac{\partial^2 f(\sigma_1(s_1), \ldots, \sigma_{j-1}(s_{j-1}), s_j, \sigma_{j+1}(s_{j+1}), \ldots, \sigma_{k-1}(s_{k-1}), s_k, \sigma_{k+1}(s_{k+1}), \ldots, \sigma_n(s_n))}{\Delta_k s_k \Delta_j s_j}$$

$$\left. \cdot \Delta_n s_n \ldots \Delta_1 s_1 \right)_{(l_1)}$$

$$+ \sum_{\substack{l_2 = 1 \\ j < k < r}}^{\binom{n}{3}} \left(\int_{\prod_{i=1}^n [a_i, b_i]} p_j(x_j, s_j) p_k(x_k, s_k) p_r(x_r, s_r) \right.$$

$$\partial^3 f\left(\sigma_1\left(s_1\right),\ldots,\sigma_{j-1}\left(s_{j-1}\right),s_j,\sigma_{j+1}\left(s_{j+1}\right),\ldots,\sigma_{k-1}\left(s_{k-1}\right),s_k,\sigma_{k+1}\left(s_{k+1}\right)\right.$$
$$\frac{\left.,\ldots,\sigma_{r-1}\left(s_{r-1}\right),s_r,\sigma_{r+1}\left(s_{r+1}\right),\ldots,\sigma_n\left(s_n\right)\right)}{\Delta_r s_r \Delta_k s_k \Delta_j s_j}$$
$$\cdot\Delta_n s_n\ldots\Delta_1 s_1)_{(l_2)}$$

$$+\cdots+\sum_{l=1}^{\binom{n}{n-1}}\left(\int_{\prod_{i=1}^n[a_i,b_i]} p_1\left(x_1,s_1\right)\ldots\widehat{p_l\left(x_l,s_l\right)}\ldots p_n\left(x_n,s_n\right)\right.$$

$$\left.\cdot\frac{\partial^{n-1}f\left(s_1,s_2,\ldots,\sigma_l\left(s_l\right),\ldots,s_n\right)}{\Delta_n s_n\ldots\widehat{\Delta_l s_l}\ldots\Delta_1 s_1}\Delta_n s_n\ldots\Delta_1 s_1\right)$$

$$+\int_{\prod_{i=1}^n[a_i,b_i]}\left(\prod_{i=1}^n p_i\left(x_i,s_i\right)\right)\frac{\partial^n f\left(s_1,\ldots,s_n\right)}{\Delta_n s_n\ldots\Delta_1 s_1}\Delta_n s_n\ldots\Delta_1 s_1\Bigg\}.$$

Above l_1 counts $(j,k): j<k$; $j,k\in\{1,\ldots,n\}$, also l_2 counts $(j,k,r): j<k<r$; $j,k,r\in\{1,\ldots,n\}$, etc. Also $\widehat{p_l\left(x_l,s_l\right)}$ and $\widehat{\Delta_l s_l}$ means that $p_l\left(x_l,s_l\right)$ and $\Delta_l s_l$ are missing, respectively.

Proof. Similar to Theorem 5.5. $\qquad\qquad\square$

Next we give multivariate ∇-representations

Theorem 5.7. *Let the time scales $(\mathbb{T}_i,\rho_i,\nabla_i)$, $i=1,2,3$ and $a_i<b_i$; $a_i,b_i\in\mathbb{T}_i$, $i=1,2,3$. Here ρ_i, $i=1,2,3$ are continuous.*
We consider $f\in C^3\left(\prod_{i=1}^3\left(\left[a_i,b_i\right]\cap\mathbb{T}_i\right)\right)$. Let $(x_1,x_2,x_3)\in\prod_{i=1}^3\left(\left[a_i,b_i\right]\cap\mathbb{T}_i\right)$. Define the kernels $p_i^*:\left[a_i,b_i\right]^2\to\mathbb{R}$:

$$p_i^*\left(x_i,s_i\right):=\begin{cases}s_i-a_i, & s_i\in\left[a_i,x_i\right],\\ s_i-b_i, & s_i\in\left(x_i,b_i\right],\end{cases}$$

for $i=1,2,3$.
  Denote

$$\int_{\prod_{i=1}^j[a_i,b_i]}\cdot=\int_{\prod_{i=1}^j\left(\left[a_i,b_i\right]\cap\mathbb{T}_i\right)}\cdot,$$

for $j=1,2,3$.
  Then

$$f\left(x_1,x_2,x_3\right)=\frac{1}{\prod_{i=1}^3\left(b_i-a_i\right)}$$

$$\cdot \left\{ \int_{\prod_{i=1}^{3}[a_i,b_i]} f\left(\rho_1\left(s_1\right),\rho_2\left(s_2\right),\rho_3\left(s_3\right)\right)\nabla_3 s_3\nabla_2 s_2\nabla_1 s_1 \right.$$

$$+\int_{\prod_{i=1}^{3}[a_i,b_i]} p_1^*\left(x_1,s_1\right)\frac{\partial f\left(s_1,\rho_2\left(s_2\right),\rho_3\left(s_3\right)\right)}{\nabla_1 s_1}\nabla_3 s_3\nabla_2 s_2\nabla_1 s_1$$

$$+\int_{\prod_{i=1}^{3}[a_i,b_i]} p_2^*\left(x_2,s_2\right)\frac{\partial f\left(\rho_1\left(s_1\right),s_2,\rho_3\left(s_3\right)\right)}{\nabla_2 s_2}\nabla_3 s_3\nabla_2 s_2\nabla_1 s_1$$

$$+\int_{\prod_{i=1}^{3}[a_i,b_i]} p_3^*\left(x_3,s_3\right)\frac{\partial f\left(\rho_1\left(s_1\right),\rho_2\left(s_2\right),s_3\right)}{\nabla_3 s_3}\nabla_3 s_3\nabla_2 s_2\nabla_1 s_1 \qquad (5.7)$$

$$+\int_{\prod_{i=1}^{3}[a_i,b_i]} p_1^*\left(x_1,s_1\right)p_2^*\left(x_2,s_2\right)\frac{\partial^2 f\left(s_1,s_2,\rho_3\left(s_3\right)\right)}{\nabla_2 s_2\nabla_1 s_1}\nabla_3 s_3\nabla_2 s_2\nabla_1 s_1$$

$$+\int_{\prod_{i=1}^{3}[a_i,b_i]} p_1^*\left(x_1,s_1\right)p_3^*\left(x_3,s_3\right)\frac{\partial^2 f\left(s_1,\rho_2\left(s_2\right),s_3\right)}{\nabla_3 s_3\nabla_1 s_1}\nabla_3 s_3\nabla_2 s_2\nabla_1 s_1$$

$$+\int_{\prod_{i=1}^{3}[a_i,b_i]} p_2^*\left(x_2,s_2\right)p_3^*\left(x_3,s_3\right)\frac{\partial^2 f\left(\rho_1\left(s_1\right),s_2,s_3\right)}{\nabla_3 s_3\nabla_2 s_2}\nabla_3 s_3\nabla_2 s_2\nabla_1 s_1$$

$$+\int_{\prod_{i=1}^{3}[a_i,b_i]} p_1^*\left(x_1,s_1\right)p_2^*\left(x_2,s_2\right)p^*\left(x_3,s_3\right)\frac{\partial^3 f\left(s_1,s_2,s_3\right)}{\nabla_3 s_3\nabla_2 s_2\nabla_1 s_1}\nabla_3 s_3\nabla_2 s_2\nabla_1 s_1 \left. \right\}.$$

Proof.   As in Theorem 5.5.   □

We give the generalization

Theorem 5.8. *Let the time scales* $(\mathbb{T}_i,\rho_i,\nabla_i)$, $i=1,2,\ldots,n$, $n\in\mathbb{N}$ *and* $a_i<b_i$; $a_i,b_i\in\mathbb{T}_i$, $i=1,\ldots,n$. *Here* ρ_i, $i=1,\ldots,n$ *are continuous. We consider* $f\in C^n\left(\prod_{i=1}^{n}\left([a_i,b_i]\cap\mathbb{T}_i\right)\right)$. *Let* $(x_1,\ldots,x_n)\in\prod_{i=1}^{n}\left([a_i,b_i]\cap\mathbb{T}_i\right)$. *Define the kernels* $p_i^*:[a_i,b_i]^2\to\mathbb{R}$:

$$p_i^*\left(x_i,s_i\right):=\begin{cases} s_i-a_i, & s_i\in[a_i,x_i],\\ s_i-b_i, & s_i\in(x_i,b_i], \end{cases}$$

for $i = 1, 2, \ldots, n$.

Denote

$$\int_{\prod\limits_{i=1}^{j} [a_i, b_i]} \cdot := \int_{\prod\limits_{i=1}^{j} ([a_i, b_i] \cap \mathbb{T}_i)} \cdot, \quad j = 1, \ldots, n.$$

Then

$$f(x_1, \ldots, x_n) = \frac{1}{\prod_{i=1}^{n} (b_i - a_i)}$$

$$\cdot \left\{ \int_{\prod\limits_{i=1}^{n} [a_i, b_i]} f(\rho_1(s_1), \rho_2(s_2), \ldots, \rho_n(s_n)) \nabla_n s_n \nabla_{n-1} s_{n-1} \ldots \nabla_1 s_1 \right.$$

$$+ \sum_{j=1}^{n} \int_{\prod\limits_{i=1}^{n} [a_i, b_i]} p_j^*(x_j, s_j) \frac{\partial f(\rho_1(s_1), \ldots, \rho_{j-1}(s_{j-1}), s_j, \rho_{j+1}(s_{j+1}), \ldots, \rho_n(s_n))}{\nabla_j s_j}$$

$$\cdot \nabla_n s_n \ldots \nabla_1 s_1$$

$$+ \sum_{\substack{l_1=1 \\ j<k}}^{\binom{n}{2}} \left(\int_{\prod\limits_{i=1}^{n} [a_i, b_i]} p_j^*(x_j, s_j) p_k^*(x_k, s_k) \right. \tag{5.8}$$

$$\cdot \frac{\partial^2 f(\rho_1(s_1), \ldots, \rho_{j-1}(s_{j-1}), s_j, \rho_{j+1}(s_{j+1}), \ldots, \rho_{k-1}(s_{k-1}), s_k, \rho_{k+1}(s_{k+1}), \ldots, \rho_n(s_n))}{\nabla_k s_k \nabla_j s_j}$$

$$\left. \cdot \nabla_n s_n \ldots \nabla_1 s_1 \right)_{(l_1)}$$

$$+ \sum_{\substack{l_2=1 \\ j<k<r}}^{\binom{n}{3}} \left(\int_{\prod\limits_{i=1}^{n} [a_i, b_i]} p_j^*(x_j, s_j) p_k^*(x_k, s_k) p_r^*(x_r, s_r) \right.$$

$$\cdot \frac{\begin{aligned}\partial^3 f(\rho_1(s_1), \ldots, \rho_{j-1}(s_{j-1}), s_j, \rho_{j+1}(s_{j+1}), \ldots, \rho_{k-1}(s_{k-1}), s_k, \rho_{k+1}(s_{k+1}) \\ , \ldots, \rho_{r-1}(s_{r-1}), s_r, \rho_{r+1}(s_{r+1}), \ldots, \rho_n(s_n))\end{aligned}}{\nabla_r s_r \nabla_k s_k \nabla_j s_j}$$

$$\left. \cdot \nabla_n s_n \ldots \nabla_1 s_1 \right)_{(l_2)}$$

$$+ \ldots + \sum_{l=1}^{\binom{n}{n-1}} \left(\int_{\prod\limits_{i=1}^{n} [a_i, b_i]} p_1^*(x_1, s_1) \ldots \widehat{p_l^*(x_l, s_l)} \ldots p_n^*(x_n, s_n) \right.$$

$$\cdot \frac{\partial^{n-1} f\left(s_1, s_2, \ldots, \rho_l\left(s_l\right), \ldots, s_n\right)}{\nabla_n s_n \ldots \widehat{\nabla_l s_l} \ldots \nabla_1 s_1} \nabla_n s_n \ldots \nabla_1 s_1\Bigg)$$

$$+ \int_{\prod_{i=1}^n [a_i, b_i]} \left(\prod_{i=1}^n p_i^*\left(x_i, s_i\right)\right) \frac{\partial^n f\left(s_1, \ldots, s_n\right)}{\nabla_n s_n \ldots \nabla_1 s_1} \nabla_n s_n \ldots \nabla_1 s_1\Bigg\}.$$

Above l_1 counts $(j, k) : j < k$; $j, k \in \{1, \ldots, n\}$, also l_2 counts $(j, k, r) : j < k < r$; $j, k, r \in \{1, \ldots, n\}$, etc. Also $p_l^*\left(x_l, s_l\right)$ and $\widehat{\Delta_l s_l}$ means that $p_l^*\left(x_l, s_l\right)$ and $\Delta_l s_l$ are missing, respectively.

Proof.   Similar to Theorem 5.7.   □

We continue with a different kind of Δ, ∇-multivariate representations.

Theorem 5.9. *Let the time scales $(\mathbb{T}_i, \sigma_i, \Delta_i)$, $i = 1, 2, 3$ and $a < A$, $b < B$, $c < C$; $a, A \in \mathbb{T}_1$; $b, B \in \mathbb{T}_2$; $c, C \in \mathbb{T}_3$, σ_i are continuous, $i = 1, 2, 3$. We consider $f \in C^3\left(([a, A] \cap \mathbb{T}_1) \times ([b, B] \cap \mathbb{T}_2) \times ([c, C] \cap \mathbb{T}_3)\right)$. We denote $[a, A] \times [b, B] \times [c, C] = ([a, A] \cap \mathbb{T}_1) \times ([b, B] \cap \mathbb{T}_2) \times ([c, C] \cap \mathbb{T}_3)$. Let also $(x, y, z) \in [a, A] \times [b, B] \times [c, C]$ be fixed. We define the kernels $p : [a, A]^2 \to \mathbb{R}$, $q : [b, B]^2 \to \mathbb{R}$, and $\theta : [c, C]^2 \to \mathbb{R}$:*

$$p(x, s) := \begin{cases} s - a, & s \in [a, x), \\ s - A, & s \in [x, A], \end{cases}$$

$$q(y, t) := \begin{cases} t - b, & t \in [b, y), \\ t - B, & t \in [y, B], \end{cases}$$

and

$$\theta(z, r) := \begin{cases} r - c, & r \in [c, z), \\ r - C, & r \in [z, C]. \end{cases}$$

Then

$$\theta_{1,3} := \int_a^A \int_b^B \int_c^C p(x, s) \, q(y, t) \, \theta(z, r) \frac{\partial^3 f(s, t, r)}{\Delta_3 r \Delta_2 t \Delta_1 s} \Delta_3 r \Delta_2 t \Delta_1 s$$

$$= \{(A - a)(B - b)(C - c) f(x, y, z)\}$$

$$- \Bigg[(B - b)(C - c) \int_a^A f(\sigma_1(s), y, z) \Delta_1 s$$

$$+ (A - a)(C - c) \int_b^B f(x, \sigma_2(t), z) \Delta_2 t$$

$$+ (A - a)(B - b) \int_c^C f(x, y, \sigma_3(r)) \Delta_3 r\Bigg]$$

$$+ \left[(C - c) \int_a^A \int_b^B f\left(\sigma_1\left(s\right), \sigma_2\left(t\right), z\right) \Delta_1 s \Delta_2 t \right.$$

$$+ (B - b) \int_a^A \int_c^C f\left(\sigma_1\left(s\right), y, \sigma_3\left(r\right)\right) \Delta_1 s \Delta_3 r$$

$$+ (A - a) \int_b^B \int_c^C f\left(x, \sigma_2\left(t\right), \sigma_3\left(r\right)\right) \Delta_2 t \Delta_3 r \right] \tag{5.9}$$

$$- \int_a^A \int_b^B \int_c^C f\left(\sigma_1\left(s\right), \sigma_2\left(t\right), \sigma_3\left(r\right)\right) \Delta_1 s \Delta_2 t \Delta_3 r =: \theta_{2,3}.$$

Proof. Integrating by parts repeatedly, see (5.1), and using Fubini's theorem, see [16], [17], [18], [19], [26], we obtain the following eight equalities:

$$\int_a^x \int_b^y \int_c^z (s - a)(t - b)(r - c) \frac{\partial^3 f\left(s, t, r\right)}{\Delta_3 r \Delta_2 t \Delta_1 s} \Delta_3 r \Delta_2 t \Delta_1 s$$

$$= (x - a)(y - b)(z - c) f\left(x, y, z\right)$$

$$- (y - b)(z - c) \int_a^x f\left(\sigma_1\left(s\right), y, z\right) \Delta_1 s - (x - a)(z - c) \int_b^y f\left(x, \sigma_2\left(t\right), z\right) \Delta_2 t$$

$$- (x - a)(y - b) \int_c^z f\left(x, y, \sigma_3\left(r\right)\right) \Delta_3 r$$

$$+ (z - c) \int_a^x \int_b^y f\left(\sigma_1\left(s\right), \sigma_2\left(t\right), z\right) \Delta_1 s \Delta_2 t$$

$$+ (y - b) \int_a^x \int_c^z f\left(\sigma_1\left(s\right), y, \sigma_3\left(r\right)\right) \Delta_1 s \Delta_3 r$$

$$+ (x - a) \int_b^y \int_c^z f\left(x, \sigma_2\left(t\right), \sigma_3\left(r\right)\right) \Delta_2 t \Delta_3 r \tag{5.10}$$

$$- \int_a^x \int_b^y \int_c^z f\left(\sigma_1\left(s\right), \sigma_2\left(t\right), \sigma_3\left(r\right)\right) \Delta_1 s \Delta_2 t \Delta_3 r.$$

Similarly we get

$$\int_x^A \int_y^B \int_z^C (s - A)(t - B)(r - C) \frac{\partial^3 f\left(s, t, r\right)}{\Delta_3 r \Delta_2 t \Delta_1 s} \Delta_3 r \Delta_2 t \Delta_1 s$$

$$= (A - x)(B - y)(C - z) f\left(x, y, z\right)$$

$$- (B - y)(C - z) \int_x^A f\left(\sigma_1\left(s\right), y, z\right) \Delta_1 s$$

$$- (A - x)(C - z) \int_y^B f(x, \sigma_2(t), z) \Delta_2 t$$

$$- (A - x)(B - y) \int_x^C f(x, y, \sigma_3(r)) \Delta_3 r$$

$$+ (C - z) \int_x^A \int_y^B f(\sigma_1(s), \sigma_2(t), z) \Delta_1 s \Delta_2 t$$

$$+ (B - y) \int_x^A \int_z^C f(\sigma_1(s), y, \sigma_3(r)) \Delta_1 s \Delta_3 r$$

$$+ (A - x) \int_y^B \int_z^C f(x, \sigma_2(t), \sigma_3(r)) \Delta_2 t \Delta_3 r \tag{5.11}$$

$$- \int_x^A \int_y^B \int_z^C f(\sigma_1(s), \sigma_2(t), \sigma_3(r)) \Delta_1 s \Delta_2 t \Delta_3 r.$$

And,

$$\int_a^x \int_b^y \int_z^C (s - a)(t - b)(r - C) \frac{\partial^3 f(s, t, r)}{\Delta_3 r \Delta_2 t \Delta_1 s} \Delta_3 r \Delta_2 t \Delta_1 s$$

$$= (x - a)(y - b)(C - z) f(x, y, z)$$

$$- (y - b)(C - z) \int_a^x f(\sigma_1(s), y, z) \Delta_1 s - (x - a)(C - z) \int_b^y f(x, \sigma_2(t), z) \Delta_2 t$$

$$- (x - a)(y - b) \int_z^C f(x, y, \sigma_3(r)) \Delta_3 r$$

$$+ (C - z) \int_a^x \int_b^y f(\sigma_1(s), \sigma_2(t), z) \Delta_1 s \Delta_2 t$$

$$+ (y - b) \int_a^x \int_z^C f(\sigma_1(s), y, \sigma_3(r)) \Delta_1 s \Delta_3 r$$

$$+ (x - a) \int_b^y \int_z^C f(x, \sigma_2(t), \sigma_3(r)) \Delta_2 t \Delta_3 r \tag{5.12}$$

$$- \int_a^x \int_b^y \int_z^C f(\sigma_1(s), \sigma_2(t), \sigma_3(r)) \Delta_1 s \Delta_2 t \Delta_3 r.$$

And,

$$\int_A^x \int_B^y \int_c^z (s - A)(t - B)(r - c) \frac{\partial^3 f(s, t, r)}{\Delta_3 r \Delta_2 t \Delta_1 s} \Delta_3 r \Delta_2 t \Delta_1 s$$

$$= (A - x)(B - y)(z - c) f(x, y, z)$$

$$- (B - y)(z - c) \int_x^A f(\sigma_1(s), y, z) \Delta_1 s - (A - x)(z - c) \int_y^B f(x, \sigma_2(t), z) \Delta_2 t$$

$$- (A - x)(B - y) \int_c^z f(x, y, \sigma_3(r)) \Delta_3 r$$

$$+ (z - c) \int_x^A \int_y^B f(\sigma_1(s), \sigma_2(t), z) \Delta_1 s \Delta_2 t$$

$$+ (B - y) \int_x^A \int_c^z f(\sigma_1(s), y, \sigma_3(r)) \Delta_1 s \Delta_3 r$$

$$+ (A - x) \int_y^B \int_c^z f(x, \sigma_2(t), \sigma_3(r)) \Delta_2 t \Delta_3 r \qquad (5.13)$$

$$- \int_x^A \int_y^B \int_c^z f(\sigma_1(s), \sigma_2(t), \sigma_3(r)) \Delta_1 s \Delta_2 t \Delta_3 r.$$

Also we find,

$$\int_a^x \int_y^B \int_z^C (s - a)(t - B)(r - C) \frac{\partial^3 f(s, t, r)}{\Delta_3 r \Delta_2 t \Delta_1 s} \Delta_3 r \Delta_2 t \Delta_1 s$$

$$= (x - a)(B - y)(C - z) f(x, y, z)$$

$$- (B - y)(C - z) \int_a^x f(\sigma_1(s), y, z) \Delta_1 s - (x - a)(C - z) \int_y^B f(x, \sigma_2(t), z) \Delta_2 t$$

$$- (x - a)(B - y) \int_z^C f(x, y, \sigma_3(r)) \Delta_3 r$$

$$+ (C - z) \int_a^x \int_y^B f(\sigma_1(s), \sigma_2(t), z) \Delta_1 s \Delta_2 t$$

$$+ (B - y) \int_a^x \int_z^C f(\sigma_1(s), y, \sigma_3(r)) \Delta_1 s \Delta_3 r$$

$$+ (x - a) \int_y^B \int_z^C f(x, \sigma_2(t), \sigma_3(r)) \Delta_2 t \Delta_3 r \qquad (5.14)$$

$$- \int_a^x \int_y^B \int_z^C f(\sigma_1(s), \sigma_2(t), \sigma_3(r)) \Delta_1 s \Delta_2 t \Delta_3 r.$$

And,

$$\int_x^A \int_b^y \int_c^z (s-A)(t-b)(r-c) \frac{\partial^3 f(s,t,r)}{\Delta_3 r \Delta_2 t \Delta_1 s} \Delta_3 r \Delta_2 t \Delta_1 s$$

$$= (A-x)(y-b)(z-c) f(x,y,z)$$

$$- (y-b)(z-c) \int_x^A f(\sigma_1(s),y,z) \Delta_1 s - (A-x)(z-c) \int_b^y f(x,\sigma_2(t),z) \Delta_2 t$$

$$- (A-x)(y-b) \int_c^z f(x,y,\sigma_3(r)) \Delta_3 r$$

$$+ (z-c) \int_x^A \int_b^y f(\sigma_1(s),\sigma_2(t),z) \Delta_1 s \Delta_2 t$$

$$+ (y-b) \int_x^A \int_c^z f(\sigma_1(s),y,\sigma_3(r)) \Delta_1 s \Delta_3 r$$

$$+ (A-x) \int_b^y \int_c^z f(x,\sigma_2(t),\sigma_3(r)) \Delta_2 t \Delta_3 r \qquad (5.15)$$

$$- \int_x^A \int_b^y \int_c^z f(\sigma_1(s),\sigma_2(t),\sigma_3(r)) \Delta_1 s \Delta_2 t \Delta_3 r.$$

And,

$$\int_a^x \int_y^B \int_c^z (s-a)(t-B)(r-c) \frac{\partial^3 f(s,t,r)}{\Delta_3 r \Delta_2 t \Delta_1 s} \Delta_3 r \Delta_2 t \Delta_1 s$$

$$= (x-a)(B-y)(z-c) f(x,y,z)$$

$$- (B-y)(z-c) \int_a^x f(\sigma_1(s),y,z) \Delta_1 s - (x-a)(z-c) \int_y^B f(x,\sigma_2(t),z) \Delta_2 t$$

$$- (x-a)(B-y) \int_c^z f(x,y,\sigma_3(r)) \Delta_3 r$$

$$+ (z-c) \int_a^x \int_y^B f(\sigma_1(s),\sigma_2(t),z) \Delta_1 s \Delta_2 t$$

$$+ (B-y) \int_a^x \int_c^z f(\sigma_1(s),y,\sigma_3(r)) \Delta_1 s \Delta_3 r$$

$$+ (x-a) \int_y^B \int_c^z f(x,\sigma_2(t),\sigma_3(r)) \Delta_2 t \Delta_3 r \qquad (5.16)$$

$$-\int_a^x \int_y^B \int_c^z f\left(\sigma_1\left(s\right),\sigma_2\left(t\right),\sigma_3\left(r\right)\right)\Delta_1 s\Delta_2 t\Delta_3 r.$$

Finally we have,

$$\int_x^A \int_b^y \int_z^C \left(s-A\right)\left(t-b\right)\left(r-C\right)\frac{\partial^3 f\left(s,t,r\right)}{\Delta_3 r\Delta_2 t\Delta_1 s}\Delta_3 r\Delta_2 t\Delta_1 s$$

$$=\left(A-x\right)\left(y-b\right)\left(C-z\right)f\left(x,y,z\right)$$

$$-\left(y-b\right)\left(C-z\right)\int_x^A f\left(\sigma_1\left(s\right),y,z\right)\Delta_1 s - \left(A-x\right)\left(C-z\right)\int_b^y f\left(x,\sigma_2\left(t\right),z\right)\Delta_2 t$$

$$-\left(A-x\right)\left(y-b\right)\int_z^C f\left(x,y,\sigma_3\left(r\right)\right)\Delta_3 r$$

$$+\left(C-z\right)\int_x^A \int_b^y f\left(\sigma_1\left(s\right),\sigma_2\left(t\right),z\right)\Delta_1 s\Delta_2 t$$

$$+\left(y-b\right)\int_x^A \int_z^C f\left(\sigma_1\left(s\right),y,\sigma_3\left(r\right)\right)\Delta_1 s\Delta_3 r$$

$$+\left(A-x\right)\int_b^y \int_z^C f\left(x,\sigma_2\left(t\right),\sigma_3\left(r\right)\right)\Delta_2 t\Delta_3 r \qquad (5.17)$$

$$-\int_x^A \int_b^y \int_z^C f\left(\sigma_1\left(s\right),\sigma_2\left(t\right),\sigma_3\left(r\right)\right)\Delta_1 s\Delta_2 t\Delta_3 r.$$

Adding all the right-hand sides of (5.10)-(5.17), we obtain

$$\textit{Big R.H.S.} = \left(A-a\right)\left(B-b\right)\left(C-c\right)f\left(x,y,z\right)$$

$$-\left(B-b\right)\left(C-c\right)\int_a^A f\left(\sigma_1\left(s\right),y,z\right)\Delta_1 s - \left(A-a\right)\left(C-c\right)\int_b^B f\left(x,\sigma_2\left(t\right),z\right)\Delta_2 t$$

$$-\left(A-a\right)\left(B-b\right)\int_c^C f\left(x,y,\sigma_3\left(r\right)\right)\Delta_3 r$$

$$+\left(C-c\right)\int_a^A \int_b^B f\left(\sigma_1\left(s\right),\sigma_2\left(t\right),z\right)\Delta_1 s\Delta_2 t$$

$$+\left(B-b\right)\int_a^A \int_c^C f\left(\sigma_1\left(s\right),y,\sigma_3\left(r\right)\right)\Delta_1 s\Delta_3 r$$

$$+\left(A-a\right)\int_b^B \int_c^C f\left(x,\sigma_2\left(t\right),\sigma_3\left(r\right)\right)\Delta_2 t\Delta_3 r$$

$$-\int_a^A \int_b^B \int_c^C f\left(\sigma_1\left(s\right),\sigma_2\left(t\right),\sigma_3\left(r\right)\right)\Delta_1 s\Delta_2 t\Delta_3 r.$$

Adding all the left-hand sides of (5.10)-(5.17), we get

$$\textit{Big L.H.S.} = \int_a^A \int_b^B \int_c^C p\left(x,s\right)q\left(y,t\right)\theta\left(z,r\right)\frac{\partial^3 f\left(s,t,r\right)}{\Delta_3 r\Delta_2 t\Delta_1 s}\Delta_3 r\Delta_2 t\Delta_1 s.$$

Clearly we have

$$\textit{Big L.H.S.} = \textit{Big R.H.S,}$$

proving the claim. $\qquad\qquad\qquad\qquad\qquad\qquad\qquad\qquad\qquad\qquad\square$

A generalization follows

Theorem 5.10. *Let the time scales* $(\mathbb{T}_i, \sigma_i, \Delta_i)$, $i = 1, \ldots, n$, $n \in \mathbb{N}$, $a_i < b_i$, $a_i, b_i \in \mathbb{T}_i$; $i = 1, \ldots, n$, *Here* σ_i *are continuous,* $i = 1, \ldots, n$. *We consider* $f \in C^n \left(\prod_{i=1}^n \left([a_i, b_i] \cap \mathbb{T}_i \right) \right)$. *We denote* $\prod_{i=1}^n [a_i, b_i] = \prod_{i=1}^n \left([a_i, b_i] \cap \mathbb{T}_i \right)$. *Let also* $(x_1, \ldots, x_n) \in \prod_{i=1}^n [a_i, b_i]$ *be fixed. We define the kernels* $p_i : [a_i, b_i]^2 \to \mathbb{R}$:

$$
p_i(x_i, s_i) := \begin{cases} s_i - a_i, & s_i \in [a_i, x_i), \\ s_i - b_i, & s_i \in [x_i, b_i], \end{cases}
$$

for all $i = 1, \ldots, n$.

Then

$$
\theta_{1,n} := \int_{\prod_{i=1}^n [a_i, b_i]} \prod_{i=1}^n p_i(x_i, s_i) \frac{\partial^n f(s_1, \ldots, s_n)}{\Delta_1 s_1 \ldots \Delta_n s_n} \Delta_1 s_1 \ldots \Delta_n s_n
$$

$$
= \left\{ \left(\prod_{i=1}^n (b_i - a_i) \right) f(x_1, \ldots, x_n) \right\}
$$

$$
- \left[\sum_{i=1}^{\binom{n}{1}} \left(\prod_{\substack{j=1 \\ j \neq i}}^n (b_j - a_j) \right) \int_{a_i}^{b_i} f(x_1, \ldots, \sigma_i(s_i), \ldots, x_n) \Delta_i s_i \right]
$$

$$
+ \left[\sum_{l=1}^{\binom{n}{2}} \left(\prod_{\substack{k=1 \\ k \neq i,j}}^n (b_k - a_k) \right) \right.
$$

$$
\left. \cdot \left(\int_{a_i}^{b_i} \int_{a_j}^{b_j} f(x_1, \ldots, \sigma_i(s_i), \ldots, \sigma_j(s_j), \ldots, x_n) \Delta_i s_i \Delta_j s_j \right)_{(l)} \right]
$$

$$
- + \ldots - + \ldots + (-1)^{n-1} . \tag{5.18}
$$

$$
+ \left[\sum_{\substack{j=1}}^{\binom{n}{n-1}} (b_j - a_j) \int_{\prod_{\substack{i=1 \\ i \neq j}}^n} f(\sigma_1(s_1), \ldots, x_j, \ldots, \sigma_n(s_n)) \Delta_1 s_1 \ldots \widehat{\Delta_j s_j} \ldots \Delta_n s_n \right]
$$

$$
+ (-1)^n \int_{\prod_{i=1}^n [a_i, b_i]} f(\sigma_1(s_1), \ldots, \sigma_n(s_n)) \Delta_1 s_1 \ldots \Delta_n s_n =: \theta_{2,n}.
$$

The above l *counts all the* (i, j)*'s,* $i < j$ *and* $i, j = 1, \ldots, n$. *Also* $\widehat{\Delta_j s_j}$ *means* $\Delta_j s_j$ *is missing.*

Proof. Similar to Theorem 5.9. □

We continue with

Theorem 5.11. *Let the time scales* $(\mathbb{T}_i, \rho_i, \nabla_i)$, $i = 1, 2, 3$ *and* $a < A$, $b < B$, $c < C$; $a, A \in \mathbb{T}_1$; $b, B \in \mathbb{T}_2$; $c, C \in \mathbb{T}_3$, ρ_i *arc continuous,* $i = 1, 2, 3$. *We consider* $f \in C^3(([a, A] \cap \mathbb{T}_1) \times ([b, B] \cap \mathbb{T}_2) \times ([c, C] \cap \mathbb{T}_3))$. *We denote* $[a, A] \times [b, B] \times [c, C] = ([a, A] \cap \mathbb{T}_1) \times ([b, B] \cap \mathbb{T}_2) \times ([c, C] \cap \mathbb{T}_3)$. *Let also* $(x, y, z) \in [a, A] \times [b, B] \times [c, C]$ *be fixed. We define the kernels* $p^* : [a, A]^2 \to \mathbb{R}$, $q^* : [b, B]^2 \to \mathbb{R}$, *and* $\theta^* : [c, C]^2 \to \mathbb{R}$:

$$p^*(x, s) := \begin{cases} s - a, & s \in [a, x], \\ s - A, & s \in (x, A], \end{cases}$$

$$q^*(y, t) := \begin{cases} t - b, & t \in [b, y], \\ t - B, & t \in (y, B], \end{cases}$$

and

$$\theta^*(z, r) := \begin{cases} r - c, & r \in [c, z], \\ r - C, & r \in (z, C]. \end{cases}$$

Then

$$\theta^*_{1,3} := \int_a^A \int_b^B \int_c^C p^*(x, s) q^*(y, t) \theta^*(z, r) \frac{\partial^3 f(s, t, r)}{\nabla_3 r \nabla_2 t \nabla_1 s} \nabla_3 r \nabla_2 t \nabla_1 s$$

$$= \{(A - a)(B - b)(C - c) f(x, y, z)\}$$

$$- \left[(B - b)(C - c) \int_a^A f(\rho_1(s), y, z) \nabla_1 s \right.$$

$$+ (A - a)(C - c) \int_b^B f(x, \rho_2(t), z) \nabla_2 t$$

$$\left. + (A - a)(B - b) \int_c^C f(x, y, \rho_3(r)) \nabla_3 r \right]$$

$$+ \left[(C - c) \int_a^A \int_b^B f(\rho_1(s), \rho_2(t), z) \nabla_1 s \nabla_2 t \right.$$

$$+ (B - b) \int_a^A \int_c^C f(\rho_1(s), y, \rho_3(r)) \nabla_1 s \nabla_3 r$$

$$\left. + (A - a) \int_b^B \int_c^C f(x, \rho_2(t), \rho_3(r)) \nabla_2 t \nabla_3 r \right] \tag{5.19}$$

$$- \int_a^A \int_b^B \int_c^C f(\rho_1(s), \rho_2(t), \rho_3(r)) \nabla_1 s \nabla_2 t \nabla_3 r =: \theta^*_{2,3}.$$

Proof.   As in Theorem 5.9.   □

A generalization follows

Theorem 5.12. *Let the time scales* $(\mathbb{T}_i, \rho_i, \nabla_i)$, $i = 1, \ldots, n$, $n \in \mathbb{N}$, $a_i < b_i$, $a_i, b_i \in \mathbb{T}_i$; $i = 1, \ldots, n$, *Here* ρ_i *are continuous,* $i = 1, \ldots, n$. *We consider* $f \in C^n \left(\prod_{i=1}^n ([a_i, b_i] \cap \mathbb{T}_i) \right)$. *We denote* $\prod_{i=1}^n [a_i, b_i] = \prod_{i=1}^n ([a_i, b_i] \cap \mathbb{T}_i)$. *Let also* $(x_1, \ldots, x_n) \in \prod_{i=1}^n [a_i, b_i]$ *be fixed. We define the kernels* $p_i^* : [a_i, b_i]^2 \to \mathbb{R}$:

$$p_i^* (x_i, s_i) := \begin{cases} s_i - a_i, & s_i \in [a_i, x_i], \\ s_i - b_i, & s_i \in (x_i, b_i], \end{cases}$$

for all $i = 1, \ldots, n$.

Then

$$\theta_{1,n}^* := \int_{\prod_{i=1}^n [a_i, b_i]} \prod_{i=1}^n p_i^* (x_i, s_i) \frac{\partial^n f (s_1, \ldots, s_n)}{\nabla_1 s_1 \ldots \nabla_n s_n} \nabla_1 s_1 \ldots \nabla_n s_n$$

$$= \left\{ \left(\prod_{i=1}^n (b_i - a_i) \right) f (x_1, \ldots, x_n) \right\}$$

$$- \left[\sum_{i=1}^{\binom{n}{1}} \left(\prod_{\substack{j=1 \\ j \neq i}}^n (b_j - a_j) \right) \int_{a_i}^{b_i} f (x_1, \ldots, \rho_i (s_i), \ldots, x_n) \nabla_i s_i \right]$$

$$+ \left[\sum_{l=1}^{\binom{n}{2}} \left(\prod_{\substack{k=1 \\ k \neq i,j}}^n (b_k - a_k) \right) \right.$$

$$\left. \cdot \left(\int_{a_i}^{b_i} \int_{a_j}^{b_j} f (x_1, \ldots, \rho_i (s_i), \ldots, \rho_j (s_j), \ldots, x_n) \nabla_i s_i \nabla_j s_j \right) \right]_{(l)}$$

$$+ - + \ldots - + \ldots + (-1)^{n-1} \tag{5.20}$$

$$\cdot \left[\sum_{j=1}^{\binom{n}{n-1}} (b_j - a_j) \int_{\substack{\prod_{i=1}^n \\ i \neq j}} f (\rho_1 (s_1), \ldots, x_j, \ldots, \rho_n (s_n)) \nabla_1 s_1 \ldots \widehat{\nabla_j s_j} \ldots \nabla_n s_n \right]$$

$$+ (-1)^n \int_{\prod_{i=1}^n [a_i, b_i]} f (\rho_1 (s_1), \ldots, \rho_n (s_n)) \nabla_1 s_1 \ldots \nabla_n s_n =: \theta_{2,n}^*.$$

The above l *counts all the* (i, j)'s, $i < j$ *and* $i, j = 1, \ldots, n$. *Also* $\widehat{\Delta_j s_j}$ *means* $\Delta_j s_j$ *is missing.*

Proof. Similar to Theorem 5.10. □

We call

$$E_1 (f) := f (x) - \frac{1}{b-a} \int_a^b f (\sigma (s_1)) \Delta s_1 \tag{5.21}$$

$$-\frac{1}{b-a} \left(\sum_{k=1}^{n-1} \int_{[a,b]^{k+1}} P (x, s_1) \prod_{i=1}^{k-1} P (s_i, s_{i+1}) f^{\Delta^k} (\sigma (s_{k+1})) \Delta s_{k+1} \Delta s_k ... \Delta s_1 \right),$$

and

$$E_2 (f) := f (x) - \frac{1}{b-a} \int_a^b f (\rho (s_1)) \nabla s_1 \tag{5.22}$$

$$-\frac{1}{b-a} \left(\sum_{k=1}^{n-1} \int_{[a,b]^{k+1}} P^* (x, s_1) \prod_{i=1}^{k-1} P^* (s_i, s_{i+1}) f^{\nabla^k} (\rho (s_{k+1})) \nabla s_{k+1} \nabla s_k ... \nabla s_1 \right).$$

By Theorem 5.3 we have that

$$E_1 (f) = \int_{[a,b]^n} P (x, s_1) \prod_{i=1}^{n-1} P (s_i, s_{i+1}) f^{\Delta^n} (s_n) \Delta s_n ... \Delta s_1, \tag{5.23}$$

and

$$E_2 (f) = \int_{[a,b]^n} P^* (x, s_1) \prod_{i=1}^{n-1} P^* (s_i, s_{i+1}) f^{\nabla^n} (s_n) \nabla s_n ... \nabla s_1. \tag{5.24}$$

We present the following Ostrowski type inequalities

Theorem 5.13. *Under the assumptions of Theorem 5.3 and additional conventions* $s_0 = x, \Delta s_0 = 1, \int_{[a,b]^0} * = *, \prod_{i=1}^{-1} * = 1,$ *we have that*

$$|E_1 (f)| \leq \min \begin{cases} \|f^{\Delta^n}\|_{\infty,([a,b] \cap \mathbb{T})} \int_{[a,b]^n} |P (x, s_1)| \prod_{i=1}^{n-1} |P (s_i, s_{i+1})| \Delta s_n ... \Delta s_1, \\ \|f^{\Delta^n}\|_{p,([a,b] \cap \mathbb{T})} \int_{[a,b]^{n-1}} |P (x, s_1)| \prod_{i=1}^{n-2} |P (s_i, s_{i+1})| \\ \quad \cdot \|P (s_{n-1}, \cdot)\|_{q,([a,b] \cap \mathbb{T})} \Delta s_{n-1} ... \Delta s_1, \\ \quad where \ p, q > 1 : \frac{1}{p} + \frac{1}{q} = 1, \\ \|f^{\Delta^n}\|_{1,([a,b] \cap \mathbb{T})} \int_{[a,b]^{n-1}} |P (x, s_1)| \prod_{i=1}^{n-2} |P (s_i, s_{i+1})| \\ \quad \cdot \|P (s_{n-1}, \cdot)\|_{\infty,([a,b] \cap \mathbb{T})} \Delta s_{n-1} ... \Delta s_1. \end{cases} \tag{5.25}$$

Proof. By (5.23) and basic properties of Δ-integration and Hölder's inequality. □

We continue with

Theorem 5.14. *Under the assumptions of Theorem 5.4 and additional conventions* $s_0 = x, \nabla s_0 = 1, \int_{[a,b]^0} * = *, \prod_{i=1}^{-1} * = 1,$ *we have that*

$$
|E_2(f)| \leq \min
\begin{cases}
\left\|f^{\nabla^n}\right\|_{\infty,([a,b]\cap\mathbb{T})} \int_{[a,b]^n} |P^*(x,s_1)| \prod_{i=1}^{n-1} |P^*(s_i, s_{i+1})| \\
\quad \cdot \nabla s_n \ldots \nabla s_1, \\
\left\|f^{\nabla^n}\right\|_{p,([a,b]\cap\mathbb{T})} \int_{[a,b]^{n-1}} |P^*(x,s_1)| \prod_{i=1}^{n-2} |P^*(s_i, s_{i+1})| \\
\quad \cdot \left\|P^*(s_{n-1},\cdot)\right\|_{q,([a,b]\cap\mathbb{T})} \nabla s_{n-1} \ldots \nabla s_1, \\
\quad where\ p, q > 1 : \frac{1}{p} + \frac{1}{q} = 1, \\
\left\|f^{\nabla^n}\right\|_{1,([a,b]\cap\mathbb{T})} \int_{[a,b]^{n-1}} |P^*(x,s_1)| \prod_{i=1}^{n-2} |P^*(s_i, s_{i+1})| \\
\quad \cdot \left\|P^*(s_{n-1},\cdot)\right\|_{\infty,([a,b]\cap\mathbb{T})} \nabla s_{n-1} \ldots \nabla s_1.
\end{cases}
\tag{5.26}
$$

Proof.   As in Theorem 5.13 using (5.24). $\square$

We make

Remark 5.1. (to Theorems 5.5-5.8) We denote

$$
D_3(f)(x_1, x_2, x_3) := f(x_1, x_2, x_3) - \frac{1}{\prod_{i=1}^3 (b_i - a_i)}
$$

$$
\cdot \left\{ \int_{\prod_{i=1}^3 [a_i, b_i]} f(\sigma_1(s_1), \sigma_2(s_2), \sigma_3(s_3)) \Delta_3 s_3 \Delta_2 s_2 \Delta_1 s_1 \right.
$$

$$
+ \int_{\prod_{i=1}^3 [a_i, b_i]} p_1(x_1, s_1) \frac{\partial f(s_1, \sigma_2(s_2), \sigma_3(s_3))}{\Delta_1 s_1} \Delta_3 s_3 \Delta_2 s_2 \Delta_1 s_1
$$

$$
+ \int_{\prod_{i=1}^3 [a_i, b_i]} p_2(x_2, s_2) \frac{\partial f(\sigma_1(s_1), s_2, \sigma_3(s_3))}{\Delta_2 s_2} \Delta_3 s_3 \Delta_2 s_2 \Delta_1 s_1
$$

$$
+ \int_{\prod_{i=1}^3 [a_i, b_i]} p_3(x_3, s_3) \frac{\partial f(\sigma_1(s_1), \sigma_2(s_2), s_3)}{\Delta_3 s_3} \Delta_3 s_3 \Delta_2 s_2 \Delta_1 s_1
\tag{5.27}
$$

$$
+ \int_{\prod_{i=1}^3 [a_i, b_i]} p_1(x_1, s_1) p_2(x_2, s_2) \frac{\partial^2 f(s_1, s_2, \sigma_3(s_3))}{\Delta_2 s_2 \Delta_1 s_1} \Delta_3 s_3 \Delta_2 s_2 \Delta_1 s_1
$$

$$
+ \int_{\prod_{i=1}^3 [a_i, b_i]} p_1(x_1, s_1) p_3(x_3, s_3) \frac{\partial^2 f(s_1, \sigma_2(s_2), s_3)}{\Delta_3 s_3 \Delta_1 s_1} \Delta_3 s_3 \Delta_2 s_2 \Delta_1 s_1
$$

$$+ \int_{\prod_{i=1}^{3}[a_i,b_i]} p_2(x_2, s_2)\, p_3(x_3, s_3)\, \frac{\partial^2 f(\sigma_1(s_1), s_2, s_3)}{\Delta_3 s_3 \Delta_2 s_2} \Delta_3 s_3 \Delta_2 s_2 \Delta_1 s_1 \Bigg\}$$

$$= \frac{1}{\prod_{i=1}^{3}(b_i - a_i)} \int_{\prod_{i=1}^{3}[a_i,b_i]} p_1(x_1, s_1)\, p_2(x_2, s_2)\, p_3(x_3, s_3)$$

$$\cdot \frac{\partial^3 f(s_1, s_2, s_3)}{\Delta_3 s_3 \Delta_2 s_2 \Delta_1 s_1} \Delta_3 s_3 \Delta_2 s_2 \Delta_1 s_1,$$

by Theorem 5.5.

We denote (see Theorem 5.6)

$$D_n(f)(x_1, \ldots, x_n) := f(x_1, \ldots, x_n) - \frac{1}{\prod_{i=1}^{n}(b_i - a_i)}$$

$$\cdot \left\{ \int_{\prod_{i=1}^{n}[a_i,b_i]} f(\sigma_1(s_1), \sigma_2(s_2), \ldots, \sigma_n(s_n))\, \Delta_n s_n \Delta_{n-1} s_{n-1} \ldots \Delta_1 s_1 \right.$$

$$+ \sum_{j=1}^{n} \int_{\prod_{i=1}^{n}[a_i,b_i]} p_j(x_j, s_j)\, \frac{\partial f(\sigma_1(s_1), \ldots, \sigma_{j-1}(s_{j-1}), s_j, \sigma_{j+1}(s_{j+1}), \ldots, \sigma_n(s_n))}{\Delta_j s_j}$$

$$\cdot \Delta_n s_n \ldots \Delta_1 s_1$$

$$+ \sum_{\substack{l_1=1 \\ j<k}}^{\binom{n}{2}} \left(\int_{\prod_{i=1}^{n}[a_i,b_i]} p_j(x_j, s_j)\, p_k(x_k, s_k) \right. \tag{5.28}$$

$$\cdot \frac{\partial^2 f(\sigma_1(s_1), \ldots, \sigma_{j-1}(s_{j-1}), s_j, \sigma_{j+1}(s_{j+1}), \ldots, \sigma_{k-1}(s_{k-1}), s_k, \sigma_{k+1}(s_{k+1}), \ldots, \sigma_n(s_n))}{\Delta_k s_k \Delta_j s_j}$$

$$\left. \Delta_n s_n \ldots \Delta_1 s_1 \right)_{(l_1)}$$

$$+ \sum_{\substack{l_2=1 \\ j<k<r}}^{\binom{n}{3}} \left(\int_{\prod_{i=1}^{n}[a_i,b_i]} p_j(x_j, s_j)\, p_k(x_k, s_k)\, p_r(x_r, s_r) \right.$$

$$\frac{\partial^3 f\left(\sigma_1\left(s_1\right),\ldots,\sigma_{j-1}\left(s_{j-1}\right),s_j,\sigma_{j+1}\left(s_{j+1}\right),\ldots,\sigma_{k-1}\left(s_{k-1}\right),s_k,\sigma_{k+1}\left(s_{k+1}\right)}{\Delta_r s_r \Delta_k s_k \Delta_j s_j}$$
$$,\ldots,\sigma_{r-1}\left(s_{r-1}\right),s_r,\sigma_{r+1}\left(s_{r+1}\right),\ldots,\sigma_n\left(s_n\right))$$

$$\Delta_n s_n ... \Delta_1 s_1)_{(l_2)}$$

$$+\ldots+\sum_{l=1}^{\binom{n}{n-1}}\left(\int_{\prod\limits_{i=1}^{n}[a_i,b_i]} p_1\left(x_1,s_1\right)...\widehat{p_l\left(x_l,s_l\right)}...p_n\left(x_n,s_n\right)\right.$$

$$\frac{\partial^{n-1}f\left(s_1,s_2,\ldots,\sigma_l\left(s_l\right),\ldots,s_n\right)}{\Delta_n s_n ... \widehat{\Delta_l s_l} ... \Delta_1 s_1}\left.\Delta_n s_n ... \Delta_1 s_1\right)\right\}$$

$$=\frac{1}{\prod\limits_{i=1}^{n}\left(b_i-a_i\right)}\int_{\prod\limits_{i=1}^{n}[a_i,b_i]}\left(\prod_{i=1}^{n}p_i\left(x_i,s_i\right)\right)\frac{\partial^n f\left(s_1,\ldots,s_n\right)}{\Delta_n s_n ... \Delta_1 s_1}\Delta_n s_n ... \Delta_1 s_1,$$

by Theorem 5.6.

We also call

$$D_3^*\left(f\right)\left(x_1,x_2,x_3\right):=\frac{1}{\prod_{i=1}^{3}\left(b_i-a_i\right)}\tag{5.29}$$

$$\cdot\int_{\prod\limits_{i=1}^{3}[a_i,b_i]} p_1^*\left(x_1,s_1\right)p_2^*\left(x_2,s_2\right)p_3^*\left(x_3,s_3\right)\frac{\partial^3 f\left(s_1,s_2,s_3\right)}{\nabla_3 s_3 \nabla_2 s_2 \nabla_1 s_1}\nabla_3 s_3 \nabla_2 s_2 \nabla_1 s_1,$$

the remainder of (5.7).

Similarly we set

$$D_n^*\left(f\right)\left(x_1,x_2,\ldots,x_n\right):=\frac{1}{\prod_{i=1}^{n}\left(b_i-a_i\right)}\tag{5.30}$$

$$\cdot\int_{\prod\limits_{i=1}^{n}[a_i,b_i]}\left(\prod_{i=1}^{n}p_i^*\left(x_i,s_i\right)\right)\frac{\partial^n f\left(s_1,\ldots,s_n\right)}{\nabla_n s_n ... \nabla_1 s_1}\nabla_n s_n ... \nabla_1 s_1,$$

the remainder of (5.8).

By Tietze's extension theorem of General Topology we easily derive that a continuous function f on $\prod_{i=1}^{n}\left([a_i,b_i]\cap\mathbb{T}_i\right)$ is bounded, since its continuous extension F on $\prod_{i=1}^{n}[a_i,b_i]$ is bounded.

We need the general Taylor monomials

$h_0^{(i)}(t,s) = 1$, $h_{k+1}^{(i)}(t,s) = \int_s^t h_k^{(i)}(\tau,s)\,\Delta_i\tau$, $k \in \mathbb{N}_0 = \mathbb{N} \cup \{0\}$, and $\widehat{h_0^{(i)}}(t,s) = 1$,

and $\widehat{h_{k+1}^{(i)}}(t,s) = \int_s^t \widehat{h_k^{(i)}}(\tau,s)\,\nabla_i\tau$, $k \in \mathbb{N}_0$, where $t,s \in \mathbb{T}_i$, a time scale, $i \in \{1,\ldots,n\}$, see [13], [22].

Notice that $h_1^{(i)}(t,s) = t - s$, and $\widehat{h_1^{(i)}}(t,s) = t - s$, so that $h_2^{(i)}(t,s) = \int_s^t (\tau - s)\,\Delta_i\tau$, and $\widehat{h_2^{(i)}}(t,s) = \int_s^t (\tau - s)\,\nabla_i\tau$.

We know that $h_2^{(i)}(t,s)$, $\widehat{h_2^{(i)}}(t,s) \geq 0$, $\forall\, s,t \in \mathbb{T}_i$, see [7], [10].

We present the following multivariate Δ-Ostrowski type inequalities.

Theorem 5.15. *All as in Theorem 5.5 and (5.27). We have that*

$$|D_3(f)(x_1,x_2,x_3)| \leq \frac{1}{\prod_{i=1}^3 (b_i - a_i)} \tag{5.31}$$

$$\cdot \min \begin{cases} \left\| \dfrac{\partial^3 f}{\Delta_3 s_3 \Delta_2 s_2 \Delta_1 s_1} \right\|_{\infty,\, \prod_{i=1}^3 ([a_i,b_i] \cap \mathbb{T}_i)} \left(\displaystyle\prod_{i=1}^3 \left(h_2^{(i)}(x_i,a_i) + h_2^{(i)}(x_i,b_i) \right) \right), \\[2em] \left\| \dfrac{\partial^3 f}{\Delta_3 s_3 \Delta_2 s_2 \Delta_1 s_1} \right\|_{p,\, \prod_{i=1}^3 ([a_i,b_i] \cap \mathbb{T}_i)} \left(\displaystyle\prod_{i=1}^3 \|p_i(x_i,\cdot)\|_{q,[a_i,b_i] \cap \mathbb{T}_i} \right), \\[1em] \text{where } p,q > 1 : \frac{1}{p} + \frac{1}{q} = 1, \\[1em] \left\| \dfrac{\partial^3 f}{\Delta_3 s_3 \Delta_2 s_2 \Delta_1 s_1} \right\|_{1,\, \prod_{i=1}^3 ([a_i,b_i] \cap \mathbb{T}_i)} \left(\displaystyle\prod_{i=1}^3 \left(\frac{b_i - a_i}{2} + \left| x_i - \left(\frac{a_i + b_i}{2} \right) \right| \right) \right). \end{cases}$$

Proof. We have that

$$D_3(f)(x_1,x_2,x_3) = \frac{1}{\prod_{i=1}^3 (b_i - a_i)}$$

$$\cdot \int_{\prod_{i=1}^3 [a_i,b_i]} \prod_{i=1}^3 p_i(x_i,s_i) \frac{\partial^3 f(s_1,s_2,s_3)}{\Delta_3 s_3 \Delta_2 s_2 \Delta_1 s_1} \Delta_3 s_3 \Delta_2 s_2 \Delta_1 s_1,$$

and

$$|D_3(f)(x_1,x_2,x_3)| \leq \frac{1}{\prod_{i=1}^3 (b_i - a_i)}$$

$$\cdot \left\| \frac{\partial^3 f(s_1,s_2,s_3)}{\Delta_3 s_3 \Delta_2 s_2 \Delta_1 s_1} \right\|_{\infty,\, \prod_{i=1}^3 ([a_i,b_i] \cap \mathbb{T}_i)} \prod_{i=1}^3 \left(\int_{a_i}^{b_i} |p_i(x_i,s_i)|\,\Delta_i s_i \right)$$

$$= \frac{1}{\prod_{i=1}^{3}(b_i - a_i)} \left\| \frac{\partial^3 f(s_1, s_2, s_3)}{\Delta_3 s_3 \Delta_2 s_2 \Delta_1 s_1} \right\|_{\infty, \prod_{i=1}^{3}([a_i, b_i] \cap \mathbb{T}_i)}$$

$$\cdot \prod_{i=1}^{3} \left[\int_{a_i}^{x_i} (s_i - a_i) \, \Delta_i s_i + \int_{x_i}^{b_i} (b_i - s_i) \, \Delta_i s_i \right]$$

$$= \frac{1}{\prod_{i=1}^{3}(b_i - a_i)} \left\| \frac{\partial^3 f(s_1, s_2, s_3)}{\Delta_3 s_3 \Delta_2 s_2 \Delta_1 s_1} \right\|_{\infty, \prod_{i=1}^{3}([a_i, b_i] \cap \mathbb{T}_i)}$$

$$\cdot \left(\prod_{i=1}^{3} \left(h_2^{(i)}(x_i, a_i) + h_2^{(i)}(x_i, b_i) \right) \right).$$

Also by Hölder's inequality (see [2]) we get that

$$|D_3(f)(x_1, x_2, x_3)| \leq \frac{1}{\prod_{i=1}^{3}(b_i - a_i)}$$

$$\cdot \left\| \frac{\partial^3 f(s_1, s_2, s_3)}{\Delta_3 s_3 \Delta_2 s_2 \Delta_1 s_1} \right\|_{p, \prod_{i=1}^{3}([a_i, b_i] \cap \mathbb{T}_i)} \prod_{i=1}^{3} \left(\int_{a_i}^{b_i} |p_i(x_i, s_i)|^q \, \Delta_i s_i \right)^{\frac{1}{q}}.$$

Finally we have

$$|D_3(f)(x_1, x_2, x_3)| \leq \frac{1}{\prod_{i=1}^{3}(b_i - a_i)}$$

$$\cdot \left\| \frac{\partial^3 f(s_1, s_2, s_3)}{\Delta_3 s_3 \Delta_2 s_2 \Delta_1 s_1} \right\|_{1, \prod_{i=1}^{3}([a_i, b_i] \cap \mathbb{T}_i)} \left(\prod_{i=1}^{3} \sup_{s_i \in ([a_i, b_i] \cap \mathbb{T}_i)} |p_i(x_i, s_i)| \right)$$

$$\leq \left(\frac{\left\| \frac{\partial^3 f(s_1, s_2, s_3)}{\Delta_3 s_3 \Delta_2 s_2 \Delta_1 s_1} \right\|_{1, \prod_{i=1}^{3}([a_i, b_i] \cap \mathbb{T}_i)}}{\prod_{i=1}^{3}(b_i - a_i)} \right) \left(\prod_{i=1}^{3} \max\{x_i - a_i, b_i - x_i\} \right)$$

$$= \left(\frac{\left\| \frac{\partial^3 f(s_1, s_2, s_3)}{\Delta_3 s_3 \Delta_2 s_2 \Delta_1 s_1} \right\|_{1, \prod_{i=1}^{3}([a_i, b_i] \cap \mathbb{T}_i)}}{\prod_{i=1}^{3}(b_i - a_i)} \right) \left(\prod_{i=1}^{3} \left(\frac{b_i - a_i}{2} + \left| x_i - \left(\frac{a_i + b_i}{2} \right) \right| \right) \right).$$

$\square$

We continue with the generalization

Theorem 5.16. *All as in Theorem 5.6 and (5.28). Then*

$$|D_n(f)(x_1,\ldots,x_n)| \le \frac{1}{\prod_{i=1}^{n}(b_i - a_i)} \tag{5.32}$$

$$\cdot \min \begin{cases} \left\| \frac{\partial^n f}{\Delta_n s_n \ldots \Delta_1 s_1} \right\|_{\infty, \prod\limits_{i=1}^{n}([a_i,b_i]\cap \mathbb{T}_i)} \left(\prod_{i=1}^{n} \left(h_2^{(i)}(x_i, a_i) + h_2^{(i)}(x_i, b_i) \right) \right), \\ \left\| \frac{\partial^n f}{\Delta_n s_n \ldots \Delta_1 s_1} \right\|_{p, \prod\limits_{i=1}^{n}([a_i,b_i]\cap \mathbb{T}_i)} \left(\prod_{i=1}^{n} \| p_i(x_i, \cdot) \|_{q, [a_i,b_i]\cap \mathbb{T}_i} \right), \\ \quad \text{where } p, q > 1 : \frac{1}{p} + \frac{1}{q} = 1, \\ \left\| \frac{\partial^n f}{\Delta_n s_n \ldots \Delta_1 s_1} \right\|_{1, \prod\limits_{i=1}^{n}([a_i,b_i]\cap \mathbb{T}_i)} \left(\prod_{i=1}^{n} \left(\frac{b_i - a_i}{2} + \left| x_i - \left(\frac{a_i + b_i}{2} \right) \right| \right) \right). \end{cases}$$

Proof. Similar to Theorem 5.15. □

Next we give multivariate ∇-Ostrowski type inequalities.

Theorem 5.17. *All as in Theorem 5.7 and (5.29). Then*

$$|D_3^*(f)(x_1, x_2, x_3)| \le \frac{1}{\prod_{i=1}^{3}(b_i - a_i)} \tag{5.33}$$

$$\cdot \min \begin{cases} \left\| \frac{\partial^3 f}{\nabla_3 s_3 \nabla_2 s_2 \nabla_1 s_1} \right\|_{\infty, \prod\limits_{i=1}^{3}([a_i,b_i]\cap \mathbb{T}_i)} \left(\prod_{i=1}^{3} \left(\widehat{h_2^{(i)}}(x_i, a_i) + \widehat{h_2^{(i)}}(x_i, b_i) \right) \right), \\ \left\| \frac{\partial^3 f}{\nabla_3 s_3 \nabla_2 s_2 \nabla_1 s_1} \right\|_{p, \prod\limits_{i=1}^{3}([a_i,b_i]\cap \mathbb{T}_i)} \left(\prod_{i=1}^{3} \| p_i^*(x_i, \cdot) \|_{q, [a_i,b_i]\cap \mathbb{T}_i} \right), \\ \quad \text{where } p, q > 1 : \frac{1}{p} + \frac{1}{q} = 1, \\ \left\| \frac{\partial^3 f}{\nabla_3 s_3 \nabla_2 s_2 \nabla_1 s_1} \right\|_{1, \prod\limits_{i=1}^{3}([a_i,b_i]\cap \mathbb{T}_i)} \left(\prod_{i=1}^{3} \left(\frac{b_i - a_i}{2} + \left| x_i - \left(\frac{a_i + b_i}{2} \right) \right| \right) \right). \end{cases}$$

Proof. As in Theorem 5.15. □

We continue with the generalization of (5.33).

Theorem 5.18. *All as in Theorem 5.8 and (5.30). Then*

$$|D_n^* (f) (x_1, \ldots, x_n)| \leq \frac{1}{\prod_{i=1}^n (b_i - a_i)} \qquad (5.34)$$

$$\cdot \min \begin{cases} \left\| \frac{\partial^n f}{\nabla_n s_n \ldots \nabla_1 s_1} \right\|_{\infty, \prod_{i=1}^n ([a_i, b_i] \cap \mathbb{T}_i)} \left(\prod_{i=1}^n \left(\widehat{h_2^{(i)}} (x_i, a_i) + \widehat{h_2^{(i)}} (x_i, b_i) \right) \right), \\ \left\| \frac{\partial^n f}{\nabla_n s_n \ldots \nabla_1 s_1} \right\|_{p, \prod_{i=1}^n ([a_i, b_i] \cap \mathbb{T}_i)} \left(\prod_{i=1}^n \| p_i^* (x_i, \cdot) \|_{q, [a_i, b_i] \cap \mathbb{T}_i} \right), \\ \text{where } p, q > 1 : \frac{1}{p} + \frac{1}{q} = 1, \\ \left\| \frac{\partial^n f}{\nabla_n s_n \ldots \nabla_1 s_1} \right\|_{1, \prod_{i=1}^n ([a_i, b_i] \cap \mathbb{T}_i)} \left(\prod_{i=1}^n \left(\frac{b_i - a_i}{2} + \left| x_i - \left(\frac{a_i + b_i}{2} \right) \right| \right) \right). \end{cases}$$

Proof.   As in Theorem 5.15.   □

We make

Remark 5.2. (to Theorems 5.9-5.12) We rewrite (5.9) as follows:

$$u_{1,3} := \frac{1}{(A - a)(B - b)(C - c)}$$

$$\cdot \int_a^A \int_b^B \int_c^C p(x, s) q(y, t) \theta(z, r) \frac{\partial^3 f(s, t, r)}{\Delta_3 r \Delta_2 t \Delta_1 s} \Delta_3 r \Delta_2 t \Delta_1 s$$

$$= f(x, y, z) - \left[\frac{1}{(A - a)} \int_a^A f(\sigma_1(s), y, z) \Delta_1 s + \frac{1}{(B - b)} \int_b^B f(x, \sigma_2(t), z) \Delta_2 t \right.$$

$$\left. + \frac{1}{(C - c)} \int_c^C f(x, y, \sigma_3(r)) \Delta_3 r \right]$$

$$+ \left[\frac{1}{(A - a)(B - b)} \int_a^A \int_b^B f(\sigma_1(s), \sigma_2(t), z) \Delta_1 s \Delta_2 t \right.$$

$$+ \frac{1}{(A - a)(C - c)} \int_a^A \int_c^C f(\sigma_1(s), y, \sigma_3(r)) \Delta_1 s \Delta_3 r$$

$$\left. + \frac{1}{(B - b)(C - c)} \int_b^B \int_c^C f(x, \sigma_2(t), \sigma_3(r)) \Delta_2 t \Delta_3 r \right] \qquad (5.35)$$

$$-\frac{1}{(A-a)(B-b)(C-c)}\int_a^A\int_b^B\int_c^C f(\sigma_1(s),\sigma_2(t),\sigma_3(r))\,\Delta_1 s\Delta_2 t\Delta_3 r =: u_{2,3}.$$

We rewrite (5.18) as follows

$$u_{1,n} := \frac{1}{\prod_{i=1}^n (b_i - a_i)} \int_{\prod_{i=1}^n [a_i,b_i]} \prod_{i=1}^n p_i(x_i,s_i) \frac{\partial^n f(s_1,\ldots,s_n)}{\Delta_1 s_1 \ldots \Delta_n s_n} \Delta_1 s_1 \ldots \Delta_n s_n$$

$$= f(x_1,\ldots,x_n) - \left[\sum_{i=1}^{\binom{n}{1}} \left(\frac{1}{(b_i - a_i)} \int_{a_i}^{b_i} f(x_1,\ldots,\sigma_i(s_i),\ldots,x_n)\,\Delta_i s_i \right) \right]$$

$$+ \left[\sum_{l=1}^{\binom{n}{2}} \left(\frac{1}{(b_i - a_i)(b_j - a_j)} \right. \right.$$

$$\left. \left. \cdot \int_{a_i}^{b_i} \int_{a_j}^{b_j} f(x_1,\ldots,\sigma_i(s_i),\ldots,\sigma_j(s_j),\ldots,x_n)\,\Delta_i s_i \Delta_j s_j \right)_{(l)} \right]$$

$$+ - + \ldots - + \ldots + (-1)^{n-1} \left[\sum_{j=1}^{\binom{n}{n-1}} \frac{1}{\left(\prod_{\substack{i=1 \\ i \neq j}}^n (b_i - a_i) \right)} \right. \qquad (5.36)$$

$$\left. \cdot \int_{\substack{\prod_{i=1}^n \\ i \neq j}} f(\sigma_1(s_1),\ldots,x_j,\ldots,\sigma_n(s_n))\,\Delta_1 s_1 \ldots \widehat{\Delta_j s_j} \ldots \Delta_n s_n \right]$$

$$+ \frac{(-1)^n}{\left(\prod_{i=1}^n (b_i - a_i)\right)} \int_{\prod_{i=1}^n [a_i,b_i]} f(\sigma_1(s_1),\ldots,\sigma_n(s_n))\,\Delta_1 s_1 \ldots \Delta_n s_n =: u_{2,n}.$$

We rewrite (5.19) as follows

$$u_{1,3}^* := \frac{1}{(A-a)(B-b)(C-c)}$$

$$\cdot \int_a^A \int_b^B \int_c^C p^*(x,s)\,q^*(y,t)\,\theta^*(z,r) \frac{\partial^3 f(s,t,r)}{\nabla_3 r \nabla_2 t \nabla_1 s} \nabla_3 r \nabla_2 t \nabla_1 s$$

$$= f(x, y, z) - \left[\frac{1}{(A-a)} \int_a^A f(\rho_1(s), y, z) \nabla_1 s + \frac{1}{(B-b)} \int_b^B f(x, \rho_2(t), z) \nabla_2 t \right.$$

$$\left. + \frac{1}{(C-c)} \int_c^C f(x, y, \rho_3(r)) \nabla_3 r \right]$$

$$+ \left[\frac{1}{(A-a)(B-b)} \int_a^A \int_b^B f(\rho_1(s), \rho_2(t), z) \nabla_1 s \nabla_2 t \right.$$

$$+ \frac{1}{(A-a)(C-c)} \int_a^A \int_c^C f(\rho_1(s), y, \rho_3(r)) \nabla_1 s \nabla_3 r$$

$$\left. + \frac{1}{(B-b)(C-c)} \int_b^B \int_c^C f(x, \rho_2(t), \rho_3(r)) \nabla_2 t \nabla_3 r \right] \tag{5.37}$$

$$- \frac{1}{(A-a)(B-b)(C-c)} \int_a^A \int_b^B \int_c^C f(\rho_1(s), \rho_2(t), \rho_3(r)) \nabla_1 s \nabla_2 t \nabla_3 r =: u_{2,3}^*.$$

We finally rewrite (5.20) as follows

$$u_{1,n}^* := \frac{1}{\prod_{i=1}^n (b_i - a_i)} \int_{\prod_{i=1}^n [a_i, b_i]} \prod_{i=1}^n p_i^*(x_i, s_i) \frac{\partial^n f(s_1, \ldots, s_n)}{\nabla_1 s_1 \ldots \nabla_n s_n} \nabla_1 s_1 \ldots \nabla_n s_n$$

$$= f(x_1, \ldots, x_n) - \left[\sum_{i=1}^{\binom{n}{1}} \frac{1}{(b_i - a_i)} \int_{a_i}^{b_i} f(x_1, \ldots, \rho_i(s_i), \ldots, x_n) \nabla_i s_i \right]$$

$$+ \left[\sum_{l=1}^{\binom{n}{2}} \frac{1}{(b_i - a_i)(b_j - a_j)} \right.$$

$$\left. \cdot \left(\int_{a_i}^{b_i} \int_{a_j}^{b_j} f(x_1, \ldots, \rho_i(s_i), \ldots, \rho_j(s_j), \ldots, x_n) \nabla_i s_i \nabla_j s_j \right)_{(l)} \right]$$

$$+ - + \ldots - + \ldots + (-1)^{n-1} \left[\sum_{j=1}^{\binom{n}{n-1}} \frac{1}{\left(\prod_{\substack{i=1 \\ i \neq j}}^n (b_i - a_i) \right)} \right. \tag{5.38}$$

$$\cdot \int_{\substack{\prod_{i=1}^{n}[a_i,b_i] \\ i \neq j}} f\left(\rho_1\left(s_1\right),\ldots,x_j,\ldots,\rho_n\left(s_n\right)\right) \nabla_1 s_1 \ldots \widehat{\nabla_j s_j} \ldots \nabla_n s_n \Bigg]$$

$$+ \frac{(-1)^n}{\prod_{i=1}^{n}\left(b_i - a_i\right)} \int_{\prod_{i=1}^{n}[a_i,b_i]} f\left(\rho_1\left(s_1\right),\ldots,\rho_n\left(s_n\right)\right) \nabla_1 s_1 \ldots \nabla_n s_n =: u_{2,n}^*.$$

We notice that $u_{1,3}$ is the same as $D_3\left(f\right)\left(x_1,x_2,x_3\right)$, see (5.27), $u_{1,n} = D_n\left(f\right)\left(x_1,\ldots,x_n\right)$, see (5.28), $u_{1,3}^*$ is the same as $D_3^*\left(f\right)\left(x_1,x_2,x_3\right)$, see (5.29), and $u_{1,n}^* = D_n^*\left(f\right)\left(x_1,\ldots,x_n\right)$, see (5.30).

Therefore Theorems 5.15-5.18 apply the same way for $u_{1,3}$, $u_{1,n}$, $u_{1,3}^*$ and $u_{1,n}^*$.

Remark 5.3. Formulae (5.35)-(5.38) simplify a lot as follows:

If $f\left(x,\cdot,\cdot\right) = f\left(\cdot,y,\cdot\right) = f\left(\cdot,\cdot,z\right) = 0$, then (5.35) collapses to

$$\int_a^A \int_b^B \int_c^C p\left(x,s\right) q\left(y,t\right) \theta\left(z,r\right) \frac{\partial^3 f\left(s,t,r\right)}{\Delta_3 r \Delta_2 t \Delta_1 s} \Delta_3 r \Delta_2 t \Delta_1 s \qquad (5.39)$$

$$= -\int_a^A \int_b^B \int_c^C f\left(\sigma_1\left(s\right),\sigma_2\left(t\right),\sigma_3\left(r\right)\right) \Delta_1 s \Delta_2 t \Delta_3 r,$$

also it holds (by (5.37))

$$\int_a^A \int_b^B \int_c^C p^*\left(x,s\right) q^*\left(y,t\right) \theta^*\left(z,r\right) \frac{\partial^3 f\left(s,t,r\right)}{\nabla_3 r \nabla_2 t \nabla_1 s} \nabla_3 r \nabla_2 t \nabla_1 s \qquad (5.40)$$

$$= -\int_a^A \int_b^B \int_c^C f\left(\rho_1\left(s\right),\rho_2\left(t\right),\rho_3\left(r\right)\right) \nabla_1 s \nabla_2 t \nabla_3 r.$$

If we assume $f\left(x_1,\cdot,\cdot,\ldots,\cdot\right) = f\left(\cdot,x_2,\cdot,\ldots,\cdot\right) = \ldots = f\left(\cdot,\cdot,\ldots,\cdot,x_n\right) = 0$, then (5.36) collapses to

$$\int_{\prod_{i=1}^{n}[a_i,b_i]} \prod_{i=1}^{n} p_i\left(x_i,s_i\right) \frac{\partial^n f\left(s_1,\ldots,s_n\right)}{\Delta_1 s_1 \ldots \Delta_n s_n} \Delta_1 s_1 \ldots \Delta_n s_n \qquad (5.41)$$

$$= (-1)^n \int_{\prod_{i=1}^{n}[a_i,b_i]} f\left(\sigma_1\left(s_1\right),\ldots,\sigma_n\left(s_n\right)\right) \Delta_1 s_1 \ldots \Delta_n s_n,$$

and (5.38) collapses to

$$\int_{\prod_{i=1}^{n}[a_i,b_i]} \prod_{i=1}^{n} p_i^*\left(x_i,s_i\right) \frac{\partial^n f\left(s_1,\ldots,s_n\right)}{\nabla_1 s_1 \ldots \nabla_n s_n} \nabla_1 s_1 \ldots \nabla_n s_n \qquad (5.42)$$

$$= (-1)^n \int_{\prod_{i=1}^{n}[a_i,b_i]} f\left(\rho_1\left(s_1\right),\ldots,\rho_n\left(s_n\right)\right) \nabla_1 s_1 \ldots \nabla_n s_n.$$

Another simplification of (5.35)-(5.38) works as follows:

In (5.35) assume that all marginal integrals are zero, then

$$\frac{1}{(A-a)(B-b)(C-c)}$$

$$\cdot \int_a^A \int_b^B \int_c^C p(x,s)\, q(y,t)\, \theta(z,r)\, \frac{\partial^3 f(s,t,r)}{\Delta_3 r \Delta_2 t \Delta_1 s}\, \Delta_3 r \Delta_2 t \Delta_1 s$$

$$= f(x,y,z) - \frac{1}{(A-a)(B-b)(C-c)}$$

$$\cdot \int_a^A \int_b^B \int_c^C f(\sigma_1(s), \sigma_2(t), \sigma_3(r))\, \Delta_1 s \Delta_2 t \Delta_3 r. \qquad (5.43)$$

If in (5.36) all marginal integrals are zero then

$$\frac{1}{\prod_{i=1}^n (b_i - a_i)} \int_{\prod_{i=1}^n [a_i,b_i]} \prod_{i=1}^n p_i(x_i,s_i)\, \frac{\partial^n f(s_1,\dots,s_n)}{\Delta_1 s_1 \dots \Delta_n s_n}\, \Delta_1 s_1 \dots \Delta_n s_n \qquad (5.44)$$

$$= f(x_1,\dots,x_n) + \frac{(-1)^n}{\prod_{i=1}^n (b_i - a_i)} \int_{\prod_{i=1}^n [a_i,b_i]} f(\sigma_1(s_1),\dots,\sigma_n(s_n))\, \Delta_1 s_1 \dots \Delta_n s_n.$$

If in (5.37) all marginal integrals are zero then

$$\frac{1}{(A-a)(B-b)(C-c)}$$

$$\cdot \int_a^A \int_b^B \int_c^C p^*(x,s)\, q^*(y,t)\, \theta^*(z,r)\, \frac{\partial^3 f(s,t,r)}{\nabla_3 r \nabla_2 t \nabla_1 s}\, \nabla_3 r \nabla_2 t \nabla_1 s$$

$$= f(x,y,z) - \frac{1}{(A-a)(B-b)(C-c)}$$

$$\cdot \int_a^A \int_b^B \int_c^C f(\rho_1(s), \rho_2(t), \rho_3(r))\, \nabla_1 s \nabla_2 t \nabla_3 r. \qquad (5.45)$$

Finally if in (5.38) all marginal integrals are zero then

$$\frac{1}{\prod_{i=1}^n (b_i - a_i)} \int_{\prod_{i=1}^n [a_i,b_i]} \prod_{i=1}^n p_i^*(x_i,s_i)\, \frac{\partial^n f(s_1,\dots,s_n)}{\nabla_1 s_1 \dots \nabla_n s_n}\, \nabla_1 s_1 \dots \nabla_n s_n \qquad (5.46)$$

$$= f(x_1,\dots,x_n) + \frac{(-1)^n}{\prod_{i=1}^n (b_i - a_i)} \int_{\prod_{i=1}^n [a_i,b_i]} f(\rho_1(s_1),\dots,\rho_n(s_n))\, \nabla_1 s_1 \dots \nabla_n s_n.$$

Based on the simplification formulae (5.39)-(5.42) and (5.43)-(5.46) the left hand sides of (5.31)-(5.34) simplify a lot and take very interesting forms.

5.3 Applications

When $\mathbb{T}_i = \mathbb{R}$, $i = 1, \ldots, n$, this chapter's results were published in [4], [5], [6], [8], [9], so this work is their generalization to time scales.

When $\mathbb{T}_i = \mathbb{Z}$, $i = 1, \ldots, n$, we have (see [22], pp. 13-14)

$f^{\Delta^n}(t) = \sum_{k=0}^{n} \binom{n}{k} (-1)^{n-k} f(t+k)$, $\sigma(t) = t + 1$, $\rho(t) = t - 1$, $h_k(t,s) = \frac{(t-s)^{(k)}}{k!}$, $\forall\ k \in \mathbb{N}_0$, $\forall\ t, s \in \mathbb{Z}$, where $t^{(0)} = 1$, $t^{(k)} = \prod_{i=0}^{k-1}(t-i)$, $k \in \mathbb{N}$, $\int_a^b f(t)\,\Delta t = \sum_{t=a}^{b-1} f(t)$, $a < b$, and an rd-continuous or a continuous function f corresponds to any f.

Also, see [22], p. 333 and [12], pp. 652, 653, that

$f^{\nabla^n}(t) = \sum_{m=0}^{n} (-1)^m \binom{n}{m} f(t-m)$, $\widehat{h}_k(t,s) = \frac{(t-s)^{\overline{k}}}{k!}$, for all $s, t \in \mathbb{Z}$, $k \in \mathbb{N}_0$,

where $t^{\overline{k}} = t(t+1)\ldots(t+k-1)$, $k \in \mathbb{N}$; $t^{\overline{0}} := 1$, and an ld-continuous function f corresponds to any f.

Furthermore we have $\int_a^b f(t)\,\nabla t = \sum_{t=a+1}^{b} f(t)$, $a < b$.

When $\mathbb{T}_i = \mathbb{R}$, then $\sigma(t) = \rho(t) = t$, and integrals and derivatives coincide with the usual numerical ones.

To keep chapter's size limited we restrict ourselves only on $\mathbb{Z}$ or combinations of $\mathbb{Z}$ and $\mathbb{R}$. One can give many other interesting applications on many other time scales or combinations of them.

We make

Remark 5.4. Here we write $E_1(f)$ of (5.21) and $E_2(f)$ of (5.22) for the time scale $\mathbb{Z}$. We have

$$\overline{E_1}(f) := f(x) - \frac{1}{b-a}\left(\sum_{s_1=a}^{b-1} f(s_1+1)\right) - \frac{1}{b-a}\left(\sum_{k=1}^{n-1}\left(\sum_{s_{k+1}=a}^{b-1}\sum_{s_k=a}^{b-1}\right.\right. \tag{5.47}$$

$$\cdots \left(\sum_{s_1=a}^{b-1} P(x,s_1)\prod_{i=1}^{k-1} P(s_i,s_{i+1})\left(\sum_{m=0}^{k}\binom{k}{m}(-1)^{k-m} f(s_{k+1}+m+1)\right)\right)\Bigg)\Bigg),$$

and

$$\overline{E_2}(f) := f(x) - \frac{1}{b-a}\left(\sum_{s_1=a+1}^{b} f(s_1-1)\right) - \frac{1}{b-a}\left(\sum_{k=1}^{n-1}\left(\sum_{s_{k+1}=a+1}^{b}\sum_{s_k=a+1}^{b}\right.\right. \tag{5.48}$$

$$\cdots \left(\sum_{s_1=a+1}^{b} P^*(x,s_1)\prod_{i=1}^{k-1} P^*(s_i,s_{i+1})\left(\sum_{m=0}^{k}(-1)^m\binom{k}{m} f(s_{k+1}-m-1)\right)\right)\Bigg)\Bigg).$$

By Theorem 5.3 applied on $\mathbb{Z}$ we have that

$$\overline{E_1}(f) = \sum_{s_n=a}^{b-1}\left(\sum_{s_{n-1}=a}^{b-1}\cdots\left(\sum_{s_1=a}^{b-1} P(x,s_1)\prod_{i=1}^{n-1} P(s_i,s_{i+1})\right.\right. \tag{5.49}$$

$$\left(\sum_{m=0}^{n} \binom{n}{m} (-1)^{n-m} f\left(s_n + m\right) \right) \right) ,$$

and by Theorem 5.6 it holds

$$\overline{E_2}\left(f\right) = \sum_{s_n=a+1}^{b} \left(\sum_{s_{n-1}=a+1}^{b} \cdots \left(\sum_{s_1=a+1}^{b} P^*\left(x, s_1\right) \prod_{i=1}^{n-1} P^*\left(s_i, s_{i+1}\right) \right. \right. \tag{5.50}$$

$$\left. \left. \left(\sum_{m=0}^{n} (-1)^m \binom{n}{m} f\left(s_n - m\right) \right) \right) \right) .$$

Here P as in Theorem 5.3 and P^* as in Theorem 5.4; $a, b \in \mathbb{Z} : a < b$, $x \in [a, b] \cap \mathbb{Z}$, $n \in \mathbb{N}$.

We have

Theorem 5.19. *It holds*

$$|\overline{E_1}\left(f\right)| \leq \min \begin{cases} \left\|f^{\Delta^n}\right\|_{\infty,([a,b]\cap\mathbb{Z})} \left(\sum_{s_1,s_2,\ldots,s_n=a}^{b-1} |P\left(x,s_1\right)| \prod_{i=1}^{n-1} |P\left(s_i,s_{i+1}\right)| \right), \\ \left\|f^{\Delta^n}\right\|_{p,([a,b]\cap\mathbb{Z})} \left(\sum_{s_1,\ldots,s_{n-1}=a}^{b-1} |P\left(x,s_1\right)| \left(\prod_{i=1}^{n-2} |P\left(s_i,s_{i+1}\right)| \right) \right. \\ \qquad \left. \cdot \left\|P\left(s_{n-1},\cdot\right)\right\|_{q,([a,b]\cap\mathbb{Z})} \right), \\ \qquad where \ p, q > 1 : \frac{1}{p} + \frac{1}{q} = 1, \\ \left\|f^{\Delta^n}\right\|_{1,([a,b]\cap\mathbb{Z})} \sum_{s_1,\ldots,s_{n-1}=a}^{b-1} |P\left(x,s_1\right)| \left(\prod_{i=1}^{n-2} |P\left(s_i,s_{i+1}\right)| \right) \\ \qquad \cdot \left\|P\left(s_{n-1},\cdot\right)\right\|_{\infty,([a,b]\cap\mathbb{Z})} . \end{cases}$$

$$\tag{5.51}$$

Proof. By Theorem 5.13 applied on $\mathbb{Z}$. □

We continue with

Theorem 5.20. *It holds*

$$|\overline{E_2}\left(f\right)| \leq \min \begin{cases} \left\|f^{\nabla^n}\right\|_{\infty,([a,b]\cap\mathbb{Z})} \left(\sum_{s_1,\ldots,s_n=a+1}^{b} |P^*\left(x,s_1\right)| \prod_{i=1}^{n-1} |P^*\left(s_i,s_{i+1}\right)| \right), \\ \left\|f^{\nabla^n}\right\|_{p,([a,b]\cap\mathbb{Z})} \left(\sum_{s_1,\ldots,s_{n-1}=a+1}^{b} |P^*\left(x,s_1\right)| \left(\prod_{i=1}^{n-2} |P^*\left(s_i,s_{i+1}\right)| \right) \right. \\ \qquad \left. \cdot \left\|P^*\left(s_{n-1},\cdot\right)\right\|_{q,([a,b]\cap\mathbb{Z})} \right), \\ \qquad where \ p, q > 1 : \frac{1}{p} + \frac{1}{q} = 1, \\ \left\|f^{\nabla^n}\right\|_{1,([a,b]\cap\mathbb{Z})} \sum_{s_1,\ldots,s_{n-1}=a+1}^{b} |P^*\left(x,s_1\right)| \left(\prod_{i=1}^{n-2} |P^*\left(s_i,s_{i+1}\right)| \right) \\ \qquad \cdot \left\|P^*\left(s_{n-1},\cdot\right)\right\|_{\infty,([a,b]\cap\mathbb{Z})} . \end{cases}$$

$$\tag{5.52}$$

Proof. By Theorem 5.14. □

We make

Remark 5.5. We apply (5.36) for $n = 3$ to get

$$u_{1,3} = \frac{1}{\prod_{i=1}^{3}(b_i - a_i)} \int_{\prod_{i=1}^{3}[a_i,b_i]} \prod_{i=1}^{3} p_i(x_i, s_i) \frac{\partial^3 f(s_1, s_2, s_3)}{\Delta_1 s_1 \Delta_2 s_2 \Delta_3 s_3} \Delta_1 s_1 \Delta_2 s_2 \Delta_3 s_3$$

$$= f(x_1, x_2, x_3) - \left[\sum_{i=1}^{3} \left(\frac{1}{(b_i - a_i)} \int_{a_i}^{b_i} f(x_1, \sigma_i(s_i), x_3) \Delta_i s_i \right) \right] \qquad (5.53)$$

$$+ \left[\sum_{l=1}^{3} \left(\frac{1}{(b_i - a_i)(b_j - a_j)} \int_{a_i}^{b_i} \int_{a_j}^{b_j} f(x_1, \sigma_i(s_i), \sigma_j(s_j)) \Delta_i s_i \Delta_j s_j \right)_{(l)} \right]$$

$$- \frac{1}{\prod_{i=1}^{3}(b_i - a_i)} \int_{\prod_{i=1}^{3}[a_i,b_i]} f(\sigma_1(s_1), \sigma_2(s_2), \sigma_3(s_3)) \Delta_1 s_1 \Delta_2 s_2 \Delta_3 s_3$$

$$= u_{2,3} = D_3(f)(x_1, x_2, x_3),$$

see (5.27).

Also we apply (5.38) for $n = 3$ to get

$$u_{1,3}^* := \frac{1}{\prod_{i=1}^{3}(b_i - a_i)} \int_{\prod_{i=1}^{3}[a_i,b_i]} \prod_{i=1}^{3} p_i^*(x_i, s_i) \frac{\partial^3 f(s_1, s_2, s_3)}{\nabla_1 s_1 \nabla_2 s_2 \nabla_3 s_3} \nabla_1 s_1 \nabla_2 s_2 \nabla_3 s_3$$

$$= f(x_1, x_2, x_3) - \left[\sum_{i=1}^{3} \frac{1}{(b_i - a_i)} \int_{a_i}^{b_i} f(x_1, \rho_i(s_i), x_3) \nabla_i s_i \right] \qquad (5.54)$$

$$+ \left[\sum_{l=1}^{3} \frac{1}{(b_i - a_i)(b_j - a_j)} \left(\int_{a_i}^{b_i} \int_{a_j}^{b_j} f(x_1, \rho_i(s_i), \rho_j(s_j)) \nabla_i s_i \nabla_j s_j \right)_{(l)} \right]$$

$$- \frac{1}{\prod_{i=1}^{3}(b_i - a_i)} \int_{\prod_{i=1}^{3}[a_i,b_i]} f(\rho_1(s_1), \rho_2(s_2), \rho_3(s_3)) \nabla_1 s_1 \nabla_2 s_2 \nabla_3 s_3$$

$$= u_{2,3}^* = D_3^*(f)(x_1, x_2, x_3),$$

see (5.29).

Next we will apply Theorems 5.5 and 5.7 for $\mathbb{T}_1 = \mathbb{Z}$, $\mathbb{T}_2 = \mathbb{R}$, $\mathbb{T}_3 = \mathbb{Z}$. Here $a_i < b_i$, $i = 1, 2, 3$; $a_1, b_1 \in \mathbb{Z}$; $a_2, b_2 \in \mathbb{R}$; $a_3, b_3 \in \mathbb{Z}$, and $f \in C^3\left(\prod_{i=1}^{3}([a_i, b_i] \cap \mathbb{T}_i)\right)$. Let $(x_1, x_2, x_3) \in (([a_1, b_1] \cap \mathbb{Z}) \times [a_2, b_2] \times ([a_3, b_3] \cap \mathbb{Z}))$, p_i as in Theorem 5.5, $i = 1, 2, 3$, and p_i^* as in Theorem 5.7, $i = 1, 2, 3$.

For $\mathbb{T}_2 = \mathbb{R}$ we have $\sigma_2(t) = \rho_2(t) = t$, $\forall\, t \in \mathbb{R}$, and $\widehat{h_k^{(2)}}(t, s) = \widehat{h_k^{(2)}}(t, s) = \frac{(t-s)^k}{k!}$, $\forall\, s, t \in \mathbb{R}$, $k \in \mathbb{N}_0$.

Thus we apply the general formulae (5.53) and (5.54) to our specific setting $(\mathbb{Z} \times \mathbb{R} \times \mathbb{Z})$. We have

$$\overline{u_{1,3}} := \frac{1}{\prod_{i=1}^{3}(b_i - a_i)}$$

$$\cdot \left(\sum_{s_1=a_1}^{b_1-1} \left(\sum_{s_3=a_3}^{b_3-1} \left(\int_{a_2}^{b_2} \left(\prod_{i=1}^{3} p_i(x_i, s_i) \right) \frac{\partial^3 f(s_1, s_2, s_3)}{\Delta_1 s_1 \partial s_2 \Delta_3 s_3} ds_2 \right) \right) \right)$$

$$= f(x_1, x_2, x_3) - \left[\frac{1}{b_1 - a_1} \sum_{s_1=a_1}^{b_1-1} f(s_1 + 1, x_2, x_3) \right. \tag{5.55}$$

$$\left. + \frac{1}{b_2 - a_2} \int_{a_2}^{b_2} f(x_1, s_2, x_3) ds_2 + \frac{1}{b_3 - a_3} \sum_{s_3=a_3}^{b_3-1} f(x_1, x_2, s_3 + 1) \right]$$

$$+ \left[\frac{1}{(b_1 - a_1)(b_2 - a_2)} \left(\sum_{s_1=a_1}^{b_1-1} \left(\int_{a_2}^{b_2} f(s_1 + 1, s_2, x_3) ds_2 \right) \right) \right.$$

$$+ \frac{1}{(b_2 - a_2)(b_3 - a_3)} \left(\sum_{s_3=a_3}^{b_3-1} \left(\int_{a_2}^{b_2} f(x_1, s_2, s_3 + 1) ds_2 \right) \right)$$

$$\left. + \frac{1}{(b_1 - a_1)(b_3 - a_3)} \left(\sum_{s_1=a_1}^{b_1-1} \sum_{s_3=a_3}^{b_3-1} f(s_1 + 1, x_2, s_3 + 1) \right) \right]$$

$$- \frac{1}{\prod_{i=1}^{3}(b_i - a_i)} \left(\sum_{s_1=a_1}^{b_1-1} \left(\sum_{s_3=a_3}^{b_3-1} \left(\int_{a_2}^{b_2} f(s_1 + 1, s_2, s_3 + 1) ds_2 \right) \right) \right)$$

$$=: \overline{u_{2,3}} =: \overline{D_3}(f)(x_1, x_2, x_3).$$

Equality (5.55) is valid by Theorem 5.5.

We also have

$$u_{1,3}^* := \frac{1}{\prod_{i=1}^{3}(b_i - a_i)}$$

$$\cdot \left(\sum_{s_1=a_1+1}^{b_1} \left(\sum_{s_3=a_3+1}^{b_3} \left(\int_{a_2}^{b_2} \left(\prod_{i=1}^{3} p_i^*(x_i, s_i) \right) \frac{\partial^3 f(s_1, s_2, s_3)}{\nabla_1 s_1 \partial s_2 \nabla_3 s_3} ds_2 \right) \right) \right)$$

$$= f(x_1, x_2, x_3) - \left[\frac{1}{b_1 - a_1} \sum_{s_1=a_1+1}^{b_1} f(s_1 - 1, x_2, x_3) \right. \tag{5.56}$$

$$+ \frac{1}{b_2 - a_2} \int_{a_2}^{b_2} f(x_1, s_2, x_3) \, ds_2 + \frac{1}{b_3 - a_3} \sum_{s_3 = a_3 + 1}^{b_3} f(x_1, x_2, s_3 - 1) \Bigg]$$

$$+ \left[\frac{1}{(b_1 - a_1)(b_2 - a_2)} \left(\sum_{s_1 = a_1 + 1}^{b_1} \left(\int_{a_2}^{b_2} f(s_1 - 1, s_2, x_3) \, ds_2 \right) \right) \right.$$

$$+ \frac{1}{(b_2 - a_2)(b_3 - a_3)} \left(\sum_{s_3 = a_3 + 1}^{b_3} \left(\int_{a_2}^{b_2} f(x_1, s_2, s_3 - 1) \, ds_2 \right) \right)$$

$$+ \frac{1}{(b_1 - a_1)(b_3 - a_3)} \left(\sum_{s_1 = a_1 + 1}^{b_1} \sum_{s_3 = a_3 + 1}^{b_3} f(s_1 - 1, x_2, s_3 - 1) \right) \Bigg]$$

$$- \frac{1}{\prod_{i=1}^{3} (b_i - a_i)} \left(\sum_{s_1 = a_1 + 1}^{b_1} \left(\sum_{s_3 = a_3 + 1}^{b_3} \left(\int_{a_2}^{b_2} f(s_1 - 1, s_2, s_3 - 1) \, ds_2 \right) \right) \right)$$

$$=: \overline{u_{2,3}^*} =: \overline{D_3^*}(f)(x_1, x_2, x_3).$$

Equality (5.56) is valid by Theorem 5.7.

We give the following special Δ-multivariate Ostrowski type inequality.

Theorem 5.21. *It holds*

$$\left| \overline{D_3}(f)(x_1, x_2, x_3) \right| \leq \frac{1}{\prod_{i=1}^{3} (b_i - a_i)} \tag{5.57}$$

$$\cdot \min \left\{ \begin{array}{l} \frac{1}{8} \left\| \frac{\partial^3 f}{\Delta_1 s_1 \partial s_2 \Delta_3 s_3} \right\|_{\infty, \prod_{i=1}^{3} ([a_i, b_i] \cap \mathbb{T}_i)} \left\{ \left((x_1 - a_1)^{(2)} + (x_1 - b_1)^{(2)} \right) \right. \\ \left. \left((x_2 - a_2)^2 + (x_2 - b_2)^2 \right) \left((x_3 - a_3)^{(2)} + (x_3 - b_3)^{(2)} \right) \right\}, \\ \left\| \frac{\partial^3 f}{\Delta_1 s_1 \partial s_2 \Delta_3 s_3} \right\|_{p, \prod_{i=1}^{3} ([a_i, b_i] \cap \mathbb{T}_i)} \left(\prod_{i=1}^{3} \| p_i(x_i, \cdot) \|_{q, [a_i, b_i] \cap \mathbb{T}_i} \right), \\ \textit{where } p, q > 1 : \frac{1}{p} + \frac{1}{q} = 1, \\ \left\| \frac{\partial^3 f}{\Delta_1 s_1 \partial s_2 \Delta_3 s_3} \right\|_{1, \prod_{i=1}^{3} ([a_i, b_i] \cap \mathbb{T}_i)} \left(\prod_{i=1}^{3} \left(\frac{b_i - a_i}{2} + \left| x_i - \left(\frac{a_i + b_i}{2} \right) \right| \right) \right). \end{array} \right.$$

Proof. By Theorem 5.15 applied on $\mathbb{Z} \times \mathbb{R} \times \mathbb{Z}$. $\qquad \qquad \square$

We finish with the next specific ∇-multivariate Ostrowski type inequality.

Theorem 5.22. *It holds*

$$\left| \overline{D_3^*} (f) (x_1, x_2, x_3) \right| \le \frac{1}{\prod_{i=1}^{3} (b_i - a_i)} \tag{5.58}$$

$$\cdot \min \left\{ \begin{array}{l} \frac{1}{8} \left\| \frac{\partial^3 f}{\nabla_1 s_1 \partial s_2 \nabla_3 s_3} \right\|_{\infty, \prod_{i=1}^{3} ([a_i, b_i] \cap \mathbb{T}_i)} \left\{ \left((x_1 - a_1)^{\overline{2}} + (x_1 - b_1)^{\overline{2}} \right) \right. \\ \left. \left((x_2 - a_2)^2 + (x_2 - b_2)^2 \right) \left((x_3 - a_3)^{\overline{2}} + (x_3 - b_3)^{\overline{2}} \right) \right\}, \\ \left\| \frac{\partial^3 f}{\nabla_1 s_1 \partial s_2 \nabla_3 s_3} \right\|_{p, \prod_{i=1}^{3} ([a_i, b_i] \cap \mathbb{T}_i)} \left(\prod_{i=1}^{3} \| p_i^* (x_i, \cdot) \|_{q, [a_i, b_i] \cap \mathbb{T}_i} \right), \\ where \ p, q > 1 : \frac{1}{p} + \frac{1}{q} = 1, \\ \left\| \frac{\partial^3 f}{\nabla_1 s_1 \partial s_2 \nabla_3 s_3} \right\|_{1, \prod_{i=1}^{3} ([a_i, b_i] \cap \mathbb{T}_i)} \left(\prod_{i=1}^{3} \left(\frac{b_i - a_i}{2} + \left| x_i - \left(\frac{a_i + b_i}{2} \right) \right| \right) \right) \end{array} \right).$$

Proof. By Theorem 5.17 applied on $\mathbb{Z} \times \mathbb{R} \times \mathbb{Z}$. □

Bibliography

1. R. Agarwal, M. Bohner, *Basic calculus on time scales and some of its applications*, Results Math. 35 (1-2) (1999), 3-22.
2. R. Agarwal, M. Bohner, A. Peterson, *Inequalities on time scales: a survey*, Math. Inequalities Appl. 4 (4) (2001), 535-557.
3. G.A. Anastassiou, *Multivariate Ostrowski type inequalities*, Acta Math. Hungarica 76 (4) (1997), 267-278.
4. G. Anastassiou, *Univariate Ostrowski inequalities, revisited*, Monatsh Math. 135 (2002), 175-189.
5. G.A. Anastassiou, *Multivariate Montgomery identities and Ostrowski inequalities*, Numer. Funct. Anal. Opt. 23 (3-4) (2002), 247-263.
6. G.A. Anastassiou, *Multidimensional Ostrowski inequalities, revisited*, Acta Math. Hungarica 97 (4) (2002), 339-353.
7. G.A. Anastassiou, *Time scales inequalities*, Intern. J. Difference Equations 5 (1) (2010), 1-23.
8. G.A. Anastassiou, *Probabilistic Inequalities*, World Scientific, Singapore, 2010.
9. G.A. Anastassiou, *Advanced Inequalities*, World Scientific, Singapore, 2011.
10. G.A. Anastassiou, *Nabla Time Scales Inequalities*, Editor Al. Paterson, special issue on Time Scales, Internat. J. Dynam. Syst. Difference Equations 3 (1/2) (2011), 59-83.
11. G.A. Anastassiou, *Representations and Ostrowski type inequalities on time scales*, Comput. Math. 62 (2011), 3933-3958.
12. G.A. Anastassiou, *Intelligent Mathematics: Computational Analysis*, Springer, New York, 2011.
13. D.R. Anderson, *Taylor Polynomials for nabla dynamic equations on time scales*, Panamer. Math. J. 12 (4) (2002), 17-27.
14. D. Anderson, J. Bullock, L. Erbe, A. Peterson, H. Tran, *Nabla dynamic equations on time scales*, Panamer. Math. J. 13 (1) (2003), 1-47.
15. F. Atici, D. Biles, A. Lebedinsky, *An application of time scales to economics*, Math. Comput. Modelling 43 (2006), 718-726.
16. M. Bohner and G. Sh. Guseinov, *Partial differentiation on time scales*, Dynamic Syst. Appl. 13 (3-4) (2004), 351-379.
17. M. Bohner, G. Guseinov, *Multiple integration on time scales*, Dynamic Syst. Appl. 14 (3-4) (2005), 579-606.
18. M. Bohner, G.S. Guseinov, *Multiple Lebesgue integration on time scales*, Adv. Difference Equations 2006 (2006), Article ID 26391, 1-12.

19. M. Bohner, G. Guseinov, *Double integral calculus of variations on times scales*, Comput. Math. Appl. 54 (2007), 45-57.
20. M. Bohner, H. Luo, *Singular second-order multipoint dynamic boundary value problems with mixed derivatives*, Adv. Difference Equations 2006 (2006), Article ID 54989, 1-15.
21. M. Bohner, T. Matthews, *Ostrowski inequalities on time scales*, JIPAM. J. Inequal. Pure Appl. Math. 9 (1) (2008), Article 6, 8 pp.
22. M. Bohner, A. Peterson, *Dynamic equations on Time Scales: An Introduction with Applications*, Birkhaüser, Boston (2001).
23. G. Guseinov, *Integration on time scales*, J. Math. Anal. Appl. 285 (2003), 107-127.
24. R. Higgins, A. Peterson, *Cauchy functions and Taylor's formula for time scales* $\mathbb{T}$, in B. Aulbach, S. Elaydi, G. Ladas (eds.), Proc. Sixth Internat. Conf. Difference Equations, New Progress in Difference Equations, Augsburg, Germany, 2001, pp. 299-308, Chapman & Hall/CRC (2004).
25. S. Hilger, *Ein Maßketten kalkül mit Anwendung auf Zentrumsmannig-faltigkeiten*, Ph.D. thesis, Universität Würzburg, Germany (1988).
26. W.J. Liu, Q.A. Ngo, W.B. Chen, *Ostrowski type inequalities on time scales for double integrals*, Acta Appl. Math. 110 (1) (2010), 477-497.
27. N. Martins, D. Torres, *Calculus of variations on time scales with nabla derivatives*, Nonlinear Anal. 71 (12) (2009), 763-773.
28. A. Ostrowski, *Über die Absolutabweichung einer differentiebaren Funktion von ihrem Integralmittelwert*, Comment. Math. Helv. 10 (1938), 226-227.

Chapter 6

Landau Inequalities on Time Scales

Here we prove some new type Landau inequalities on Time Scales. Both delta and nabla cases are presented. We give applications. It follows [14].

6.1 Introduction

We are motivated by [28], where E. Landau in 1913 proved the following two theorems:

Theorem 6.1. *Let $f \in C^2(\mathbb{R}_+, \mathbb{R})$ with $\|f\|_{\infty,\mathbb{R}_+}$, $\|f''\|_{\infty,\mathbb{R}_+} < \infty$. Then*

$$\|f'\|_{\infty,\mathbb{R}_+} \leq 2\sqrt{\|f\|_{\infty,\mathbb{R}_+} \|f''\|_{\infty,\mathbb{R}_+}},$$

with 2 the best constant.

Theorem 6.2. *Let $f \in C^2(\mathbb{R}, \mathbb{R})$ with $\|f\|_{\infty,\mathbb{R}}$, $\|f''\|_{\infty,\mathbb{R}} < \infty$. Then*

$$\|f'\|_{\infty,\mathbb{R}} \leq \sqrt{2\|f\|_{\infty,\mathbb{R}} \|f''\|_{\infty,\mathbb{R}}},$$

with $\sqrt{2}$ the best constant.

In this chapter we establish a new type of Landau inequalities on time scales. We treat both delta and nabla cases with applications. To keep paper short for basics on time scales we refer the reader to the following sources: [1], [2], [9], [12], [15], [16], [17], [20], [21], [22], [24], [25], [26], [27], [29], [30].

6.2 Background

Let $f \in C^1_{rd}(\mathbb{T})$, $s, t \in \mathbb{T}$, where $\mathbb{T}$ is a time scale, then

$$\int_s^t f^\Delta(\tau)\, \Delta\tau = f(t) - f(s).$$

Here we use $h_0(t, s) = 1$, $\forall\ s, t \in \mathbb{T}$, $k \in \mathbb{N}_0 = \mathbb{N} \cup \{0\}$, and $h_{k+1}(t, s) = \int_s^t h_k(\tau, s)\, \Delta\tau$, $\forall\ s, t \in \mathbb{T}$.

Notice that

$$h_1(r, s) = \int_s^r 1 \Delta \tau = r - s, \ \forall \, r, s \in \mathbb{T},$$

and

$$h_2(r, s) = \int_s^r (\tau - s) \, \Delta \tau, \ \forall \, r, s \in \mathbb{T}.$$

If $r \geq s$, then $0 \leq \tau - s \leq r - s$ and $0 \leq h_2(r, s) \leq (r - s)^2$.

If $r \leq s$, then

$$0 \leq h_2(r, s) = \int_r^s (s - \tau) \, \Delta \tau \leq (s - r)^2.$$

Hence in general

$$0 \leq h_2(r, s) \leq (r - s)^2, \ \forall \, r, s \in \mathbb{T}.$$

Also $h_2(r, s) \neq h_2(s, r)$.

Similarly we define and use

$$\widehat{h_0}(t, s) = 1, \ \forall \, s, t \in \mathbb{T},$$

and

$$\widehat{h_{k+1}}(t, s) = \int_s^t \widehat{h_k}(\tau, s) \, \nabla \tau, \ \forall \, s, t \in \mathbb{T}, \ k \in \mathbb{N}_0.$$

We have that

$$\widehat{h_1}(t, s) = t - s,$$

and

$$\widehat{h_2}(t, s) = \int_s^t (\tau - s) \, \nabla \tau.$$

If $t \geq s$, then

$$0 \leq \widehat{h_2}(t, s) \leq (t - s)^2.$$

If $t \leq s$, then

$$0 \leq \widehat{h_2}(t, s) = \int_t^s (s - \tau) \, \nabla \tau \leq (s - t)^2.$$

Therefore

$$0 \leq \widehat{h_2}(t, s) \leq (t - s)^2, \ \forall \, t, s \in \mathbb{T}.$$

Let $f \in C^1_{ld}(\mathbb{T})$, then

$$\int_s^t f^\nabla(\tau) \, \nabla \tau = f(t) - f(s).$$

We need

Theorem 6.3. *(see [9], [13], p. 634) Here* $\mathbb{T} = \mathbb{T}^k$. *Let* $f \in C^1_{rd}(\mathbb{T})$, $a, b, c \in \mathbb{T}$:
$a \leq c \leq b$. *Then*

$$\left| \frac{1}{b - a} \int_a^b f(t) \, \Delta t - f(c) \right| \leq \left(\frac{h_2(a, c) + h_2(b, c)}{b - a} \right) \|f^\Delta\|_{\infty, [a, b] \cap \mathbb{T}}. \qquad (6.1)$$

The last is a basic Δ-Ostrowski inequality on time scales, see for basics on $\mathbb{R}$ [31].

Remark 6.1. We notice that

$$R.H.S.\,(6.1) \leq \left(\frac{(c-a)^2 + (b-c)^2}{b-a}\right) \left\|f^\Delta\right\|_{\infty,[a,b]\cap\mathbb{T}}$$

$$\leq \frac{(b-a)^2}{(b-a)} \left\|f^\Delta\right\|_{\infty,[a,b]\cap\mathbb{T}} = (b-a) \left\|f^\Delta\right\|_{\infty,[a,b]\cap\mathbb{T}}.$$

Therefore

$$\left|\frac{1}{b-a}\int_a^b f(t)\,\Delta t - f(c)\right| \leq (b-a)\left\|f^\Delta\right\|_{\infty,[a,b]\cap\mathbb{T}}. \tag{6.2}$$

We need

Theorem 6.4. *(see [12], [13], p. 659) Here* $\mathbb{T} = \mathbb{T}^k = \mathbb{T}_k$. *Let* $f \in C_{ld}^1(\mathbb{T})$, $a,b,c \in \mathbb{T} : a \leq c \leq b$. *Then*

$$\left|\frac{1}{b-a}\int_a^b f(t)\,\nabla t - f(c)\right| \leq \left(\frac{\widehat{h_2}(a,c) + \widehat{h_2}(b,c)}{b-a}\right)\left\|f^\nabla\right\|_{\infty,[a,b]\cap\mathbb{T}}. \tag{6.3}$$

The last is a basic ∇-Ostrowski inequality on time scales.

Remark 6.2. We notice that

$$R.H.S.\,(6.3) \leq \left(\frac{(c-a)^2 + (b-c)^2}{b-a}\right) \left\|f^\nabla\right\|_{\infty,[a,b]\cap\mathbb{T}}$$

$$\leq \frac{(b-a)^2}{(b-a)} \left\|f^\nabla\right\|_{\infty,[a,b]\cap\mathbb{T}} = (b-a) \left\|f^\nabla\right\|_{\infty,[a,b]\cap\mathbb{T}}.$$

Therefore

$$\left|\frac{1}{b-a}\int_a^b f(t)\,\nabla t - f(c)\right| \leq (b-a)\left\|f^\nabla\right\|_{\infty,[a,b]\cap\mathbb{T}}. \tag{6.4}$$

6.3 Main Results

We give the following.

Theorem 6.5. *Here* $\mathbb{T} = \mathbb{T}^k$. *Let* $f \in C_{rd}^2(\mathbb{T})$, $a,b \in \mathbb{T}$, $a < b$. *Then*

$$\left\|f^\Delta\right\|_{\infty,[a,b]\cap\mathbb{T}} \leq \left(\frac{2}{b-a}\right)\|f\|_{\infty,[a,b]\cap\mathbb{T}} + (b-a)\left\|f^{\Delta^2}\right\|_{\infty,[a,b]\cap\mathbb{T}}. \tag{6.5}$$

Proof. By (6.2) we have

$$\left| \frac{1}{b-a} \int_a^b f^\Delta(t)\, \Delta t - f'(c) \right| \le (b-a) \left\| f^{\Delta^2} \right\|_{\infty, [a,b] \cap \mathbb{T}},$$

for any $c \in \mathbb{T} : a \le c \le b$.

Therefore

$$\left| \frac{1}{b-a} \left(f(b) - f(a) \right) - f^\Delta(c) \right| \le (b-a) \left\| f^{\Delta^2} \right\|_{\infty, [a,b] \cap \mathbb{T}}.$$

Hence

$$\left| f^\Delta(c) \right| - \frac{1}{b-a} \left| f(b) - f(a) \right| \le (b-a) \left\| f^{\Delta^2} \right\|_{\infty, [a,b] \cap \mathbb{T}}$$

and

$$\left| f^\Delta(c) \right| \le \left(\frac{2}{b-a} \right) \left\| f \right\|_{\infty, [a,b] \cap \mathbb{T}} + (b-a) \left\| f^{\Delta^2} \right\|_{\infty, [a,b] \cap \mathbb{T}},$$

for any $c \in [a, b] \cap \mathbb{T}$.

The last implies (6.5). $\qquad\square$

Next we present a ∇-Landau type inequality.

Theorem 6.6. *Here* $\mathbb{T} = \mathbb{T}^k = \mathbb{T}_k$. *Let* $f \in C_{ld}^2(\mathbb{T})$, $a, b \in \mathbb{T}$, $a < b$. *Then*

$$\left\| f^\nabla \right\|_{\infty, [a,b] \cap \mathbb{T}} \le \frac{2 \left\| f \right\|_{\infty, [a,b] \cap \mathbb{T}}}{(b-a)} + (b-a) \left\| f^{\nabla^2} \right\|_{\infty, [a,b] \cap \mathbb{T}}. \qquad (6.6)$$

Proof. By (6.4) we get

$$\left| \frac{1}{b-a} \int_a^b f^\nabla(t)\, \nabla t - f^\nabla(c) \right| \le (b-a) \left\| f^{\nabla^2} \right\|_{\infty, [a,b] \cap \mathbb{T}},$$

for any $c \in \mathbb{T} : a \le c \le b$.

That is

$$\left| \left(\frac{f(b) - f(a)}{b-a} \right) - f^\nabla(c) \right| \le (b-a) \left\| f^{\nabla^2} \right\|_{\infty, [a,b] \cap \mathbb{T}},$$

and

$$\left| f^\nabla(c) \right| - \frac{1}{b-a} \left| f(b) - f(a) \right| \le (b-a) \left\| f^{\nabla^2} \right\|_{\infty, [a,b] \cap \mathbb{T}},$$

and

$$\left| f^\nabla(c) \right| \le \frac{1}{b-a} \left| f(b) - f(a) \right| + (b-a) \left\| f^{\nabla^2} \right\|_{\infty, [a,b] \cap \mathbb{T}}.$$

Therefore

$$\left| f^\nabla(c) \right| \le \frac{2 \left\| f \right\|_{\infty, [a,b] \cap \mathbb{T}}}{(b-a)} + (b-a) \left\| f^{\nabla^2} \right\|_{\infty, [a,b] \cap \mathbb{T}},$$

for any $c \in [a, b] \cap \mathbb{T}$, proving the claim. $\qquad\square$

6.4   Applications

i) When $\mathbb{T} = \mathbb{R}$, then see [3], [4].

   ii) When $\mathbb{T} = \mathbb{Z}$, see that $\mathbb{Z} = \mathbb{Z}^k = \mathbb{Z}_k$, then

$$f^{\Delta}(t) = f(t+1) - f(t),$$
$$f^{\nabla}(t) = f(t) - f(t-1), \tag{6.7}$$
$$f^{\Delta^2}(t) = f(t) - 2f(t+1) + f(t+2),$$

and

$$f^{\nabla^2}(t) = f(t) - 2f(t-1) + f(t-2),$$

where $f : \mathbb{Z} \to \mathbb{R}$.

   Then by (6.5) we get

$$\|f(\cdot+1) - f(\cdot)\|_{\infty, [a,b] \cap \mathbb{Z}}$$

$$\leq \left(\frac{2}{b-a}\right) \|f\|_{\infty, [a,b] \cap \mathbb{Z}} + (b-a) \|f(\cdot) - 2f(\cdot+1) + f(\cdot+2)\|_{\infty, [a,b] \cap \mathbb{Z}}. \tag{6.8}$$

Also by (6.6) we have that

$$\|f(\cdot) - f(\cdot - 1)\|_{\infty, [a,b] \cap \mathbb{Z}}$$

$$\leq \frac{2 \|f\|_{\infty, [a,b] \cap \mathbb{Z}}}{(b-a)} + (b-a) \|f(\cdot) - 2f(\cdot-1) + f(\cdot-2)\|_{\infty, [a,b] \cap \mathbb{Z}}. \tag{6.9}$$

   iii) Next let $q > 1$, $q^{\mathbb{Z}} = \{q^k : k \in \mathbb{Z}\}$, and we take $\mathbb{T} = q^{\overline{\mathbb{Z}}} = q^{\mathbb{Z}} \cup \{0\}$, which is a very important time scale to q-difference equations. See that $q^{\overline{\mathbb{Z}}} = \left(q^{\overline{\mathbb{Z}}}\right)^k = \left(q^{\overline{\mathbb{Z}}}\right)_k$.

   Let $f : q^{\overline{\mathbb{Z}}} \to \mathbb{R}$.

   By [24], p. 17 we get that

$$f^{\Delta}(t) = \frac{f(qt) - f(t)}{(q-1)t}, \quad \forall\, t \in q^{\overline{\mathbb{Z}}} - \{0\}$$

and

$$f^{\Delta}(0) = \lim_{s \to 0} \frac{f(s) - f(0)}{s},$$

provided that the limit exists.

   For $t \neq 0$ we have that

$$f^{\Delta^2}(t) = \frac{f(q^2 t) - (q+1)f(qt) + qf(t)}{q(q-1)^2 t^2},$$

and

$$f^{\Delta^2}(0) = \lim_{s \to 0} \frac{f^{\Delta}(s) - f^{\Delta}(0)}{s},$$

provided that the limit exists.

Here 0 is a right-dense minimum and every other point in $q^{\overline{\mathbb{Z}}}$ is isolated, also $\sigma(t) = qt$ and $\rho(t) = \frac{t}{q}$, $t \in q^{\overline{\mathbb{Z}}}$, see [24], p. 16.

For $t \in q^{\overline{\mathbb{Z}}} - \{0\}$ we have that (see [15])

$$f^{\nabla}(t) = \frac{f(t) - f\left(\frac{t}{q}\right)}{t - \frac{t}{q}} = \frac{f\left(\frac{1}{q}t\right) - f(t)}{\left(\frac{1}{q} - 1\right)t},$$

and since zero is a left-dense point of $q^{\overline{\mathbb{Z}}}$ we have

$$f^{\nabla}(0) = \lim_{s \to 0} \frac{f(s) - f(0)}{s},$$

provided that the limit exists.

For $t \neq 0$ we have that

$$f^{\nabla^2}(t) = \frac{f\left(\frac{1}{q^2}t\right) - \left(\frac{1}{q} + 1\right)f\left(\frac{1}{q}t\right) + \frac{1}{q}f(t)}{\frac{1}{q}\left(\frac{1}{q} - 1\right)^2 t^2},$$

and

$$f^{\nabla^2}(0) = \lim_{s \to 0} \frac{f^{\nabla}(s) - f^{\nabla}(0)}{s},$$

provided that the limit exists.

Let $f \in C_{rd}^2\left(q^{\overline{\mathbb{Z}}}\right)$; $a, b \in q^{\overline{\mathbb{Z}}}$, $a < b$, then by (6.5) we get

$$\left\|f^{\Delta}\right\|_{\infty, [a,b] \cap q^{\overline{\mathbb{Z}}}} \leq \left(\frac{2}{b-a}\right) \|f\|_{\infty, [a,b] \cap q^{\overline{\mathbb{Z}}}} + (b-a)\left\|f^{\Delta^2}\right\|_{\infty, [a,b] \cap q^{\overline{\mathbb{Z}}}}. \tag{6.10}$$

If $f \in C_{ld}^2\left(q^{\overline{\mathbb{Z}}}\right)$, then by (6.6) we get

$$\left\|f^{\nabla}\right\|_{\infty, [a,b] \cap q^{\overline{\mathbb{Z}}}} \leq \frac{2\|f\|_{\infty, [a,b] \cap q^{\overline{\mathbb{Z}}}}}{(b-a)} + (b-a)\left\|f^{\nabla^2}\right\|_{\infty, [a,b] \cap q^{\overline{\mathbb{Z}}}}. \tag{6.11}$$

6.5   Addendum

Similarly as in Theorem 1.24 (i), [24], pp. 9-10, we have

Theorem 6.7. *Let α be a constant and $m \in \mathbb{N}$. For $f(t) = (t - \alpha)^m$ we have*

$$f^{\nabla}(t) = \sum_{\nu=0}^{m-1} (\rho(t) - \alpha)^{\nu} (t - \alpha)^{m-1-\nu}. \tag{6.12}$$

Proof. By mathematical induction:

If $m = 1$, then $f(t) = t - \alpha$, and $f^{\nabla}(t) = 1$, so that (6.12) is valid.

Now we assume that (6.12) is true for $m \in \mathbb{N}$, we will prove it correct for $m+1$.

So let $F(t) = (t-\alpha)^{m+1} = (t-\alpha) f(t)$. Hence by ∇-product rule (see p. 331, Theorem 8.41(iii), of [24]) we get:

$$F^{\nabla}(t) = ((t-\alpha) f(t))^{\nabla} = (f(t)(t-\alpha))^{\nabla}$$

$$= f^{\nabla}(t)(t-\alpha) + f(\rho(t))(t-\alpha)^{\nabla} = f^{\nabla}(t)(t-\alpha) + f(\rho(t))$$

$$= f(\rho(t)) + f^{\nabla}(t)(t-\alpha) = f(\rho(t)) + (t-\alpha) f^{\nabla}(t)$$

$$= (\rho(t)-\alpha)^m + (t-\alpha) \sum_{\nu=0}^{m-1} (\rho(t)-\alpha)^{\nu} (t-\alpha)^{m-1-\nu}$$

$$= (\rho(t)-\alpha)^m + \sum_{\nu=0}^{m-1} (\rho(t)-\alpha)^{\nu} (t-\alpha)^{m-\nu} = \sum_{\nu=0}^{m} (\rho(t)-\alpha)^{\nu} (t-\alpha)^{m-\nu}.$$

The claim is proved.   □

Bibliography

1. R. Agarwal, M. Bohner, *Basic calculus on time scales and some of its applications*, Results Math. 35 (1-2) (1999), 3-22.
2. R. Agarwal, M. Bohner, A. Peterson, *Inequalities on time scales: a survey*, Math. Inequalities Appl. 4 (4) (2001), 535-557.
3. A. Aglic Aljinovic, Lj. Marangunic, J. Pecaric, *On Landau type inequalities via extension of Montgomery identity, Euler and Fink identities*, Nonlinear Funct. Anal. Appl. 10 (2) (2005), 273-283.
4. A. Aglic Aljinovic, Lj. Marangunic, J. Pecaric, *On Landau type inequalities via Ostrowski inequalities*, Nonlinear Funct. Anal. Appl. 10 (4) (2005), 565-579.
5. G.A. Anastassiou, *Multivariate Ostrowski type inequalities*, Acta Math. Hungarica 76 (4) (1997), 267-278.
6. G. Anastassiou, *Univariate Ostrowski inequalities, revisited*, Monatsh. Math. 135 (2002), 175-189.
7. G.A. Anastassiou, *Multivariate Montgomery identities and Ostrowski inequalities*, Numer. Funct. Anal. Opt. 23 (3-4) (2002), 247-263.
8. G.A. Anastassiou, *Multidimensional Ostrowski inequalities, revisited*, Acta Math. Hungarica 97 (4) (2002), 339-353.
9. G.A. Anastassiou, *Time scales inequalities*, Intern. J. Difference Equations 5 (1) (2010), 1-23.
10. G.A. Anastassiou, *Probabilistic Inequalities*, World Scientific, Singapore, 2010.
11. G.A. Anastassiou, *Advanced Inequalities*, World Scientific, Singapore, 2011.
12. G.A. Anastassiou, *Nabla time scales inequalities*, Editor Al. Paterson, special issue on Time Scales, Internat. J. Dynam. Syst. Difference Equations 3 (1/2) (2011), 59-83.
13. G.A. Anastassiou, *Intelligent Mathematics: Computational Analysis*, Springer, Heidelberg, 2011.
14. G.A. Anastassiou, *Landau type inequalities on time scales*, J. Comput. Anal. Appl. 14 (6) (2012), 1130-1138.
15. D.R. Anderson, *Taylor polynomials for nabla dynamic equations on time scales*, Panamer. Math. J. 12 (4) (2002), 17-27.
16. D. Anderson, J. Bullock, L. Erbe, A. Peterson, H. Tran, *Nabla dynamic equations on time scales*, Panamer. Math. J. 13 (1) (2003), 1-47.
17. F. Atici, D. Biles, A. Lebedinsky, *An application of time scales to economics*, Math. Comput. Modelling 43 (2006), 718-726.
18. M. Bohner, G.Sh. Guseinov, *Partial differentiation on time scales*, Dynamic Syst. Appl. 13 (3-4) (2004), 351-379.

19. M. Bohner, G. Guseinov, *Multiple integration on time scales*, Dynamic Syst. Appl. 14 (3-4) (2005), 579-606.
20. M. Bohner, G.S. Guseinov, *Multiple Lebesgue integration on time scales*, Adv. Difference Equations 2006 (2006), Article ID 26391, 1-12.
21. M. Bohner, G. Guseinov, *Double integral calculus of variations on times scales*, Comput. Math. Appl. 54 (2007), 45-57.
22. M. Bohner, H. Luo, *Singular second-order multipoint dynamic boundary value problems with mixed derivatives*, Adv. Difference Equations 2006 (2006), Article ID 54989, 1-15.
23. M. Bohner, T. Matthews, *Ostrowski inequalities on time scales*, JIPAM. J. Inequal. Pure Appl. Math. 9 (1) (2008), Article 6, 8 pp.
24. M. Bohner, A. Peterson, *Dynamic Equations on Time Scales: An Introduction with Applications*, Birkhaüser, Boston (2001).
25. G. Guseinov, *Integration on time scales*, J. Math. Anal. Appl. 285 (2003), 107-127.
26. R. Higgins, A. Peterson, *Cauchy functions and Taylor's formula for time scales* $\mathbb{T}$, in B. Aulbach, S. Elaydi, G. Ladas (eds.), Proc. Sixth Internat. Conf. Difference Equations, New Progress in Difference Equations, Augsburg, Germany, 2001, pp. 299-308, Chapman & Hall/CRC (2004).
27. S. Hilger, *Ein Maßketten kalkül mit Anwendung auf Zentrumsmannig-faltigkeiten*, Ph.D. thesis, Universität Würzburg, Germany (1988).
28. E. Landau, *Einige Ungleichungen für zweimal differenzierbaren Funktionen*, Proc. London Math. Soc. (2) 13 (1913), 43-49.
29. W.J. Liu, Q.A. Ngo, W.B. Chen, *Ostrowski type inequalities on time scales for double integrals*, Acta Appl. Math. 110 (1) (2010), 477-497.
30. N. Martins, D. Torres, *Calculus of variations on time scales with nabla derivatives*, Nonlinear Anal. 71 (12) (2009), 763-773.
31. A. Ostrowski, *Über die Absolutabweichung einer differentiebaren Funktion von ihrem Integralmittelwert*, Comment. Math. Helv. 10 (1938), 226-227.

Chapter 7

Grüss and Comparison of Means Inequalities over Time Scales

Here we give univariate and multivariate delta and nabla Grüss type and comparison of means inequalities on Time Scales. The estimates involve the higher order delta and nabla derivatives and partial derivatives of the engaged functions. We finish with applications on the time scales $\mathbb{R}$ and $\mathbb{Z}$. It follows [11].

7.1 Introduction

We are motivated and inspired by the following theorems.

Theorem 7.1. *(Chebyshev, 1882, [22]). Let $f, g : [a, b] \to \mathbb{R}$ absolutely continuous functions. If $f', g' \in L_\infty([a, b])$, then*

$$\left| \frac{1}{b-a} \int_a^b f(x) g(x) \, dx - \left(\frac{1}{b-a} \int_a^b f(x) \, dx \right) \left(\frac{1}{b-a} \int_a^b g(x) \, dx \right) \right|$$

$$\leq \frac{1}{12} (b-a)^2 \|f'\|_\infty \|g'\|_\infty. \tag{7.1}$$

Theorem 7.2. *(Grüss, 1935, [23]). Let f, g integrable functions from $[a, b]$ into $\mathbb{R}$, such that $m \leq f(x) \leq M$, $\rho \leq g(x) \leq \sigma$, for all $x \in [a, b]$, where $m, M, \rho, \sigma \in \mathbb{R}$. Then*

$$\left| \frac{1}{b-a} \int_a^b f(x) g(x) \, dx - \left(\frac{1}{b-a} \int_a^b f(x) \, dx \right) \left(\frac{1}{b-a} \int_a^b g(x) \, dx \right) \right|$$

$$\leq \frac{1}{4} (M - m) (\sigma - \rho). \tag{7.2}$$

We are also inspired by [20], 2007, written by M. Bohner and T. Matthews, where they prove Grüss inequality (7.2) on Time Scales.

So here we present univariate and multivariate delta and nabla inequalities of Grüss-type and for comparison of means. The estimates are with respect to all norms $\|\cdot\|_p$, $1 \leq p \leq \infty$, and they involve the functions under consideration, as well as, their delta and nabla higher order derivatives and partial derivatives.

To keep the size of chapter limited, for the basics of Time Scales we refer to reader to [1], [2], [7], [8], [9], [12], [13], [14], [15], [16], [17], [18], [19], [21], [24], [25], [26], [27], [28].

7.2 Main Results

In [10] we proved the following representation result.

Theorem 7.3. *Let* $a, b \in \mathbb{T}$, $\mathbb{T}$ *is a time scale*, $a < b$ *and* $f \in C_{rd}^n([a,b])$, $n \in \mathbb{N}$, *where* $[a,b]$ *is* $[a,b] \cap \mathbb{T}$. *Let* $x \in [a,b] \cap \mathbb{T}$. *Define the kernel*

$$P(r,s) := \begin{cases} \frac{s-a}{b-a}, & a \le s < r, \\ \frac{s-b}{b-a}, & r \le s \le b, \end{cases}$$

where $r, s \in [a,b] \cap \mathbb{T}$. *Then*

$$f(t) = \frac{1}{b-a} \int_a^b f(\sigma(s_1)) \Delta s_1 \tag{7.3}$$

$$+ \frac{1}{b-a} \left(\sum_{k=1}^{n-1} \int_{[a,b]^{k+1}} P(x,s_1) \prod_{i=1}^{k-1} P(s_i, s_{i+1}) f^{\Delta^k}(\sigma(s_{k+1})) \Delta s_{k+1} \Delta s_k ... \Delta s_1 \right)$$

$$+ \int_{[a,b]^n} P(x,s_1) \prod_{i=1}^{n-1} P(s_i, s_{i+1}) f^{\Delta^n}(s_n) \Delta s_n ... \Delta s_1.$$

We make the conventions $\sum_{k=1}^0 \cdot = 0$, $\prod_{i=1}^0 \cdot = 1$, $\int_{[a,b]^j} \cdot = \int_{([a,b] \cap \mathbb{T})^j} \cdot$, $j \in \mathbb{N}$.

We present the following Grüss type univariate Δ-inequality.

Theorem 7.4. *Let* $a, b \in \mathbb{T}$, $a < b$ *and* $f, g \in C_{rd}^n([a,b])$, $n \in \mathbb{N}$. *Define the kernel*

$$P(r,s) := \begin{cases} \frac{s-a}{b-a}, & a \le s < r, \\ \frac{s-b}{b-a}, & r \le s \le b, \end{cases} \tag{7.4}$$

where $r, s \in [a,b] \cap \mathbb{T}$. *For any* $h \in C_{rd}^n([a,b])$ *we define*

$$T_{n-1}^h(x) := \frac{1}{b-a}$$

$$\cdot \left(\sum_{k=1}^{n-1} \int_{[a,b]^{k+1}} P(x,s_1) \prod_{i=1}^{k-1} P(s_i, s_{i+1}) h^{\Delta^k}(\sigma(s_{k+1})) \Delta s_{k+1} ... \Delta s_1 \right), \tag{7.5}$$

and $T_0^h(x) := 0$, $\forall \ x \in [a,b] \cap \mathbb{T}$.
    We further define

$$D(f,g) := \int_a^b f(x) g(x) \Delta x - \frac{1}{2(b-a)}$$

$$\cdot \left[\left(\int_a^b f(x)\,\Delta x\right)\left(\int_a^b g(\sigma(x))\,\Delta x\right) + \left(\int_a^b f(\sigma(x))\,\Delta x\right)\left(\int_a^b g(x)\,\Delta x\right)\right]$$

$$-\frac{1}{2}\left[\int_a^b \left(f(x)\,T_{n-1}^g(x) + g(x)\,T_{n-1}^f(x)\right)\Delta x\right]. \tag{7.6}$$

Then

$$|D(f,g)| \le \frac{1}{2}$$

$$\cdot \min \begin{cases} \left(\|f\|_{\infty,([a,b]\cap\mathbb{T})}\left\|g^{\Delta^n}\right\|_{\infty,([a,b]\cap\mathbb{T})} + \|g\|_{\infty,([a,b]\cap\mathbb{T})}\left\|f^{\Delta^n}\right\|_{\infty,([a,b]\cap\mathbb{T})}\right) \\ \quad\cdot\left(\int_{[a,b]^{n+1}} |P(x,s_1)|\left(\prod_{i=1}^{n-1}|P(s_i,s_{i+1})|\right)\Delta s_n...\Delta s_1\Delta x\right), \\[2ex] \left(\|f\|_{\infty,([a,b]\cap\mathbb{T})}\left\|g^{\Delta^n}\right\|_{p,([a,b]\cap\mathbb{T})} + \|g\|_{\infty,([a,b]\cap\mathbb{T})}\left\|f^{\Delta^n}\right\|_{p,([a,b]\cap\mathbb{T})}\right) \\ \quad\cdot\left(\int_{[a,b]^n} |P(x,s_1)|\left(\prod_{i=1}^{n-2}|P(s_i,s_{i+1})|\right)\right. \\ \quad\left.\cdot\|P(s_{n-1},\cdot)\|_{q,([a,b]\cap\mathbb{T})}\Delta s_{n-1}...\Delta s_1\Delta x\right), \\ \textit{where }p,q>1:\frac{1}{p}+\frac{1}{q}=1, \\[2ex] \left(\|f\|_{\infty,([a,b]\cap\mathbb{T})}\left\|g^{\Delta^n}\right\|_{1,([a,b]\cap\mathbb{T})} + \|g\|_{\infty,([a,b]\cap\mathbb{T})}\left\|f^{\Delta^n}\right\|_{1,([a,b]\cap\mathbb{T})}\right) \\ \quad\cdot\left(\int_{[a,b]^n} |P(x,s_1)|\left(\prod_{i=1}^{n-2}|P(s_i,s_{i+1})|\right)\right. \\ \quad\left.\cdot\|P(s_{n-1},\cdot)\|_{\infty,([a,b]\cap\mathbb{T})}\Delta s_{n-1}...\Delta s_1\Delta x\right). \end{cases} \tag{7.7}$$

We make the conventions $s_0 = x$, $\Delta s_0 = 1$, $\sum_{k=1}^0 \cdot = 0$, $\prod_{i=1}^0 \cdot = \prod_{i=1}^{-1} \cdot = 1$, $\int_{[a,b]^j} \cdot = \int_{([a,b]\cap\mathbb{T})^j} \cdot$, $j \in \mathbb{N}$.

Proof. Let $f,g \in C_{rd}^n([a,b])$, $n \in \mathbb{N}$. We have that

$$T_{n-1}^f(x) = \frac{1}{b-a}$$

$$\cdot\left(\sum_{k=1}^{n-1}\int_{[a,b]^{k+1}} P(x,s_1)\prod_{i=1}^{k-1} P(s_i,s_{i+1})\, f^{\Delta^k}(\sigma(s_{k+1}))\,\Delta s_{k+1}...\Delta s_1\right), \tag{7.8}$$

and define

$$R_n^f(x) := \int_{[a,b]^n} P(x,s_1)\prod_{i=1}^{n-1} P(s_i,s_{i+1})\, f^{\Delta^n}(s_n)\,\Delta s_n...\Delta s_1. \tag{7.9}$$

Similarly we define $T_{n-1}^g(x)$, $R_n^g(x)$.

By Theorem 7.3 we get

$$f(x) = \frac{1}{b-a}\int_a^b f(\sigma(s_1))\,\Delta s_1 + T_{n-1}^f(x) + R_n^f(x), \tag{7.10}$$

and

$$g\left(x\right) = \frac{1}{b-a}\int_a^b g\left(\sigma\left(s_1\right)\right)\Delta s_1 + T_{n-1}^g\left(x\right) + R_n^g\left(x\right). \qquad (7.11)$$

Then

$$f\left(x\right)g\left(x\right) = \frac{g\left(x\right)}{b-a}\int_a^b f\left(\sigma\left(s_1\right)\right)\Delta s_1 + g\left(x\right)T_{n-1}^f\left(x\right) + g\left(x\right)R_n^f\left(x\right), \qquad (7.12)$$

and

$$f\left(x\right)g\left(x\right) = \frac{f\left(x\right)}{b-a}\int_a^b g\left(\sigma\left(s_1\right)\right)\Delta s_1 + f\left(x\right)T_{n-1}^g\left(x\right) + f\left(x\right)R_n^g\left(x\right). \qquad (7.13)$$

Thus by integration we obtain

$$\int_a^b f\left(x\right)g\left(x\right)\Delta x - \frac{\left(\int_a^b g\left(x\right)\Delta x\right)}{b-a}\int_a^b f\left(\sigma\left(s_1\right)\right)\Delta s_1 \qquad (7.14)$$

$$-\int_a^b g\left(x\right)T_{n-1}^f\left(x\right)\Delta x = \int_a^b g\left(x\right)R_n^f\left(x\right)\Delta x,$$

and

$$\int_a^b f\left(x\right)g\left(x\right)\Delta x - \frac{\left(\int_a^b f\left(x\right)\Delta x\right)}{b-a}\int_a^b g\left(\sigma\left(s_1\right)\right)\Delta s_1 \qquad (7.15)$$

$$-\int_a^b f\left(x\right)T_{n-1}^g\left(x\right)\Delta x = \int_a^b f\left(x\right)R_n^g\left(x\right)\Delta x.$$

By adding the last two equalities (7.14), (7.15), and divide by 2, we find

$$D\left(f,g\right) = \int_a^b f\left(x\right)g\left(x\right)\Delta x - \frac{1}{2\left(b-a\right)}$$

$$\cdot\left[\left(\int_a^b f\left(x\right)\Delta x\right)\left(\int_a^b g\left(\sigma\left(x\right)\right)\Delta x\right) + \left(\int_a^b f\left(\sigma\left(x\right)\right)\Delta x\right)\left(\int_a^b g\left(x\right)\Delta x\right)\right]$$

$$-\frac{1}{2}\left[\int_a^b \left(f\left(x\right)T_{n-1}^g\left(x\right) + g\left(x\right)T_{n-1}^f\left(x\right)\right)\Delta x\right] \qquad (7.16)$$

$$= \frac{1}{2}\left[\int_a^b \left(f\left(x\right)R_n^g\left(x\right) + g\left(x\right)R_n^f\left(x\right)\right)\Delta x\right].$$

Hence it holds

$$\left|D\left(f,g\right)\right| \le \frac{1}{2}\left[\|f\|_{\infty,\left(\left[a,b\right]\cap\mathbb{T}\right)}\int_a^b \left|R_n^g\left(x\right)\right|\Delta x\right.$$

$$+ \|g\|_{\infty,([a,b]\cap\mathbb{T})} \int_a^b \left|R_n^f(x)\right| \Delta x \right]. \tag{7.17}$$

By Theorem 13 of [10] we have

$$\left|R_n^f(x)\right|$$

$$\leq \min \begin{cases} \left\|f^{\Delta^n}\right\|_{\infty,([a,b]\cap\mathbb{T})} \int_{[a,b]^n} |P(x,s_1)| \left(\prod_{i=1}^{n-1} |P(s_i,s_{i+1})|\right) \Delta s_n...\Delta s_1, \\[3mm]
\left\|f^{\Delta^n}\right\|_{p,([a,b]\cap\mathbb{T})} \int_{[a,b]^{n-1}} |P(x,s_1)| \left(\prod_{i=1}^{n-2} |P(s_i,s_{i+1})|\right) \\[2mm]
\quad \cdot \|P(s_{n-1},\cdot)\|_{q,([a,b]\cap\mathbb{T})} \Delta s_{n-1}...\Delta s_1, \text{ where } p,q > 1: \frac{1}{p}+\frac{1}{q}=1, \\[3mm]
\left\|f^{\Delta^n}\right\|_{1,([a,b]\cap\mathbb{T})} \int_{[a,b]^{n-1}} |P(x,s_1)| \left(\prod_{i=1}^{n-2} |P(s_i,s_{i+1})|\right) \\[2mm]
\quad \cdot \|P(s_{n-1},\cdot)\|_{\infty,([a,b]\cap\mathbb{T})} \Delta s_{n-1}...\Delta s_1. \end{cases} \tag{7.18}$$

A similar estimate as in (7.18) holds for $\left|R_n^g(x)\right|$.

Therefore we obtain

$$\int_a^b \left|R_n^f(x)\right| \Delta x$$

$$\leq \min \begin{cases} \left\|f^{\Delta^n}\right\|_{\infty,([a,b]\cap\mathbb{T})} \int_{[a,b]^{n+1}} |P(x,s_1)| \left(\prod_{i=1}^{n-1} |P(s_i,s_{i+1})|\right) \Delta s_n...\Delta s_1 \Delta x, \\[3mm]
\left\|f^{\Delta^n}\right\|_{p,([a,b]\cap\mathbb{T})} \int_{[a,b]^n} |P(x,s_1)| \left(\prod_{i=1}^{n-2} |P(s_i,s_{i+1})|\right) \\[2mm]
\quad \cdot \|P(s_{n-1},\cdot)\|_{q,([a,b]\cap\mathbb{T})} \Delta s_{n-1}...\Delta s_1 \Delta x, \text{ where } p,q > 1: \frac{1}{p}+\frac{1}{q}=1, \\[3mm]
\left\|f^{\Delta^n}\right\|_{1,([a,b]\cap\mathbb{T})} \int_{[a,b]^n} |P(x,s_1)| \left(\prod_{i=1}^{n-2} |P(s_i,s_{i+1})|\right) \\[2mm]
\quad \cdot \|P(s_{n-1},\cdot)\|_{\infty,([a,b]\cap\mathbb{T})} \Delta s_{n-1}...\Delta s_1 \Delta x. \end{cases} \tag{7.19}$$

One can write a similar estimate for $\int_a^b \left|R_n^g(x)\right| \Delta x$. Now by (7.17), (7.19) the claim is proved. $\square$

Next we give the following Grüss type univariate ∇-inequality.

Theorem 7.5. *Let* $a,b \in \mathbb{T}$, $a < b$ *and* $f,g \in C_{ld}^n([a,b])$, $n \in \mathbb{N}$. *Define the kernel*

$$P^*(r,s) := \begin{cases} \frac{s-a}{b-a}, & a \leq s \leq r, \\ \frac{s-b}{b-a}, & r < s \leq b, \end{cases} \tag{7.20}$$

where $r,s \in [a,b] \cap \mathbb{T}$. *For any* $h \in C_{ld}^n([a,b])$ *we define*

$$\overline{T}_{n-1}^h(x) := \frac{1}{b-a}$$

$$\cdot \left(\sum_{k=1}^{n-1} \int_{[a,b]^{k+1}} P^*(x,s_1) \prod_{i=1}^{k-1} P^*(s_i,s_{i+1}) h^{\nabla^k}(\rho(s_{k+1})) \nabla s_{k+1}...\nabla s_1\right), \tag{7.21}$$

and $\overline{T}_0^h(x) := 0$, $\forall\, x \in [a, b] \cap \mathbb{T}$.

We further define

$$D^*(f, g) := \int_a^b f(x)\, g(x)\, \nabla x - \frac{1}{2(b-a)}$$

$$\cdot \left[\left(\int_a^b f(x)\, \nabla x \right) \left(\int_a^b g(\rho(x))\, \nabla x \right) + \left(\int_a^b f(\rho(x))\, \nabla x \right) \left(\int_a^b g(x)\, \nabla x \right) \right]$$

$$-\frac{1}{2}\left[\int_a^b \left(f(x)\, \overline{T}_{n-1}^g(x) + g(x)\, \overline{T}_{n-1}^f(x) \right) \nabla x \right]. \tag{7.22}$$

Then

$$|D^*(f, g)| \le \frac{1}{2}$$

$$\cdot \min \begin{cases} \left(\|f\|_{\infty,([a,b]\cap\mathbb{T})} \left\|g^{\nabla^n}\right\|_{\infty,([a,b]\cap\mathbb{T})} + \|g\|_{\infty,([a,b]\cap\mathbb{T})} \left\|f^{\nabla^n}\right\|_{\infty,([a,b]\cap\mathbb{T})} \right) \\ \cdot \left(\int_{[a,b]^{n+1}} |P^*(x, s_1)| \left(\prod_{i=1}^{n-1} |P^*(s_i, s_{i+1})| \right) \nabla s_n \ldots \nabla s_1 \nabla x \right), \\[12pt] \left(\|f\|_{\infty,([a,b]\cap\mathbb{T})} \left\|g^{\nabla^n}\right\|_{p,([a,b]\cap\mathbb{T})} + \|g\|_{\infty,([a,b]\cap\mathbb{T})} \left\|f^{\nabla^n}\right\|_{p,([a,b]\cap\mathbb{T})} \right) \\ \cdot \left(\int_{[a,b]^n} |P^*(x, s_1)| \left(\prod_{i=1}^{n-2} |P^*(s_i, s_{i+1})| \right) \right. \\ \left. \cdot \|P^*(s_{n-1}, \cdot)\|_{q,([a,b]\cap\mathbb{T})}\, \nabla s_{n-1} \ldots \nabla s_1 \nabla x \right), \\ \text{where } p, q > 1 : \frac{1}{p} + \frac{1}{q} = 1, \\[12pt] \left(\|f\|_{\infty,([a,b]\cap\mathbb{T})} \left\|g^{\nabla^n}\right\|_{1,([a,b]\cap\mathbb{T})} + \|g\|_{\infty,([a,b]\cap\mathbb{T})} \left\|f^{\nabla^n}\right\|_{1,([a,b]\cap\mathbb{T})} \right) \\ \cdot \left(\int_{[a,b]^n} |P^*(x, s_1)| \left(\prod_{i=1}^{n-2} |P^*(s_i, s_{i+1})| \right) \right. \\ \left. \cdot \|P^*(s_{n-1}, \cdot)\|_{\infty,([a,b]\cap\mathbb{T})}\, \nabla s_{n-1} \ldots \nabla s_1 \nabla x \right). \end{cases} \tag{7.23}$$

We make the conventions $s_0 = x$, $\nabla s_0 = 1$, $\sum_{k=1}^0 \cdot = 0$, $\prod_{i=1}^0 \cdot = \prod_{i=1}^{-1} \cdot = 1$, $\int_{[a,b]^j} \cdot = \int_{([a,b]\cap\mathbb{T})^j} \cdot$, $j \in \mathbb{N}$.

Proof. Similar to the proof of Theorem 7.4, using Theorems 4 and 14 of [10]. $\square$

We continue with a comparison of integral means Δ-inequality.

Theorem 7.6. Let $a \le c < d \le b$, where $a, b, c, d \in \mathbb{T}$ and $f \in C_{rd}^n([a,b])$, $n \in \mathbb{N}$. Let $\mathcal{P} := \mathcal{P}([c,d] \cap \mathbb{T})$ the power set of $[c,d] \cap \mathbb{T}$, and let a probability measure μ on $\mathcal{P}$, and let $T_{n-1}^f(x)$ as in (7.5). Define

$$M := \int_{[c,d]\cap\mathbb{T}} f(x)\, d\mu(x) - \frac{1}{b-a} \int_a^b f(\sigma(x))\, \Delta x - \int_{[c,d]\cap\mathbb{T}} T_{n-1}^f(x)\, d\mu(x). \tag{7.24}$$

Then

$$|M| \le \min \begin{cases} \|f^{\Delta^n}\|_{\infty,([a,b]\cap\mathbb{T})} \int_{[c,d]\cap\mathbb{T}} \left(\int_{[a,b]^n} |P(x,s_1)| \right. \\ \left. \cdot \left(\prod_{i=1}^{n-1} |P(s_i,s_{i+1})| \right) \Delta s_n...\Delta s_1 \right) d\mu(x), \\[2ex] \|f^{\Delta^n}\|_{p,([a,b]\cap\mathbb{T})} \int_{[c,d]\cap\mathbb{T}} \left(\int_{[a,b]^{n-1}} |P(x,s_1)| \left(\prod_{i=1}^{n-2} |P(s_i,s_{i+1})| \right) \right. \\ \left. \cdot \|P(s_{n-1},\cdot)\|_{q,([a,b]\cap\mathbb{T})} \Delta s_{n-1}...\Delta s_1 \right) d\mu(x), \\ \text{where } p,q > 1 : \frac{1}{p} + \frac{1}{q} = 1, \\[2ex] \|f^{\Delta^n}\|_{1,([a,b]\cap\mathbb{T})} \int_{[c,d]\cap\mathbb{T}} \left(\int_{[a,b]^{n-1}} |P(x,s_1)| \left(\prod_{i=1}^{n-2} |P(s_i,s_{i+1})| \right) \right. \\ \left. \cdot \|P(s_{n-1},\cdot)\|_{\infty,([a,b]\cap\mathbb{T})} \Delta s_{n-1}...\Delta s_1 \right) d\mu(x). \end{cases}$$

(7.25)

Proof. By Theorem 7.3 and (7.10) we get that

$$f(x) = \frac{1}{b-a} \int_a^b f(\sigma(s_1)) \Delta s_1 + T_{n-1}^f(x) + R_n^f(x), \qquad (7.26)$$

where $T_{n-1}^f(x)$ as in (7.8) and $R_n^f(x)$ as in (7.9).

Hence

$$M = \int_{[c,d]\cap\mathbb{T}} f(x)\,d\mu(x) - \frac{1}{b-a} \int_a^b f(\sigma(s_1))\,\Delta s_1 \qquad (7.27)$$

$$- \int_{[c,d]\cap\mathbb{T}} T_{n-1}^f(x)\,d\mu(x) = \int_{[c,d]\cap\mathbb{T}} R_n^f(x)\,d\mu(x).$$

Therefore

$$|M| \le \int_{[c,d]\cap\mathbb{T}} \left| R_n^f(x) \right| d\mu(x) \qquad (7.28)$$

and by (7.18) the claim is proved. □

We give also a comparison of integral means ∇-inequality.

Theorem 7.7. *Let $a \le c < d \le b$, where $a,b,c,d \in \mathbb{T}$ and $f \in C_{ld}^n([a,b])$, $n \in \mathbb{N}$. Let $\mathcal{P} := \mathcal{P}([c,d] \cap \mathbb{T})$ the power set of $[c,d] \cap \mathbb{T}$, and let a probability measure μ on $\mathcal{P}$, and let $\overline{T}_{n-1}^f(x)$ as in (7.21). Define*

$$M^* := \int_{[c,d]\cap\mathbb{T}} f(x)\,d\mu(x) - \frac{1}{b-a} \int_a^b f(\rho(x))\,\nabla x - \int_{[c,d]\cap\mathbb{T}} \overline{T}_{n-1}^f(x)\,d\mu(x). \quad (7.29)$$

Then

$$|M^*| \leq \min \begin{cases} \left\| f^{\nabla^n} \right\|_{\infty,([a,b]\cap\mathbb{T})} \int_{[c,d]\cap\mathbb{T}} \left(\int_{[a,b]^n} |P^*(x,s_1)| \right. \\ \left. \cdot \left(\prod_{i=1}^{n-1} |P^*(s_i,s_{i+1})| \right) \nabla s_n ... \nabla s_1 \right) d\mu(x), \\ \\ \left\| f^{\nabla^n} \right\|_{p,([a,b]\cap\mathbb{T})} \int_{[c,d]\cap\mathbb{T}} \left(\int_{[a,b]^{n-1}} |P^*(x,s_1)| \left(\prod_{i=1}^{n-2} |P^*(s_i,s_{i+1})| \right) \right. \\ \left. \cdot \|P^*(s_{n-1},\cdot)\|_{q,([a,b]\cap\mathbb{T})} \nabla s_{n-1} ... \nabla s_1 \right) d\mu(x), \\ where \ \ p,q > 1 : \frac{1}{p} + \frac{1}{q} = 1, \\ \\ \left\| f^{\nabla^n} \right\|_{1,([a,b]\cap\mathbb{T})} \int_{[c,d]\cap\mathbb{T}} \left(\int_{[a,b]^{n-1}} |P^*(x,s_1)| \left(\prod_{i=1}^{n-2} |P^*(s_i,s_{i+1})| \right) \right. \\ \left. \cdot \|P^*(s_{n-1},\cdot)\|_{\infty,([a,b]\cap\mathbb{T})} \nabla s_{n-1} ... \nabla s_1 \right) d\mu(x). \end{cases}$$

$$(7.30)$$

Remark 7.1. (to Theorems 7.6, 7.7) One can replace in (7.25) and (7.30) μ by $\frac{\Delta x}{d-c}$ and $\frac{\nabla x}{d-c}$, respectively.

From [10] we need the following Δ-representation result.

Theorem 7.8. *Let the time scales* $(\mathbb{T}_i, \sigma_i, \Delta_i)$, $i = 1, 2, \ldots, n$, $n \in \mathbb{N}$ *and* $a_i < b_i$; $a_i, b_i \in \mathbb{T}_i$, $i = 1, \ldots, n$. *Here* σ_i, $i = 1, \ldots, n$ *are continuous. We consider* $f \in C^n \left(\prod_{i=1}^n ([a_i, b_i] \cap \mathbb{T}_i) \right)$. *Let* $(x_1, \ldots, x_n) \in \prod_{i=1}^n ([a_i, b_i] \cap \mathbb{T}_i)$. *Define the kernels* $p_i : [a_i, b_i]^2 \to \mathbb{R}$:

$$p_i(x_i, s_i) := \begin{cases} s_i - a_i, & s_i \in [a_i, x_i), \\ s_i - b_i, & s_i \in [x_i, b_i], \end{cases} \qquad (7.31)$$

for $i = 1, 2, \ldots, n$.

Denote $\int_{\prod_{i=1}^j [a_i, b_i]} \cdot = \int_{\prod_{i=1}^j ([a_i, b_i] \cap \mathbb{T}_i)} \cdot$, $j = 1, \ldots, n$.
Then

$$f(x_1, \ldots, x_n) = \frac{1}{\prod_{i=1}^n (b_i - a_i)}$$

$$\cdot \left\{ \int_{\prod_{i=1}^n [a_i, b_i]} f(\sigma_1(s_1), \sigma_2(s_2), \ldots, \sigma_n(s_n)) \Delta_n s_n ... \Delta_1 s_1 \right.$$

$$+ \sum_{j=1}^n \int_{\prod_{i=1}^n [a_i, b_i]} p_j(x_j, s_j)$$

$$\cdot \frac{\partial f(\sigma_1(s_1), \ldots, \sigma_{j-1}(s_{j-1}), s_j, \sigma_{j+1}(s_{j+1}), \ldots, \sigma_n(s_n))}{\Delta_j s_j} \Delta_n s_n ... \Delta_1 s_1$$

$$+ \sum_{\substack{l_1=1 \\ j<k}}^{\binom{n}{2}} \left(\int_{\prod_{i=1}^n [a_i, b_i]} p_j(x_j, s_j) p_k(x_k, s_k) \cdot \right.$$

$$\cdot \frac{\partial^2 f(\sigma_1(s_1),\ldots,\sigma_{j-1}(s_{j-1}),s_j,\sigma_{j+1}(s_{j+1}),\ldots,\sigma_{k-1}(s_{k-1}),s_k,\sigma_{k+1}(s_{k+1}),\ldots,\sigma_n(s_n))}{\Delta_k s_k \Delta_j s_j}$$

$$\cdot \Delta_n s_n \ldots \Delta_1 s_1)_{(l_1)} + \sum_{\substack{l_2=1 \\ j<k<r}}^{\binom{n}{3}} \left(\int_{\prod_{i=1}^n [a_i,b_i]} p_j(x_j,s_j)\, p_k(x_k,s_k)\, p_r(x_r,s_r) \right.$$

$$\cdot \frac{\partial^3 f(\sigma_1(s_1),\ldots,\sigma_{j-1}(s_{j-1}),s_j,\sigma_{j+1}(s_{j+1}),\ldots,\sigma_{k-1}(s_{k-1}),s_k,\sigma_{k+1}(s_{k+1}),\ldots,\sigma_{r-1}(s_{r-1}),s_r,\sigma_{r+1}(s_{r+1}),\ldots,\sigma_n(s_n))}{\Delta_r s_r \Delta_k s_k \Delta_j s_j}$$

$$\cdot \Delta_n s_n \ldots \Delta_1 s_1)_{(l_2)} + \ldots + \sum_{l=1}^{\binom{n}{n-1}} \left(\int_{\prod_{i=1}^n [a_i,b_i]} p_1(x_1,s_1)\ldots \widehat{p_l(x_l,s_l)}\ldots p_n(x_n,s_n) \right.$$

$$(7.32)$$

$$\cdot \frac{\partial^{n-1} f(s_1,s_2,\ldots,\sigma_l(s_l),\ldots,s_n)}{\Delta_n s_n \ldots \widehat{\Delta_l s_l}\ldots \Delta_1 s_1} \Delta_n s_n \ldots \Delta_1 s_1 \bigg)$$

$$+ \int_{\prod_{i=1}^n [a_i,b_i]} \left(\prod_{i=1}^n p_i(x_i,s_i) \right) \frac{\partial^n f(s_1,\ldots,s_n)}{\Delta_n s_n \ldots \Delta_1 s_1} \Delta_n s_n \ldots \Delta_1 s_1 \bigg\}.$$

Above l_1 counts $(j,k): j<k$; $j,k \in \{1,\ldots,n\}$, also l_2 counts $(j,k,r): j<k<r$; $j,k,r \in \{1,\ldots,n\}$, etc. Also $\widehat{p_l(x_l,s_l)}$ and $\widehat{\Delta_l s_l}$ means that $p_l(x_l,s_l)$ and $\Delta_l s_l$ are missing, respectively.

We make

Remark 7.2. (to Theorem 7.8) we define

$$F_1(f)(x_j) := \int_{\prod_{i=1}^n [a_i,b_i]} p_j(x_j,s_j)$$

$$\cdot \frac{\partial f(\sigma_1(s_1),\ldots,\sigma_{j-1}(s_{j-1}),s_j,\sigma_{j+1}(s_{j+1}),\ldots,\sigma_n(s_n))}{\Delta_j s_j} \Delta_n s_n \ldots \Delta_1 s_1, \qquad (7.33)$$

for $j=1,\ldots,n$,

$$F_2(f)(x_j,x_k) := \int_{\prod_{i=1}^n [a_i,b_i]} p_j(x_j,s_j)\, p_k(x_k,s_k) \qquad (7.34)$$

$$\cdot \frac{\partial^2 f(\sigma_1(s_1),\ldots,\sigma_{j-1}(s_{j-1}),s_j,\sigma_{j+1}(s_{j+1}),\ldots,\sigma_{k-1}(s_{k-1}),s_k,\sigma_{k+1}(s_{k+1}),\ldots,\sigma_n(s_n))}{\Delta_k s_k \Delta_j s_j}$$

$$\cdot \Delta_n s_n \ldots \Delta_1 s_1,$$

for $j,k \in \{1,\ldots,n\}$,

$$F_3(f)(x_j,x_k,x_r) := \int_{\prod_{i=1}^n [a_i,b_i]} p_j(x_j,s_j)\, p_k(x_k,s_k)\, p_r(x_r,s_r) \qquad (7.35)$$

$$. \frac{\partial^3 f\left(\sigma_1\left(s_1\right),\ldots,\sigma_{j-1}\left(s_{j-1}\right),s_j,\sigma_{j+1}\left(s_{j+1}\right),\ldots,\sigma_{k-1}\left(s_{k-1}\right),s_k,\sigma_{k+1}\left(s_{k+1}\right)}{\Delta_r s_r \Delta_k s_k \Delta_j s_j}$$

$$\cdot \Delta_n s_n ... \Delta_1 s_1,$$

for $j,k,r \in \{1,\ldots,n\}$,

$$\vdots$$

$$F_{n-1}\left(f\right)\left(x_1,\ldots,\widehat{x_l},\ldots,x_n\right) := \int_{\prod_{i=1}^n [a_i,b_i]} p_1\left(x_1,s_1\right)...\widehat{p_l\left(x_l,s_l\right)}...p_n\left(x_n,s_n\right)$$

$$\cdot \frac{\partial^{n-1} f\left(s_1,s_2,\ldots,\sigma_l\left(s_l\right),\ldots,s_n\right)}{\Delta_n s_n ... \widehat{\Delta_l s_l} ... \Delta_1 s_1} \Delta_n s_n ... \Delta_1 s_1, \tag{7.36}$$

for $l = 1,\ldots,n$; $\widehat{x_l}$ means x_l is missing, and

$$R_n\left(f\right)\left(x_1,\ldots,x_n\right) := \int_{\prod_{i=1}^n [a_i,b_i]} \left(\prod_{i=1}^n p_i\left(x_i,s_i\right)\right) \frac{\partial^n f\left(s_1,\ldots,s_n\right)}{\Delta_n s_n ... \Delta_1 s_1} \Delta_n s_n ... \Delta_1 s_1. \tag{7.37}$$

So one can rewrite (7.32) as follows

$$f\left(x_1,\ldots,x_n\right)$$

$$= \frac{1}{\prod_{i=1}^n \left(b_i - a_i\right)} \left\{ \int_{\prod_{i=1}^n [a_i,b_i]} f\left(\sigma_1\left(s_1\right),\ldots,\sigma_n\left(s_n\right)\right) \Delta_n s_n ... \Delta_1 s_1 \right.$$

$$+ \sum_{j=1}^n F_1\left(f\right)\left(x_j\right) + \sum_{\substack{l_1=1 \\ j<k}}^{\binom{n}{2}} F_2\left(f\right)\left(x_j,x_k\right) + \sum_{\substack{l_2=1 \\ j<k<r}}^{\binom{n}{3}} F_3\left(f\right)\left(x_j,x_k,x_r\right)$$

$$\left. +...+ \sum_{l=1}^{\binom{n}{n-1}} F_{n-1}\left(f\right)\left(x_1,\ldots,\widehat{x_l},\ldots,x_n\right) + R_n\left(f\right)\left(x_1,\ldots,x_n\right) \right\}, \tag{7.38}$$

where l_1 counts $(j,k) : j < k$; $j,k \in \{1,\ldots,n\}$, l_2 counts $(j,k,r) : j < k < r$; $j,k,r \in \{1,\ldots,n\}$, etc, also $\widehat{p_l\left(x_l,s_l\right)}$, $\widehat{\Delta_l s_l}$, $\widehat{x_l}$ are missing elements.

Next we set

$$F\left(f\right)\left(x_1,\ldots,x_n\right) := \sum_{j=1}^n F_1\left(f\right)\left(x_j\right) + \sum_{\substack{l_1=1 \\ j<k}}^{\binom{n}{2}} F_2\left(f\right)\left(x_j,x_k\right) \tag{7.39}$$

$$+\sum_{\substack{l_2=1 \\ j<k<r}}^{\binom{n}{3}} F_3\left(f\right)\left(x_j,x_k,x_r\right)+...+\sum_{l=1}^{\binom{n}{n-1}} F_{n-1}\left(f\right)\left(x_1,\ldots,\widehat{x_l},\ldots,x_n\right).$$

So finally (7.32) is rewritten as

$$f\left(x_1,\ldots,x_n\right)=\frac{1}{\prod_{i=1}^n\left(b_i-a_i\right)}\int_{\prod_{i=1}^n[a_i,b_i]} f\left(\sigma_1\left(s_1\right),\ldots,\sigma_n\left(s_n\right)\right)\Delta_n s_n...\Delta_1 s_1$$

$$+\frac{F\left(f\right)\left(x_1,\ldots,x_n\right)}{\prod_{i=1}^n\left(b_i-a_i\right)}+\frac{R_n\left(f\right)\left(x_1,\ldots,x_n\right)}{\prod_{i=1}^n\left(b_i-a_i\right)}, \tag{7.40}$$

and

$$f\left(x_1,\ldots,x_n\right)-\frac{1}{\prod_{i=1}^n\left(b_i-a_i\right)}\int_{\prod_{i=1}^n[a_i,b_i]} f\left(\sigma_1\left(s_1\right),\ldots,\sigma_n\left(s_n\right)\right)\Delta_n s_n...\Delta_1 s_1$$

$$-\frac{F\left(f\right)\left(x_1,\ldots,x_n\right)}{\prod_{i=1}^n\left(b_i-a_i\right)}=\frac{R_n\left(f\right)\left(x_1,\ldots,x_n\right)}{\prod_{i=1}^n\left(b_i-a_i\right)}. \tag{7.41}$$

We need

Definition 7.1. The general Taylor monomials are defined as follows

$$h_0^{(i)}\left(t,s\right)=1,\ h_{k+1}^{(i)}\left(t,s\right)=\int_s^t h_k^{(i)}\left(\tau,s\right)\Delta_i\tau,\ k\in\mathbb{N}_0=\mathbb{N}\cup\{0\}, \tag{7.42}$$

and

$$\widehat{h}_0^{(i)}\left(t,s\right)=1,\ \widehat{h}_{k+1}^{(i)}\left(t,s\right)=\int_s^t \widehat{h}_k^{(i)}\left(\tau,s\right)\nabla_i\tau,\ k\in\mathbb{N}_0,$$

where $t,s\in\mathbb{T}_i$ a time scale, $i\in\{1,\ldots,n\}$, see [12], [21].

Notice that $h_1^{(i)}\left(t,s\right)=t-s$, and $\widehat{h}_1^{(i)}\left(t,s\right)=t-s$, so that

$$h_2^{(i)}\left(t,s\right)=\int_s^t\left(\tau-s\right)\Delta_i\tau,\ \text{and}\ \widehat{h}_2^{(i)}\left(t,s\right)=\int_s^t\left(\tau-s\right)\nabla_i\tau. \tag{7.43}$$

By [7], [8], we get that $h_2^{(i)}\left(t,s\right),\widehat{h}_2^{(i)}\left(t,s\right)\geq 0,\ \forall\ s,t\in\mathbb{T}_i$.

From [10], we also need the estimate

Theorem 7.9. *(all as in Theorem 7.8 and (7.37)) Then*

$$\left|R_n\left(f\right)\left(x_1,\ldots,x_n\right)\right|$$

$$\leq\min\begin{cases}\left\|\frac{\partial^n f}{\Delta_n s_n...\Delta_1 s_1}\right\|_{\infty,\prod_{i=1}^n([a_i,b_i]\cap\mathbb{T}_i)}\left(\prod_{i=1}^n\left(h_2^{(i)}\left(x_i,a_i\right)+h_2^{(i)}\left(x_i,b_i\right)\right)\right),\\ \left\|\frac{\partial^n f}{\Delta_n s_n...\Delta_1 s_1}\right\|_{p,\prod_{i=1}^n([a_i,b_i]\cap\mathbb{T}_i)}\left(\prod_{i=1}^n\|p_i\left(x_i,\cdot\right)\|_{q,\prod_{i=1}^n([a_i,b_i]\cap\mathbb{T}_i)}\right),\\ \text{where } p,q>1:\frac{1}{p}+\frac{1}{q}=1,\\ \left\|\frac{\partial^n f}{\Delta_n s_n...\Delta_1 s_1}\right\|_{1,\prod_{i=1}^n([a_i,b_i]\cap\mathbb{T}_i)}\left(\prod_{i=1}^n\left(\frac{b_i-a_i}{2}+\left|x_i-\left(\frac{a_i+b_i}{2}\right)\right|\right)\right).\end{cases}$$

$$\tag{7.44}$$

Next we present the following general Grüss type Δ-multivariate inequality.

Theorem 7.10. *Let the time scales* $(\mathbb{T}_i, \sigma_i, \Delta_i)$, $i = 1, 2, \ldots, n$, $n \in \mathbb{N}$ *and* $a_i < b_i$; $a_i, b_i \in \mathbb{T}_i$, $i = 1, \ldots, n$. *Here* σ_i, $i = 1, \ldots, n$ *are continuous. We consider* $f, g \in C^n \left(\prod_{i=1}^n ([a_i, b_i] \cap \mathbb{T}_i) \right)$. *Here* p_i *is as in (7.31), and* $F(f)$ *and* $F(g)$ *as in (7.39). Define*

$$E(f, g) := \int_{\prod_{i=1}^n [a_i, b_i]} f(x_1, \ldots, x_n) \, g(x_1, \ldots, x_n) \, \Delta_n x_n \ldots \Delta_1 x_1$$

$$- \frac{1}{2 \left(\prod_{i=1}^n (b_i - a_i) \right)} \left[\left(\int_{\prod_{i=1}^n [a_i, b_i]} f(x_1, \ldots, x_n) \, \Delta_n x_n \ldots \Delta_1 x_1 \right) \right.$$

$$\cdot \left(\int_{\prod_{i=1}^n [a_i, b_i]} g(\sigma_1(x_1), \ldots, \sigma_n(x_n)) \, \Delta_n x_n \ldots \Delta_1 x_1 \right)$$

$$+ \left(\int_{\prod_{i=1}^n [a_i, b_i]} f(\sigma_1(x_1), \ldots, \sigma_n(x_n)) \, \Delta_n x_n \ldots \Delta_1 x_1 \right)$$

$$\left. \cdot \left(\int_{\prod_{i=1}^n [a_i, b_i]} g(x_1, \ldots, x_n) \, \Delta_n x_n \ldots \Delta_1 x_1 \right) \right]$$

$$- \frac{1}{2 \left(\prod_{i=1}^n (b_i - a_i) \right)} \left[\int_{\prod_{i=1}^n [a_i, b_i]} [F(f)(x_1, \ldots, x_n) \, g(x_1, \ldots, x_n) \right.$$

$$\left. + F(g)(x_1, \ldots, x_n) \, f(x_1, \ldots, x_n)] \, \Delta_n x_n \ldots \Delta_1 x_1 \right]. \tag{7.45}$$

Then

$$|E(f, g)| \le \frac{1}{\prod_{i=1}^n (b_i - a_i)} \cdot$$

$$\cdot \min \begin{cases} \left[\|f\|_{\infty, \prod_{i=1}^n ([a_i, b_i] \cap \mathbb{T}_i)} \left\| \frac{\partial^n g}{\Delta_n s_n \ldots \Delta_1 s_1} \right\|_{\infty, \prod_{i=1}^n ([a_i, b_i] \cap \mathbb{T}_i)} \right. \\ \left. + \|g\|_{\infty, \prod_{i=1}^n ([a_i, b_i] \cap \mathbb{T}_i)} \left\| \frac{\partial^n f}{\Delta_n s_n \ldots \Delta_1 s_1} \right\|_{\infty, \prod_{i=1}^n ([a_i, b_i] \cap \mathbb{T}_i)} \right] \\ \cdot \left(\prod_{i=1}^n \int_{a_i}^{b_i} \left[h_2^{(i)}(x_i, a_i) + h_2^{(i)}(x_i, b_i) \right] \Delta_i x_i \right), \\[2mm] \left[\|f\|_{\infty, \prod_{i=1}^n ([a_i, b_i] \cap \mathbb{T}_i)} \left\| \frac{\partial^n g}{\Delta_n s_n \ldots \Delta_1 s_1} \right\|_{p, \prod_{i=1}^n ([a_i, b_i] \cap \mathbb{T}_i)} \right. \\ \left. + \|g\|_{\infty, \prod_{i=1}^n ([a_i, b_i] \cap \mathbb{T}_i)} \left\| \frac{\partial^n f}{\Delta_n s_n \ldots \Delta_1 s_1} \right\|_{p, \prod_{i=1}^n ([a_i, b_i] \cap \mathbb{T}_i)} \right] \\ \cdot \left(\prod_{i=1}^n \left(\int_{a_i}^{b_i} \|p_i(x_i, \cdot)\|_{q, ([a_i, b_i] \cap \mathbb{T}_i)} \Delta_i x_i \right) \right), \\ \text{where } p, q > 1 : \frac{1}{p} + \frac{1}{q} = 1. \\[2mm] \left[\|f\|_{\infty, \prod_{i=1}^n ([a_i, b_i] \cap \mathbb{T}_i)} \left\| \frac{\partial^n g}{\Delta_n s_n \ldots \Delta_1 s_1} \right\|_{1, \prod_{i=1}^n ([a_i, b_i] \cap \mathbb{T}_i)} \right. \\ \left. + \|g\|_{\infty, \prod_{i=1}^n ([a_i, b_i] \cap \mathbb{T}_i)} \left\| \frac{\partial^n f}{\Delta_n s_n \ldots \Delta_1 s_1} \right\|_{1, \prod_{i=1}^n ([a_i, b_i] \cap \mathbb{T}_i)} \right] \\ \cdot \left(\prod_{i=1}^n \int_{a_i}^{b_i} \left(\frac{b_i - a_i}{2} + \left| x_i - \left(\frac{a_i + b_i}{2} \right) \right| \right) \Delta_i x_i \right). \end{cases} \tag{7.46}$$

Proof. Let $f, g \in C^n \left(\prod_{i=1}^{n} \left([a_i, b_i] \cap \mathbb{T}_i \right) \right)$, $n \in \mathbb{N}$. Then by Theorem 7.8 and Remark 7.2, in particular (7.40), we have that

$$f(x_1, \ldots, x_n) = \frac{1}{\prod_{i=1}^{n}(b_i - a_i)} \int_{\prod_{i=1}^{n}[a_i, b_i]} f(\sigma_1(s_1), \ldots, \sigma_n(s_n)) \, \Delta_n s_n \ldots \Delta_1 s_1$$

$$+ \frac{F(f)(x_1, \ldots, x_n)}{\prod_{i=1}^{n}(b_i - a_i)} + \frac{R_n(f)(x_1, \ldots, x_n)}{\prod_{i=1}^{n}(b_i - a_i)}, \qquad (7.47)$$

and

$$g(x_1, \ldots, x_n) = \frac{1}{\prod_{i=1}^{n}(b_i - a_i)} \int_{\prod_{i=1}^{n}[a_i, b_i]} g(\sigma_1(s_1), \ldots, \sigma_n(s_n)) \, \Delta_n s_n \ldots \Delta_1 s_1$$

$$+ \frac{F(g)(x_1, \ldots, x_n)}{\prod_{i=1}^{n}(b_i - a_i)} + \frac{R_n(g)(x_1, \ldots, x_n)}{\prod_{i=1}^{n}(b_i - a_i)}. \qquad (7.48)$$

Then we obtain

$$f(x_1, \ldots, x_n) g(x_1, \ldots, x_n)$$

$$= \frac{g(x_1, \ldots, x_n)}{\prod_{i=1}^{n}(b_i - a_i)} \int_{\prod_{i=1}^{n}[a_i, b_i]} f(\sigma_1(s_1), \ldots, \sigma_n(s_n)) \, \Delta_n s_n \ldots \Delta_1 s_1$$

$$+ \frac{F(f)(x_1, \ldots, x_n) g(x_1, \ldots, x_n)}{\prod_{i=1}^{n}(b_i - a_i)} + \frac{g(x_1, \ldots, x_n) R_n(f)(x_1, \ldots, x_n)}{\prod_{i=1}^{n}(b_i - a_i)}, \qquad (7.49)$$

and

$$f(x_1, \ldots, x_n) g(x_1, \ldots, x_n)$$

$$= \frac{f(x_1, \ldots, x_n)}{\prod_{i=1}^{n}(b_i - a_i)} \int_{\prod_{i=1}^{n}[a_i, b_i]} g(\sigma_1(s_1), \ldots, \sigma_n(s_n)) \, \Delta_n s_n \ldots \Delta_1 s_1$$

$$+ \frac{F(g)(x_1, \ldots, x_n) f(x_1, \ldots, x_n)}{\prod_{i=1}^{n}(b_i - a_i)} + \frac{f(x_1, \ldots, x_n) R_n(g)(x_1, \ldots, x_n)}{\prod_{i=1}^{n}(b_i - a_i)}. \qquad (7.50)$$

Thus by integration we get

$$\int_{\prod_{i=1}^{n}[a_i, b_i]} f(x_1, \ldots, x_n) g(x_1, \ldots, x_n) \, \Delta_n x_n \ldots \Delta_1 x_1$$

$$\frac{\left(\int_{\prod_{i=1}^{n}[a_i, b_i]} g(x_1, \ldots, x_n) \, \Delta_n x_n \ldots \Delta_1 x_1 \right)}{\prod_{i=1}^{n}(b_i - a_i)}$$

$$\cdot \left(\int_{\prod_{i=1}^{n}[a_i, b_i]} f(\sigma_1(s_1), \ldots, \sigma_n(s_n)) \, \Delta_n s_n \ldots \Delta_1 s_1 \right)$$

$$\frac{\int_{\prod_{i=1}^{n}[a_i, b_i]} F(f)(x_1, \ldots, x_n) g(x_1, \ldots, x_n) \, \Delta_n x_n \ldots \Delta_1 x_1}{\prod_{i=1}^{n}(b_i - a_i)} \qquad (7.51)$$

$$= \frac{\int_{\prod_{i=1}^{n}[a_i,b_i]} g\left(x_1,\ldots,x_n\right) R_n\left(f\right)\left(x_1,\ldots,x_n\right) \Delta_n x_n\ldots\Delta_1 x_1}{\prod_{i=1}^{n}\left(b_i-a_i\right)},$$

and

$$\int_{\prod_{i=1}^{n}[a_i,b_i]} f\left(x_1,\ldots,x_n\right) g\left(x_1,\ldots,x_n\right) \Delta_n x_n\ldots\Delta_1 x_1$$

$$-\frac{\left(\int_{\prod_{i=1}^{n}[a_i,b_i]} f\left(x_1,\ldots,x_n\right) \Delta_n x_n\ldots\Delta_1 x_1\right)}{\prod_{i=1}^{n}\left(b_i-a_i\right)}$$

$$\cdot\left(\int_{\prod_{i=1}^{n}[a_i,b_i]} g\left(\sigma_1\left(s_1\right),\ldots,\sigma_n\left(s_n\right)\right) \Delta_n s_n\ldots\Delta_1 s_1\right)$$

$$-\frac{\int_{\prod_{i=1}^{n}[a_i,b_i]} F\left(g\right)\left(x_1,\ldots,x_n\right) f\left(x_1,\ldots,x_n\right) \Delta_n x_n\ldots\Delta_1 x_1}{\prod_{i=1}^{n}\left(b_i-a_i\right)} \tag{7.52}$$

$$= \frac{\int_{\prod_{i=1}^{n}[a_i,b_i]} f\left(x_1,\ldots,x_n\right) R_n\left(g\right)\left(x_1,\ldots,x_n\right) \Delta_n x_n\ldots\Delta_1 x_1}{\prod_{i=1}^{n}\left(b_i-a_i\right)}.$$

Adding (7.51) and (7.52) and dividing by 2 we derive

$$E\left(f,g\right) = \int_{\prod_{i=1}^{n}[a_i,b_i]} f\left(x_1,\ldots,x_n\right) g\left(x_1,\ldots,x_n\right) \Delta_n x_n\ldots\Delta_1 x_1$$

$$-\frac{1}{2\left(\prod_{i=1}^{n}\left(b_i-a_i\right)\right)}\left[\left(\int_{\prod_{i=1}^{n}[a_i,b_i]} f\left(x_1,\ldots,x_n\right) \Delta_n x_n\ldots\Delta_1 x_1\right)\right.$$

$$\cdot\left(\int_{\prod_{i=1}^{n}[a_i,b_i]} g\left(\sigma_1\left(x_1\right),\ldots,\sigma_n\left(x_n\right)\right) \Delta_n x_n\ldots\Delta_1 x_1\right)$$

$$+\left(\int_{\prod_{i=1}^{n}[a_i,b_i]} f\left(\sigma_1\left(x_1\right),\ldots,\sigma_n\left(x_n\right)\right) \Delta_n x_n\ldots\Delta_1 x_1\right)$$

$$\left.\cdot\left(\int_{\prod_{i=1}^{n}[a_i,b_i]} g\left(x_1,\ldots,x_n\right) \Delta_n x_n\ldots\Delta_1 x_1\right)\right]$$

$$-\frac{1}{2\left(\prod_{i=1}^{n}\left(b_i-a_i\right)\right)}\left[\left(\int_{\prod_{i=1}^{n}[a_i,b_i]} F\left(f\right)\left(x_1,\ldots,x_n\right) g\left(x_1,\ldots,x_n\right) \Delta_n x_n\ldots\Delta_1 x_1\right)\right.$$

$$\left.+\int_{\prod_{i=1}^{n}[a_i,b_i]} F\left(g\right)\left(x_1,\ldots,x_n\right) f\left(x_1,\ldots,x_n\right) \Delta_n x_n\ldots\Delta_1 x_1\right] \tag{7.53}$$

$$= \frac{\int_{\prod_{i=1}^n [a_i,b_i]} f(x_1,\ldots,x_n) R_n(g)(x_1,\ldots,x_n) \Delta_n x_n \ldots \Delta_1 x_1}{\prod_{i=1}^n (b_i - a_i)}$$

$$+ \frac{\int_{\prod_{i=1}^n [a_i,b_i]} g(x_1,\ldots,x_n) R_n(f)(x_1,\ldots,x_n) \Delta_n x_n \ldots \Delta_1 x_1}{\prod_{i=1}^n (b_i - a_i)}.$$

Hence

$$|E(f,g)| \le \frac{1}{\prod_{i=1}^n (b_i - a_i)} \tag{7.54}$$

$$\cdot \left[\|f\|_{\infty, \prod_{i=1}^n ([a_i,b_i] \cap \mathbb{T}_i)} \int_{\prod_{i=1}^n [a_i,b_i]} |R_n(g)(x_1,\ldots,x_n)| \Delta_n x_n \ldots \Delta_1 x_1 \right.$$

$$\left. + \|g\|_{\infty, \prod_{i=1}^n ([a_i,b_i] \cap \mathbb{T}_i)} \int_{\prod_{i=1}^n [a_i,b_i]} |R_n(f)(x_1,\ldots,x_n)| \Delta_n x_n \ldots \Delta_1 x_1 \right]$$

By Tietze's extension theorem for the continuous functions f,g we get that they are bounded, so that the right hand side of (7.54) makes sense.

By (7.44) we get that

$$\int_{\prod_{i=1}^n [a_i,b_i]} |R_n(f)(x_1,\ldots,x_n)| \Delta_n x_n \ldots \Delta_1 x_1$$

$$\le \min \begin{cases} \left\| \frac{\partial^n f}{\Delta_n s_n \ldots \Delta_1 s_1} \right\|_{\infty, \prod_{i=1}^n ([a_i,b_i] \cap \mathbb{T}_i)} \\ \quad \cdot \left(\prod_{i=1}^n \int_{a_i}^{b_i} \left[h_2^{(i)}(x_i, a_i) + h_2^{(i)}(x_i, b_i) \right] \Delta_i x_i \right), \\ \left\| \frac{\partial^n f}{\Delta_n s_n \ldots \Delta_1 s_1} \right\|_{p, \prod_{i=1}^n ([a_i,b_i] \cap \mathbb{T}_i)} \\ \quad \cdot \left(\prod_{i=1}^n \left(\int_{a_i}^{b_i} \|p_i(x_i, \cdot)\|_{q, ([a_i,b_i] \cap \mathbb{T}_i)} \Delta_i x_i \right) \right), \\ \quad \text{where } p, q > 1 : \frac{1}{p} + \frac{1}{q} = 1, \\ \left\| \frac{\partial^n f}{\Delta_n s_n \ldots \Delta_1 s_1} \right\|_{1, \prod_{i=1}^n ([a_i,b_i] \cap \mathbb{T}_i)} \\ \quad \cdot \left(\prod_{i=1}^n \left(\int_{a_i}^{b_i} \left(\frac{b_i - a_i}{2} + \left| x_i - \left(\frac{a_i + b_i}{2} \right) \right| \right) \Delta_i x_i \right) \right). \end{cases} \tag{7.55}$$

Using Theorem 7.9, a similar estimate to (7.55) can be written for $\int_{\prod_{i=1}^n [a_i,b_i]} |R_n(g)(x_1,\ldots,x_n)| \Delta_n x_n \ldots \Delta_1 x_1$.

Now by (7.54) and (7.55) the claim is proved. $\qquad\square$

We make

Remark 7.3. Let the time scales $(\mathbb{T}_i, \rho_i, \nabla_i)$, $i = 1, 2, \ldots, n$, $n \in \mathbb{N}$ and $a_i < b_i$; $a_i, b_i \in \mathbb{T}_i$, $i = 1, \ldots, n$. Here ρ_i, $i = 1, \ldots, n$ are continuous. We consider $f, g \in C^n \left(\prod_{i=1}^n ([a_i, b_i] \cap \mathbb{T}_i) \right)$. Define the kernels $p_i^* : [a_i, b_i]^2 \to \mathbb{R}$:

$$p_i^*(x_i, s_i) := \begin{cases} s_i - a_i, & s_i \in [a_i, x_i], \\ s_i - b_i, & s_i \in (x_i, b_i], \end{cases} \tag{7.56}$$

for $i = 1, 2, \ldots, n$.

Denote $\int_{\prod_{i=1}^{j}[a_i,b_i]} \cdot := \int_{\prod_{i=1}^{j}([a_i,b_i]\cap\mathbb{T}_i)} \cdot$, $j = 1, \ldots, n$.

We define

$$F_1^*(f)(x_j) := \int_{\prod_{i=1}^{n}[a_i,b_i]} p_j^*(x_j, s_j)$$

$$\cdot \frac{\partial f(\rho_1(s_1), \ldots, \rho_{j-1}(s_{j-1}), s_j, \rho_{j+1}(s_{j+1}), \ldots, \rho_n(s_n))}{\nabla_j s_j} \nabla_n s_n \ldots \nabla_1 s_1, \quad (7.57)$$

for $j = 1, \ldots, n$,

$$F_2^*(f)(x_j, x_k) := \int_{\prod_{i=1}^{n}[a_i,b_i]} p_j^*(x_j, s_j) p_k^*(x_k, s_k) \quad (7.58)$$

$$\cdot \frac{\partial^2 f(\rho_1(s_1), \ldots, \rho_{j-1}(s_{j-1}), s_j, \rho_{j+1}(s_{j+1}), \ldots, \rho_{k-1}(s_{k-1}), s_k, \rho_{k+1}(s_{k+1}), \ldots, \rho_n(s_n))}{\nabla_k s_k \nabla_j s_j}$$

$$\cdot \nabla_n s_n \ldots \nabla_1 s_1,$$

for $j, k \in \{1, \ldots, n\}$,

$$F_3^*(f)(x_j, x_k, x_r) := \int_{\prod_{i=1}^{n}[a_i,b_i]} p_j^*(x_j, s_j) p_k^*(x_k, s_k) p_r^*(x_r, s_r) \quad (7.59)$$

$$\cdot \frac{\partial^3 f(\rho_1(s_1), \ldots, \rho_{j-1}(s_{j-1}), s_j, \rho_{j+1}(s_{j+1}), \ldots, \rho_{k-1}(s_{k-1}), s_k, \rho_{k+1}(s_{k+1})}{\nabla_r s_r \nabla_k s_k \nabla_j s_j}$$
$$\underset{}{\cdot, \ldots, \rho_{r-1}(s_{r-1}), s_r, \rho_{r+1}(s_{r+1}), \ldots, \rho_n(s_n))}$$

$$\cdot \nabla_n s_n \ldots \nabla_1 s_1,$$

for $j, k, r \in \{1, \ldots, n\}$,

$$\vdots$$

$$F_{n-1}^*(f)(x_1, \ldots, \widehat{x_l}, \ldots, x_n) := \int_{\prod_{i=1}^{n}[a_i,b_i]} p_1^*(x_1, s_1) \ldots \widehat{p_l^*(x_l, s_l)} \ldots p_n^*(x_n, s_n)$$

$$\cdot \frac{\partial^{n-1} f(s_1, s_2, \ldots, \rho_l(s_l), \ldots, s_n)}{\nabla_n s_n \ldots \widehat{\nabla_l s_l} \ldots \nabla_1 s_1} \nabla_n s_n \ldots \nabla_1 s_1, \quad (7.60)$$

for $l = 1, \ldots, n$; $\widehat{x_l}$ means x_l is missing, and

$$R_n^*(f)(x_1, \ldots, x_n) := \int_{\prod_{i=1}^{n}[a_i,b_i]} \left(\prod_{i=1}^{n} p_i^*(x_i, s_i)\right) \frac{\partial^n f(s_1, \ldots, s_n)}{\nabla_n s_n \ldots \nabla_1 s_1} \nabla_n s_n \ldots \nabla_1 s_1. \quad (7.61)$$

We set

$$F^*(f)(x_1, \ldots, x_n) := \sum_{j=1}^{n} F_1^*(f)(x_j) + \sum_{\substack{l_1=1 \\ j<k}}^{\binom{n}{2}} F_2^*(f)(x_j, x_k) \quad (7.62)$$

$$+ \sum_{\substack{l_2=1 \\ j<k<r}}^{\binom{n}{3}} F_3^* (f)(x_j, x_k, x_r) + \ldots + \sum_{l=1}^{\binom{n}{n-1}} F_{n-1}^* (f)(x_1, \ldots, \widehat{x_l}, \ldots, x_n),$$

where l_1 counts $(j,k) : j < k;$ $j,k \in \{1, \ldots, n\}$, l_2 counts $(j,k,r) : j < k < r;$ $j, k, r \in \{1, \ldots, n\}$, etc. Also $p_l^* \, \widehat{(x_l, s_l)}$, $\widehat{\nabla_l s_l}$ are missing elements.

Similarly one can define $F_1^* (g)$, $F_2^* (g)$, $F_3^* (g)$, $\ldots$, $F_{n-1}^* (g)$, $R_n^* (g)$ and $F^* (g)$.

By Theorem 8 of [10] we obtain the ∇-representation of f:

$$f(x_1, \ldots, x_n) = \frac{1}{\prod_{i=1}^n (b_i - a_i)} \int_{\prod_{i=1}^n [a_i, b_i]} f(\rho_1(s_1), \ldots, \rho_n(s_n)) \nabla_n s_n \ldots \nabla_1 s_1$$

$$+ \frac{F^*(f)(x_1, \ldots, x_n)}{\prod_{i=1}^n (b_i - a_i)} + \frac{R_n^*(f)(x_1, \ldots, x_n)}{\prod_{i=1}^n (b_i - a_i)}, \qquad (7.63)$$

for any $(x_1, \ldots, x_n) \in \prod_{i=1}^n ([a_i, b_i] \cap \mathbb{T}_i)$.

A similar representation holds for g.

From [10], we also need the estimate

Theorem 7.11. *(all as in Remark 7.3) Then*

$$|R_n^* (f)(x_1, \ldots, x_n)|$$

$$\leq \min \begin{cases} \left\| \frac{\partial^n f}{\nabla_n s_n \ldots \nabla_1 s_1} \right\|_{\infty, \prod_{i=1}^n ([a_i, b_i] \cap \mathbb{T}_i)} \left(\prod_{i=1}^n \left(\widehat{h}_2^{(i)} (x_i, a_i) + \widehat{h}_2^{(i)} (x_i, b_i) \right) \right), \\ \left\| \frac{\partial^n f}{\nabla_n s_n \ldots \nabla_1 s_1} \right\|_{p, \prod_{i=1}^n ([a_i, b_i] \cap \mathbb{T}_i)} \left(\prod_{i=1}^n \| p_i^* (x_i, \cdot) \|_{q, ([a_i, b_i] \cap \mathbb{T}_i)} \right), \\ \text{where } p, q > 1 : \frac{1}{p} + \frac{1}{q} = 1, \\ \left\| \frac{\partial^n f}{\nabla_n s_n \ldots \nabla_1 s_1} \right\|_{1, \prod_{i=1}^n ([a_i, b_i] \cap \mathbb{T}_i)} \left(\prod_{i=1}^n \left(\frac{b_i - a_i}{2} + \left| x_i - \left(\frac{a_i + b_i}{2} \right) \right| \right) \right). \end{cases}$$

$$(7.64)$$

So based on Remark 7.3 and Theorem 7.11 and acting, similarly, as in the proof of Theorem 7.10, we present the following general Grüss type ∇-multivariate inequality.

Theorem 7.12. *Let the time scales* $(\mathbb{T}_i, \rho_i, \nabla_i)$, $i = 1, 2, \ldots, n$, $n \in \mathbb{N}$ *and* $a_i < b_i$; $a_i, b_i \in \mathbb{T}_i$, $i = 1, \ldots, n$. *Here* ρ_i, $i = 1, \ldots, n$ *are continuous. We consider* $f, g \in C^n (\prod_{i=1}^n ([a_i, b_i] \cap \mathbb{T}_i))$. *Here* p_i^* *is as in (7.56), and* $F^* (f)$ *and* $F^* (g)$ *as in (7.62). Define*

$$E^*(f, g) := \int_{\prod_{i=1}^n [a_i, b_i]} f(x_1, \ldots, x_n) g(x_1, \ldots, x_n) \nabla_n x_n \ldots \nabla_1 x_1$$

$$- \frac{1}{2 \left(\prod_{i=1}^n (b_i - a_i) \right)} \left[\left(\int_{\prod_{i=1}^n [a_i, b_i]} f(x_1, \ldots, x_n) \nabla_n x_n \ldots \nabla_1 x_1 \right) \cdot \right.$$

$$\left(\int_{\prod_{i=1}^{n}[a_i,b_i]} g\left(\rho_1(x_1),\ldots,\rho_n(x_n)\right) \nabla_n x_n \ldots \nabla_1 x_1 \right)$$

$$+ \left(\int_{\prod_{i=1}^{n}[a_i,b_i]} f\left(\rho_1(x_1),\ldots,\rho_n(x_n)\right) \nabla_n x_n \ldots \nabla_1 x_1 \right)$$

$$\cdot \left(\int_{\prod_{i=1}^{n}[a_i,b_i]} g\left(x_1,\ldots,x_n\right) \nabla_n x_n \ldots \nabla_1 x_1 \right) \Big]$$

$$- \frac{1}{2\left(\prod_{i=1}^{n}(b_i - a_i)\right)} \left[\int_{\prod_{i=1}^{n}[a_i,b_i]} [F^*(f)(x_1,\ldots,x_n) g(x_1,\ldots,x_n) \right.$$

$$\left. + F^*(g)(x_1,\ldots,x_n) f(x_1,\ldots,x_n)] \nabla_n x_n \ldots \nabla_1 x_1 \right]. \tag{7.65}$$

Then

$$|E^*(f,g)| \leq \frac{1}{\prod_{i=1}^{n}(b_i - a_i)}$$

$$\cdot \min \begin{cases} \left[\|f\|_{\infty,\prod_{i=1}^{n}([a_i,b_i]\cap\mathbb{T}_i)} \left\| \frac{\partial^n g}{\nabla_n s_n \ldots \nabla_1 s_1} \right\|_{\infty,\prod_{i=1}^{n}([a_i,b_i]\cap\mathbb{T}_i)} \right. \\ \left. + \|g\|_{\infty,\prod_{i=1}^{n}([a_i,b_i]\cap\mathbb{T}_i)} \left\| \frac{\partial^n f}{\nabla_n s_n \ldots \nabla_1 s_1} \right\|_{\infty,\prod_{i=1}^{n}([a_i,b_i]\cap\mathbb{T}_i)} \right] \\ \cdot \left(\prod_{i=1}^{n} \int_{a_i}^{b_i} \left[\widehat{h}_2^{(i)}(x_i,a_i) + \widehat{h}_2^{(i)}(x_i,b_i) \right] \nabla_i x_i \right), \\[12pt] \left[\|f\|_{\infty,\prod_{i=1}^{n}([a_i,b_i]\cap\mathbb{T}_i)} \left\| \frac{\partial^n g}{\nabla_n s_n \ldots \nabla_1 s_1} \right\|_{p,\prod_{i=1}^{n}([a_i,b_i]\cap\mathbb{T}_i)} \right. \\ \left. + \|g\|_{\infty,\prod_{i=1}^{n}([a_i,b_i]\cap\mathbb{T}_i)} \left\| \frac{\partial^n f}{\nabla_n s_n \ldots \nabla_1 s_1} \right\|_{p,\prod_{i=1}^{n}([a_i,b_i]\cap\mathbb{T}_i)} \right] \\ \cdot \left(\prod_{i=1}^{n} \left(\int_{a_i}^{b_i} \|p_i^*(x_i,\cdot)\|_{q,([a_i,b_i]\cap\mathbb{T}_i)} \nabla_i x_i \right) \right), \\ \text{where } p,q > 1 : \frac{1}{p} + \frac{1}{q} = 1. \\[12pt] \left[\|f\|_{\infty,\prod_{i=1}^{n}([a_i,b_i]\cap\mathbb{T}_i)} \left\| \frac{\partial^n g}{\nabla_n s_n \ldots \nabla_1 s_1} \right\|_{1,\prod_{i=1}^{n}([a_i,b_i]\cap\mathbb{T}_i)} \right. \\ \left. + \|g\|_{\infty,\prod_{i=1}^{n}([a_i,b_i]\cap\mathbb{T}_i)} \left\| \frac{\partial^n f}{\nabla_n s_n \ldots \nabla_1 s_1} \right\|_{1,\prod_{i=1}^{n}([a_i,b_i]\cap\mathbb{T}_i)} \right] \\ \cdot \left(\prod_{i=1}^{n} \int_{a_i}^{b_i} \left(\frac{b_i - a_i}{2} + \left| x_i - \left(\frac{a_i + b_i}{2} \right) \right| \right) \nabla_i x_i \right). \end{cases} \tag{7.66}$$

It follows a comparison of integral means Δ-multivariate inequality.

Theorem 7.13. *Let the time scales* $(\mathbb{T}_i, \sigma_i, \Delta_i)$, $i = 1,\ldots,n$, $n \in \mathbb{N}$ *and* $a_i \leq c_i < d_i \leq b_i$; $a_i, c_i, d_i, b_i \in \mathbb{T}_i$, $i = 1,\ldots,n$. *Here* σ_i, $i = 1,\ldots,n$ *are continuous. We consider* $f \in C^n\left(\prod_{i=1}^{n}([a_i,b_i]\cap\mathbb{T}_i)\right)$. *Here* p_i *as in* (7.31) *and* $F(f)$ *as in* (7.39).

Let $\mathcal{P} := \mathcal{P}\left(\prod_{i=1}^{n}\left([c_i, d_i] \cap \mathbb{T}_i\right)\right)$ *the power set of* $\prod_{i=1}^{n}\left([c_i, d_i] \cap \mathbb{T}_i\right)$ *and* μ *be a probability measure on* $\left(\left(\prod_{i=1}^{n}\left([c_i, d_i] \cap \mathbb{T}_i\right)\right), \mathcal{P}\right)$. *We define*

$$K := \int_{\prod_{i=1}^{n}[c_i, d_i]} f(x_1, \ldots, x_n)\, d\mu(x_1, \ldots, x_n)$$

$$-\frac{1}{\prod_{i=1}^{n}(b_i - a_i)} \int_{\prod_{i=1}^{n}[a_i, b_i]} f(\sigma_1(x_1), \ldots, \sigma_n(x_n))\, \Delta_n x_n \ldots \Delta_1 x_1$$

$$-\frac{\int_{\prod_{i=1}^{n}[c_i, d_i]} F(f)(x_1, \ldots, x_n)\, d\mu(x_1, \ldots, x_n)}{\prod_{i=1}^{n}(b_i - a_i)}. \tag{7.67}$$

Then

$$|K| \leq \frac{1}{\prod_{i=1}^{n}(b_i - a_i)}$$

$$\cdot \min \begin{cases} \left\|\frac{\partial^n f}{\Delta_n s_n \ldots \Delta_1 s_1}\right\|_{\infty, \prod_{i=1}^{n}([a_i, b_i] \cap \mathbb{T}_i)} \\ \cdot \left(\int_{\prod_{i=1}^{n}[c_i, d_i]} \left(\prod_{i=1}^{n}\left(h_2^{(i)}(x_i, a_i) + h_2^{(i)}(x_i, b_i)\right)\right) d\mu(\overrightarrow{x})\right), \\[2mm] \left\|\frac{\partial^n f}{\Delta_n s_n \ldots \Delta_1 s_1}\right\|_{p, \prod_{i=1}^{n}([a_i, b_i] \cap \mathbb{T}_i)} \\ \cdot \left(\int_{\prod_{i=1}^{n}[c_i, d_i]} \left(\prod_{i=1}^{n} \|p_i(x_i, \cdot)\|_{q, ([a_i, b_i] \cap \mathbb{T}_i)}\right) d\mu(\overrightarrow{x})\right), \\ \text{where } p, q > 1 : \frac{1}{p} + \frac{1}{q} = 1, \\[2mm] \left\|\frac{\partial^n f}{\Delta_n s_n \ldots \Delta_1 s_1}\right\|_{1, \prod_{i=1}^{n}([a_i, b_i] \cap \mathbb{T}_i)} \\ \cdot \left(\int_{\prod_{i=1}^{n}[c_i, d_i]} \left(\prod_{i=1}^{n}\left(\frac{b_i - a_i}{2} + \left|x_i - \left(\frac{a_i + b_i}{2}\right)\right|\right)\right) d\mu(\overrightarrow{x})\right), \end{cases} \tag{7.68}$$

where $\overrightarrow{x} := (x_1, \ldots, x_n)$.

Proof. Integrating (7.41) against μ over $\prod_{i=1}^{n}\left([c_i, d_i] \cap \mathbb{T}_i\right)$ we obtain

$$K = \int_{\prod_{i=1}^{n}[c_i, d_i]} f(x_1, \ldots, x_n)\, d\mu(x_1, \ldots, x_n)$$

$$-\frac{1}{\prod_{i=1}^{n}(b_i - a_i)} \int_{\prod_{i=1}^{n}[a_i, b_i]} f(\sigma_1(x_1), \ldots, \sigma_n(x_n))\, \Delta_n x_n \ldots \Delta_1 x_1$$

$$-\frac{\int_{\prod_{i=1}^{n}[c_i, d_i]} F(f)(x_1, \ldots, x_n)\, d\mu(x_1, \ldots, x_n)}{\prod_{i=1}^{n}(b_i - a_i)} \tag{7.69}$$

$$= \frac{\int_{\prod_{i=1}^{n}[c_i, d_i]} R_n(f)(x_1, \ldots, x_n)\, d\mu(x_1, \ldots, x_n)}{\prod_{i=1}^{n}(b_i - a_i)}.$$

Hence

$$|K| \leq \frac{\int_{\prod_{i=1}^{n}[c_i, d_i]} |R_n(f)(x_1, \ldots, x_n)|\, d\mu(x_1, \ldots, x_n)}{\prod_{i=1}^{n}(b_i - a_i)}, \tag{7.70}$$

and by (7.44) the claim is proved. $\qquad\square$

We finish main results with a comparison of integral means ∇-multivariate inequality.

Theorem 7.14. *Let the time scales* $(\mathbb{T}_i, \rho_i, \nabla_i)$, $i = 1, \ldots, n$, $n \in \mathbb{N}$ *and* $a_i \leq c_i < d_i \leq b_i$; $a_i, c_i, d_i, b_i \in \mathbb{T}_i$, $i = 1, \ldots, n$. *Here* ρ_i, $i = 1, \ldots, n$ *are continuous. We consider* $f \in C^n \left(\prod_{i=1}^n ([a_i, b_i] \cap \mathbb{T}_i) \right)$. *Here* p_i^* *as in (7.56) and* $F^*(f)$ *as in (7.62). Let* $\mathcal{P} := \mathcal{P} \left(\prod_{i=1}^n ([c_i, d_i] \cap \mathbb{T}_i) \right)$ *the power set of* $\prod_{i=1}^n ([c_i, d_i] \cap \mathbb{T}_i)$ *and* μ *be a probability measure on* $\left(\left(\prod_{i=1}^n ([c_i, d_i] \cap \mathbb{T}_i) \right), \mathcal{P} \right)$. *We define*

$$K^* := \int_{\prod_{i=1}^n [c_i, d_i]} f(x_1, \ldots, x_n) \, d\mu(x_1, \ldots, x_n)$$

$$- \frac{1}{\prod_{i=1}^n (b_i - a_i)} \int_{\prod_{i=1}^n [a_i, b_i]} f(\rho_1(x_1), \ldots, \rho_n(x_n)) \, \nabla_n x_n \ldots \nabla_1 x_1$$

$$- \frac{\int_{\prod_{i=1}^n [c_i, d_i]} F^*(f)(x_1, \ldots, x_n) \, d\mu(x_1, \ldots, x_n)}{\prod_{i=1}^n (b_i - a_i)}. \tag{7.71}$$

Then

$$|K^*| \leq \frac{1}{\prod_{i=1}^n (b_i - a_i)}$$

$$\cdot \min \begin{cases} \left\| \dfrac{\partial^n f}{\nabla_n s_n \ldots \nabla_1 s_1} \right\|_{\infty, \prod_{i=1}^n ([a_i, b_i] \cap \mathbb{T}_i)} \\ \quad \cdot \left(\int_{\prod_{i=1}^n [c_i, d_i]} \left(\prod_{i=1}^n \left(\widehat{h}_2^{(i)}(x_i, a_i) + \widehat{h}_2^{(i)}(x_i, b_i) \right) \right) d\mu(\overrightarrow{x}) \right), \\[1.5em] \left\| \dfrac{\partial^n f}{\nabla_n s_n \ldots \nabla_1 s_1} \right\|_{p, \prod_{i=1}^n ([a_i, b_i] \cap \mathbb{T}_i)} \\ \quad \cdot \left(\int_{\prod_{i=1}^n [c_i, d_i]} \left(\prod_{i=1}^n \| p_i^*(x_i, \cdot) \|_{q, ([a_i, b_i] \cap \mathbb{T}_i)} \right) d\mu(\overrightarrow{x}) \right), \\ \quad \text{where } p, q > 1 : \frac{1}{p} + \frac{1}{q} = 1, \\[1.5em] \left\| \dfrac{\partial^n f}{\nabla_n s_n \ldots \Delta_1 s_1} \right\|_{1, \prod_{i=1}^n ([a_i, b_i] \cap \mathbb{T}_i)} \\ \quad \cdot \left(\int_{\prod_{i=1}^n [c_i, d_i]} \left(\prod_{i=1}^n \left(\frac{b_i - a_i}{2} + \left| x_i - \left(\frac{a_i + b_i}{2} \right) \right| \right) \right) d\mu(\overrightarrow{x}) \right), \end{cases} \tag{7.72}$$

where $\overrightarrow{x} := (x_1, \ldots, x_n)$.

Proof.   Similar to Theorem 7.13, using (7.63) and (7.64). $\square$

Remark 7.4. Inequalities (7.68) and (7.72) are valid when μ takes any of the forms $\frac{\Delta_1 s_1 \ldots \Delta_n s_n}{\prod_{i=1}^n (d_i - c_i)}$, $\frac{\nabla_1 s_1 \ldots \nabla_n s_n}{\prod_{i=1}^n (d_i - c_i)}$, $\frac{\Delta_1 s_1 \nabla_2 s_2 \Delta_3 s_3 \ldots \nabla_n s_n}{\prod_{i=1}^n (d_i - c_i)}$, and any combination product of Δs_j, ∇s_k divided by $\prod_{i=1}^n (d_i - c_i)$, where $j, k \in \{1, \ldots, n\}$.

7.3 Applications

When $\mathbb{T}_i = \mathbb{R}$, $i = 1, \ldots, n$, this chapter's related results were published in [3], [4], [5], [6], so this work is their generalization to the time scales setting.

When $\mathbb{T}_i = \mathbb{Z}$, $i = 1, \ldots, n$, we have (see [21], pp. 13-14)

$$f^{\Delta^n}(t) = \sum_{k=0}^{n} \binom{n}{k} (-1)^{n-k} f(t+k), \tag{7.73}$$

with $\sigma(t) = t + 1$, and $h_k(t,s) = \frac{(t-s)^{(k)}}{k!}$, $\forall\, k \in \mathbb{N}_0$, $\forall\, t, s \in \mathbb{Z}$, where $t^{(0)} = 1$, $t^{(k)} = \prod_{i=0}^{k-1}(t-i)$, $k \in \mathbb{N}$, also

$$\int_a^b f(t)\,\Delta t = \sum_{t=a}^{b-1} f(t), \quad a < b, \tag{7.74}$$

and an rd-continuous or a continuous function f corresponds to any f.

When $\mathbb{T}_i = \mathbb{R}$, then $\sigma(t) = t$, and integrals and derivatives coincide with the usual numerical ones.

To keep paper's size limited we restrict ourselves to applications on $\mathbb{Z} \times \mathbb{R} \times \mathbb{Z}$. One can give many other interesting applications, univariate and multivariate on many other time scales or their combinations, also applications for ∇-results.

So here we take $\mathbb{T}_1 = \mathbb{Z}$, $\mathbb{T}_2 = \mathbb{R}$, $\mathbb{T}_3 = \mathbb{Z}$. Here $\sigma_2(t) = t$, $\forall\, t \in \mathbb{R}$, and

$$h_k^{(2)}(t,s) = \frac{(t-s)^k}{k!}, \quad \forall\, s, t \in \mathbb{R}, \ k \in \mathbb{N}_0. \tag{7.75}$$

We take $a_i \leq c_i < d_i \leq b_i$, $i = 1, 2, 3$; $a_1, b_1, c_1, d_1 \in \mathbb{Z}$; $a_2, b_2, c_2, d_2 \in \mathbb{R}$; $a_3, b_3, c_3, d_3 \in \mathbb{Z}$, and $f, g \in C^3\left(\prod_{i=1}^{3}([a_i, b_i] \cap \mathbb{T}_i)\right)$, and $(x_1, x_2, x_3) \in (([a_1, b_1] \cap \mathbb{Z}) \times [a_2, b_2] \times ([a_3, b_3] \cap \mathbb{Z}))$, while p_i, $i = 1, 2, 3$, as in Theorem 7.8, see (7.31).

By (7.33) we have in our setting that

$$F_1(f)(x_1) = \sum_{s_1=a_1}^{b_1-1} \left(\sum_{s_3=a_3}^{b_3-1} \left(\int_{a_2}^{b_2} p_1(x_1, s_1) \frac{\partial f(s_1, s_2, s_3+1)}{\Delta_1 s_1}\,ds_2 \right) \right), \tag{7.76}$$

$$F_1(f)(x_2) = \sum_{s_1=a_1}^{b_1-1} \left(\sum_{s_3=a_3}^{b_3-1} \left(\int_{a_2}^{b_2} p_2(x_2, s_2) \frac{\partial f(s_1+1, s_2, s_3+1)}{\partial s_2}\,ds_2 \right) \right), \tag{7.77}$$

$$F_1(f)(x_3) = \sum_{s_1=a_1}^{b_1-1} \left(\sum_{s_3=a_3}^{b_3-1} \left(\int_{a_2}^{b_2} p_3(x_3, s_3) \frac{\partial f(s_1+1, s_2, s_3)}{\Delta_3 s_3}\,ds_2 \right) \right), \tag{7.78}$$

also by (7.34) we get

$$F_2(f)(x_1, x_2)$$

$$= \sum_{s_1=a_1}^{b_1-1} \left(\sum_{s_3=a_3}^{b_3-1} \left(\int_{a_2}^{b_2} p_1(x_1, s_1) p_2(x_2, s_2) \frac{\partial^2 f(s_1, s_2, s_3+1)}{\Delta_1 s_1 \partial s_2}\,ds_2 \right) \right), \tag{7.79}$$

$$F_2\left(f\right)\left(x_1, x_3\right)$$

$$= \sum_{s_1=a_1}^{b_1-1}\left(\sum_{s_3=a_3}^{b_3-1}\left(\int_{a_2}^{b_2} p_1\left(x_1, s_1\right) p_3\left(x_3, s_3\right)\frac{\partial^2 f\left(s_1, s_2, s_3\right)}{\Delta_1 s_1 \Delta_3 s_2} ds_2\right)\right), \qquad (7.80)$$

$$F_2\left(f\right)\left(x_2, x_3\right)$$

$$= \sum_{s_1=a_1}^{b_1-1}\left(\sum_{s_3=a_3}^{b_3-1}\left(\int_{a_2}^{b_2} p_2\left(x_2, s_2\right) p_3\left(x_3, s_3\right)\frac{\partial^2 f\left(s_1+1, s_2, s_3\right)}{\partial s_2 \Delta_3 s_3} ds_2\right)\right), \qquad (7.81)$$

and by (7.37) we have

$$R_3\left(f\right)\left(x_1, x_2, x_3\right)$$

$$= \sum_{s_1=a_1}^{b_1-1}\left(\sum_{s_3=a_3}^{b_3-1}\left(\int_{a_2}^{b_2} p_1\left(x_1, s_1\right) p_2\left(x_2, s_2\right) p_3\left(x_3, s_3\right)\frac{\partial^3 f\left(s_1, s_2, s_3\right)}{\Delta_3 s_3 \partial s_2 \Delta_1 s_1} ds_2\right)\right).$$
$$(7.82)$$

Next by (7.39) we have

$$F\left(f\right)\left(x_1, x_2, x_3\right) = F_1\left(f\right)\left(x_1\right) + F_1\left(f\right)\left(x_2\right) + F_1\left(f\right)\left(x_3\right) \qquad (7.83)$$

$$+F_2\left(f\right)\left(x_1, x_2\right) + F_2\left(f\right)\left(x_1, x_3\right) + F_2\left(f\right)\left(x_2, x_3\right).$$

So by (7.40) we obtain

$$f\left(x_1, x_2, x_3\right) = \frac{1}{\prod_{i=1}^{3}\left(b_i - a_i\right)}\sum_{s_1=a_1}^{b_1-1}\left(\sum_{s_3=a_3}^{b_3-1}\left(\int_{a_2}^{b_2} f\left(s_1+1, s_2, s_3+1\right) ds_2\right)\right)$$
$$(7.84)$$

$$+\frac{F\left(f\right)\left(x_1, x_2, x_3\right)}{\prod_{i=1}^{3}\left(b_i - a_i\right)} + \frac{R_3\left(f\right)\left(x_1, x_2, x_3\right)}{\prod_{i=1}^{3}\left(b_i - a_i\right)}.$$

Similarly one can write all these formulae for g.

Next following (7.45) we have in our specific case

$$E\left(f, g\right) = \sum_{x_1=a_1}^{b_1-1}\left(\sum_{x_3=a_3}^{b_3-1}\left(\int_{a_2}^{b_2} f\left(x_1, x_2, x_3\right) g\left(x_1, x_2, x_3\right) dx_2\right)\right)$$

$$-\frac{1}{2\prod_{i=1}^{3}\left(b_i - a_i\right)}\left[\left(\sum_{x_1=a_1}^{b_1-1}\left(\sum_{x_3=a_3}^{b_3-1}\left(\int_{a_2}^{b_2} f\left(x_1, x_2, x_3\right) dx_2\right)\right)\right)\right.$$

$$\cdot\left(\sum_{x_1=a_1}^{b_1-1}\left(\sum_{x_3=a_3}^{b_3-1}\left(\int_{a_2}^{b_2} g\left(x_1+1, x_2, x_3+1\right) dx_2\right)\right)\right)$$

$$+\left(\sum_{x_1=a_1}^{b_1-1}\left(\sum_{x_3=a_3}^{b_3-1}\left(\int_{a_2}^{b_2} f\left(x_1+1, x_2, x_3+1\right) dx_2\right)\right)\right)$$

$$\cdot \left(\sum_{x_1=a_1}^{b_1-1} \left(\sum_{x_3=a_3}^{b_3-1} \left(\int_{a_2}^{b_2} g\left(x_1,x_2,x_3\right)dx_2 \right) \right) \right) \right] \tag{7.85}$$

$$-\frac{1}{2\prod_{i=1}^{3}\left(b_i-a_i\right)}\left[\sum_{x_1=a_1}^{b_1-1}\left(\sum_{x_3=a_3}^{b_3-1}\left(\int_{a_2}^{b_2}\left[F\left(f\right)\left(x_1,x_2,x_3\right)g\left(x_1,x_2,x_3\right)\right.\right.\right.\right.$$

$$\left.\left.\left.\left.+F\left(g\right)\left(x_1,x_2,x_3\right)f\left(x_1,x_2,x_3\right)\right]ds_2\right)\right)\right].$$

Thus by (7.46) we obtain on $\mathbb{Z}\times\mathbb{R}\times\mathbb{Z}$ the following Grüss type Δ-inequality:

$$\left|E\left(f,g\right)\right|\leq\frac{1}{\prod_{i=1}^{3}\left(b_i-a_i\right)}$$

$$\cdot\min\begin{cases}\left[\|f\|_{\infty,\prod_{i=1}^{3}\left(\left[a_i,b_i\right]\cap\mathbb{T}_i\right)}\left\|\dfrac{\partial^3 g}{\Delta_3 s_3 \partial s_2 \Delta_1 s_1}\right\|_{\infty,\prod_{i=1}^{3}\left(\left[a_i,b_i\right]\cap\mathbb{T}_i\right)}\right.\\[2mm] \left.+\|g\|_{\infty,\prod_{i=1}^{3}\left(\left[a_i,b_i\right]\cap\mathbb{T}_i\right)}\left\|\dfrac{\partial^3 f}{\Delta_3 s_3 \partial s_2 \Delta_1 s_1}\right\|_{\infty,\prod_{i=1}^{3}\left(\left[a_i,b_i\right]\cap\mathbb{T}_i\right)}\right]\\[2mm] \cdot\left(\prod_{i=1}^{3}\int_{a_i}^{b_i}\left[h_2^{(i)}\left(x_i,a_i\right)+h_2^{(i)}\left(x_i,b_i\right)\right]\Delta_i x_i\right),\\[4mm] \left[\|f\|_{\infty,\prod_{i=1}^{3}\left(\left[a_i,b_i\right]\cap\mathbb{T}_i\right)}\left\|\dfrac{\partial^3 g}{\Delta_3 s_3 \partial s_2 \Delta_1 s_1}\right\|_{p,\prod_{i=1}^{3}\left(\left[a_i,b_i\right]\cap\mathbb{T}_i\right)}\right.\\[2mm] \left.+\|g\|_{\infty,\prod_{i=1}^{3}\left(\left[a_i,b_i\right]\cap\mathbb{T}_i\right)}\left\|\dfrac{\partial^3 f}{\Delta_3 s_3 \partial s_2 \Delta_1 s_1}\right\|_{p,\prod_{i=1}^{3}\left(\left[a_i,b_i\right]\cap\mathbb{T}_i\right)}\right]\\[2mm] \cdot\left(\prod_{i=1}^{3}\left(\int_{a_i}^{b_i}\left\|p_i\left(x_i,\cdot\right)\right\|_{q,\left(\left[a_i,b_i\right]\cap\mathbb{T}_i\right)}\Delta_i x_i\right)\right),\\[2mm] \text{where } p,q>1:\frac{1}{p}+\frac{1}{q}=1.\\[4mm] \left[\|f\|_{\infty,\prod_{i=1}^{3}\left(\left[a_i,b_i\right]\cap\mathbb{T}_i\right)}\left\|\dfrac{\partial^3 g}{\Delta_3 s_3 \partial s_2 \Delta_1 s_1}\right\|_{1,\prod_{i=1}^{3}\left(\left[a_i,b_i\right]\cap\mathbb{T}_i\right)}\right.\\[2mm] \left.+\|g\|_{\infty,\prod_{i=1}^{3}\left(\left[a_i,b_i\right]\cap\mathbb{T}_i\right)}\left\|\dfrac{\partial^3 f}{\Delta_3 s_3 \partial s_2 \Delta_1 s_1}\right\|_{1,\prod_{i=1}^{3}\left(\left[a_i,b_i\right]\cap\mathbb{T}_i\right)}\right]\\[2mm] \cdot\left(\prod_{i=1}^{3}\int_{a_i}^{b_i}\left(\frac{b_i-a_i}{2}+\left|x_i-\left(\frac{a_i+b_i}{2}\right)\right|\right)\Delta_i x_i\right).\end{cases} \tag{7.86}$$

Next time write (7.67) for $\mu=\frac{\Delta_3 x_3 \partial x_2 \Delta_1 x_1}{\prod_{i=1}^{3}\left(d_i-c_i\right)}$. We have

$$K=\frac{\left(\sum_{x_1=a_1}^{b_1-1}\left(\sum_{x_3=a_3}^{b_3-1}\left(\int_{a_2}^{b_2}f\left(x_1,x_2,x_3\right)dx_2\right)\right)\right)}{\prod_{i=1}^{3}\left(d_i-c_i\right)}$$

$$-\frac{1}{\prod_{i=1}^{3}\left(b_i-a_i\right)}\left(\sum_{x_1=a_1}^{b_1-1}\left(\sum_{x_3=a_3}^{b_3-1}\left(\int_{a_2}^{b_2}f\left(x_1+1,x_2,x_3+1\right)dx_2\right)\right)\right) \tag{7.87}$$

$$-\frac{\left(\sum_{x_1=a_1}^{b_1-1}\left(\sum_{x_3=a_3}^{b_3-1}\left(\int_{a_2}^{b_2}F\left(f\right)\left(x_1,x_2,x_3\right)dx_2\right)\right)\right)}{\left(\prod_{i=1}^{3}\left(b_i-a_i\right)\right)\left(\prod_{i=1}^{3}\left(d_i-c_i\right)\right)}.$$

Then by (7.68) we derive the specific comparison of means Δ-inequality

$$|K| \leq \frac{1}{\prod_{i=1}^{3}(b_i - a_i)}$$

$$\cdot \min \begin{cases} \left\| \frac{\partial^3 f}{\Delta_3 s_3 \partial s_2 \Delta_1 s_1} \right\|_{\infty, \prod_{i=1}^{3}([a_i,b_i] \cap \mathbb{T}_i)} \\ \cdot \frac{\left(\int_{\prod_{i=1}^{3}[c_i,d_i]} \left(\prod_{i=1}^{3} \left(h_2^{(i)}(x_i,a_i) + h_2^{(i)}(x_i,b_i) \right) \right) \Delta_3 s_3 ds_2 \Delta_1 s_1 \right)}{\prod_{i=1}^{3}(d_i - c_i)}, \\[3em] \left\| \frac{\partial^3 f}{\Delta_3 s_3 \partial s_2 \Delta_1 s_1} \right\|_{p, \prod_{i=1}^{3}([a_i,b_i] \cap \mathbb{T}_i)} \\ \cdot \frac{\left(\int_{\prod_{i=1}^{3}[c_i,d_i]} \left(\prod_{i=1}^{3} \| p_i(x_i,\cdot) \|_{q,([a_i,b_i] \cap \mathbb{T}_i)} \right) \Delta_3 x_3 dx_2 \Delta_1 x_1 \right)}{\prod_{i=1}^{3}(d_i - c_i)}, \\ \text{where } p,q > 1 : \frac{1}{p} + \frac{1}{q} = 1, \\[3em] \left\| \frac{\partial^3 f}{\Delta_3 s_3 \partial s_2 \Delta_1 s_1} \right\|_{1, \prod_{i=1}^{3}([a_i,b_i] \cap \mathbb{T}_i)} \\ \cdot \frac{\left(\int_{\prod_{i=1}^{3}[c_i,d_i]} \left(\prod_{i=1}^{3} \left(\frac{b_i - a_i}{2} + \left| x_i - \left(\frac{a_i + b_i}{2} \right) \right| \right) \right) \Delta_3 x_3 dx_2 \Delta_1 x_1 \right)}{\prod_{i=1}^{3}(d_i - c_i)}. \end{cases}$$

$$(7.88)$$

Bibliography

1. R. Agarwal, M. Bohner, *Basic calculus on time scales and some of its applications*, Results Math. 35 (1-2) (1999), 3-22.
2. R. Agarwal, M. Bohner, A. Peterson, *Inequalities on time scales: a survey*, Math. Inequalities Appl. 4 (4) (2001), 535-557.
3. G.A. Anastassiou, *On Grüss type multivariate integral inequalities*, Math. Balkanica 17 (1-2) (2003), 1-13.
4. G.A. Anastassiou, *Multivariate Chebyshev–Grüss and comparison of integral euler means type inequalities via a multivariate Euler type identity*, Demonstratio Math. 40 (3) (2007), 537-558.
5. G.A. Anastassiou, *Chebyshev–Grüss type inequalities via Euler type and Fink identities*, Math. Comput. Modelling 45 (9) (2007), 1189-1200.
6. G.A. Anastassiou, *Multivariate Fink type identity and multivariate Ostrowski, comparison of Means and Grüss type inequalities*, Math. Comput. Modelling 46 (3-4) (2007), 351-374.
7. G.A. Anastassiou, *Time scales inequalities*, Intern. J. Difference Equations 5 (1) (2010), 1-23.
8. G.A. Anastassiou, *Nabla time scales inequalities*, Editor Al. Paterson, special issue on Time Scales, Internat. J. Dynam. Syst. Difference Equations 3 (1/2) (2011), 59-83.
9. G.A. Anastassiou, *Intelligent Mathematics: Computational Analysis*, Springer, Heidelberg, 2011.
10. G.A. Anastassiou, *Representations and Ostrowski type inequalities on time scales*, Comput. Math. 62 (2011), 3933-3958.
11. G.A. Anastassiou, *Grüss type and comparison of means inequalities on time scales*, Commun. Appl. Anal. 16 (4) (2012), 541-564.
12. D.R. Anderson, *Taylor polynomials for nabla dynamic equations on time scales*, Panamer. Math. J. 12 (4) (2002), 17-27.
13. D. Anderson, J. Bullock, L. Erbe, A. Peterson, H. Tran, *Nabla dynamic equations on time scales*, Panamer. Math. J. 13 (1) (2003), 1-47.
14. F. Atici, D. Biles, A. Lebedinsky, *An application of time scales to economics*, Math. Comput. Modelling 43 (2006), 718-726.
15. M. Bohner and G. Sh. Guseinov, *Partial differentiation on time scales*, Dynamic Syst. Appl. 13 (3-4) (2004), 351-379.
16. M. Bohner, G. Guseinov, *Multiple integration on time scales*, Dynamic Syst. Appl. 14 (3-4) (2005), 579-606.
17. M. Bohner, G.S. Guseinov, *Multiple Lebesgue integration on time scales*, Adv. Difference Equations 2006 (2006), Article ID 26391, 1-12.

18. M. Bohner, G. Guseinov, *Double integral calculus of variations on times scales*, Comput. Math. Appl. 54 (2007), 45-57.
19. M. Bohner, H. Luo, *Singular second-order multipoint dynamic boundary value problems with mixed derivatives*, Adv. Difference Equations 2006 (2006), Article ID 54989, 1-15.
20. M. Bohner and T. Matthews, *The Grüss inequality on time scales*, Commun. Math. Anal. 3 (1) (2007), 1-8.
21. M. Bohner, A. Peterson, *Dynamic Equations on Time Scales: An Introduction with Applications*, Birkhaüser, Boston (2001).
22. P.L. Čebyšev, *Sur les expressions approximatives des integrales definies par les autres prises entre les mêmes limites*, Proc. Math. Soc. Charkov 2 (1882), 93-98.
23. G. Grüss, *Über das maximum des absoluten Betrages von* $\left[\left(\frac{1}{b-a} \int_a^b f(x)g(x)dx \right) - \left(\frac{1}{(b-a)^2} \int_a^b f(x)dx \int_a^b g(x)dx \right) \right]$, Math. Z. 39 (1935), 215-226.
24. G. Guseinov, *Integration on time scales*, J. Math. Anal. Appl. 285 (2003), 107-127.
25. R. Higgins, A. Peterson, *Cauchy functions and Taylor's formula for time scales* $\mathbb{T}$, in B. Aulbach, S. Elaydi, G. Ladas (eds.), Proc. Sixth Internat. Conf. Difference Equations, New Progress in Difference Equations, Augsburg, Germany, 2001, pp. 299-308, Chapman & Hall/CRC (2004).
26. S. Hilger, *Ein Maßketten kalkül mit Anwendung auf Zentrumsmannig-faltigkeiten*, Ph.D. thesis, Universität Würzburg, Germany (1988).
27. W.J. Liu, Q.A. Ngo, W.B. Chen, *Ostrowski type inequalities on time scales for double integrals*, Acta Appl. Math. 110 (1) (2010), 477-497.
28. N. Martins, D. Torres, *Calculus of variations on time scales with nabla derivatives*, Nonlinear Anal. 71 (12) (2009), 763-773.

Chapter 8

About Integral Operator Inequalities over Time Scales

Here we present a wide range integral operator general inequalities on time scales under convexity. Our treatment is combined by using the diamond-alpha integral. When that fails in the fractional setting we use the delta and nabla integrals. We give plenty of interesting applications. It follows [6].

8.1 Introduction

We start with the definition of the Riemann-Liouville fractional integrals, see [20]. Let $[a, b]$, $(-\infty < a < b < \infty)$ be a finite interval on the real axis $\mathbb{R}$. The Riemann-Liouville fractional integrals $I_{a+}^\alpha f$ and $I_{b-}^\alpha f$ of order $\alpha > 0$ are defined by

$$\left(I_{a+}^\alpha f\right)(x) = \frac{1}{\Gamma(\alpha)} \int_a^x f(t)(x-t)^{\alpha-1} \, dt, \quad (x > a), \tag{8.1}$$

$$\left(I_{b-}^\alpha f\right)(x) = \frac{1}{\Gamma(\alpha)} \int_x^b f(t)(t-x)^{\alpha-1} \, dt, \quad (x < b), \tag{8.2}$$

respectively. Here $\Gamma(\alpha)$ is the Gamma function. These integrals are called the left-sided and the right-sided fractional integrals. We mention a basic property of the operators $I_{a+}^\alpha f$ and $I_{b-}^\alpha f$ of order $\alpha > 0$, see also [23]. The result says that the fractional integral operators $I_{a+}^\alpha f$ and $I_{b-}^\alpha f$ are bounded in $L_p(a, b)$, $1 \le p \le \infty$, that is

$$\left\| I_{a+}^\alpha f \right\|_p \le K \left\| f \right\|_p, \quad \left\| I_{b-}^\alpha f \right\|_p \le K \left\| f \right\|_p \tag{8.3}$$

where

$$K = \frac{(b-a)^\alpha}{\alpha \Gamma(\alpha)}. \tag{8.4}$$

Inequality (8.3), that is the result involving the left-sided fractional integral, was proved by H. G. Hardy in one of his first papers, see [16]. He did not write down the constant, but the calculation of the constant was hidden inside his proof.

So we are motivated by (8.3), and also [5], [8], [19], [2], and we will prove analogous properties on Time Scales. But first we need some background on Time Scales, see also [12].

A time scale $\mathbb{T}$ is an arbitrary nonempty closed subset of the real numbers. The time scales calculus was initiated by S. Hilger in his PhD thesis in order to unify discrete and continuous analysis [17; 18]. Let $\mathbb{T}$ be a time scale with the topology that it inherits from the real numbers. For $t \in \mathbb{T}$, we define the forward jump operator $\sigma : \mathbb{T} \to \mathbb{T}$ by

$$\sigma(t) = \inf\{s \in \mathbb{T} : s > t\}, \tag{8.5}$$

and the backward jump operator $\rho : \mathbb{T} \to \mathbb{T}$ by

$$\rho(t) = \sup\{s \in \mathbb{T} : s < t\}. \tag{8.6}$$

If $\sigma(t) > t$ we say that t is right-scattered, while if $\rho(t) < t$ we say that t is left-scattered. Points that are simultaneously right-scattered and left-scattered are said to be isolated. If $\sigma(t) = t$, then t is called right-dense; if $\rho(t) = t$, then t is called left-dense. The mappings $\mu, \nu : \mathbb{T} \to [0, +\infty)$ defined by $\mu(t) := \sigma(t) - t$ and $\nu(t) := t - \rho(t)$ are called, respectively, the forward and backward graininess function.

Given a time scale $\mathbb{T}$, we introduce the sets $\mathbb{T}^k$, $\mathbb{T}_k$, and $\mathbb{T}_k^k$ as follows. If $\mathbb{T}$ has a left-scattered maximum t_1, then $\mathbb{T}^k = \mathbb{T} - \{t_1\}$, otherwise $\mathbb{T}^k = \mathbb{T}$. If $\mathbb{T}$ has a right-scattered minimum t_2, then $\mathbb{T}_k = \mathbb{T} - \{t_2\}$, otherwise $\mathbb{T}_k = \mathbb{T}$. Finally, $\mathbb{T}_k^k = \mathbb{T}^k \cap \mathbb{T}_k$.

Let $f : \mathbb{T} \to \mathbb{R}$ be a real valued function on a time scale $\mathbb{T}$. Then, for $t \in \mathbb{T}^k$, we define $f^{\Delta}(t)$ to be the number, if one exists, such that for all $\epsilon > 0$, there is a neighborhood U of t such that for all $s \in U$,

$$\left| f(\sigma(t)) - f(s) - f^{\Delta}(t)(\sigma(t) - s) \right| \le \epsilon |\sigma(t) - s|. \tag{8.7}$$

We say that f is delta differentiable on $\mathbb{T}^k$ provided $f^{\Delta}(t)$ exists for all $t \in \mathbb{T}^k$. Similarly, for $t \in \mathbb{T}_k$ we define $f^{\nabla}(t)$ to be the number, if one exists, such that for all $\epsilon > 0$, there is a neighborhood V of t such that for all $s \in V$

$$\left| f(\rho(t)) - f(s) - f^{\nabla}(t)(\rho(t) - s) \right| \le \epsilon |\rho(t) - s|. \tag{8.8}$$

We say that f is nabla differentiable on $\mathbb{T}_k$, provided that $f^{\nabla}(t)$ exists for all $t \in \mathbb{T}_k$.

For $f : \mathbb{T} \to \mathbb{R}$ we define the function $f^{\sigma} : \mathbb{T} \to \mathbb{R}$ by $f^{\sigma}(t) = f(\sigma(t))$ for all $t \in \mathbb{T}$, that is $f^{\sigma} = f \circ \sigma$. Similarly, we define the function $f^{\rho} : \mathbb{T} \to \mathbb{R}$ by $f^{\rho}(t) = f(\rho(t))$ for all $t \in \mathbb{T}$, that is, $f^{\rho} = f \circ \rho$.

A function $f : \mathbb{T} \to \mathbb{R}$ is called rd-continuous, provided it is continuous at all right-dense points in $\mathbb{T}$ and its left-sided limits finite at all left-dense points in $\mathbb{T}$. A function $f : \mathbb{T} \to \mathbb{R}$ is called ld-continuous, provided it is continuous at all left-dense points in $\mathbb{T}$ and its right-sided limits finite at all right-dense points in $\mathbb{T}$.

A function $F : \mathbb{T} \to \mathbb{R}$ is called a delta antiderivative of $f : \mathbb{T} \to \mathbb{R}$ provided that $F^{\Delta}(t) = f(t)$ holds for all $t \in \mathbb{T}^k$. Then the delta integral of f is defined by

$$\int_a^b f(t)\,\Delta t = F(b) - F(a). \tag{8.9}$$

A function $G : \mathbb{T} \to \mathbb{R}$ is called a nabla antiderivative of $g : \mathbb{T} \to \mathbb{R}$, provided $G^\nabla(t) = g(t)$ holds for all $t \in \mathbb{T}_k$. Then the nabla integral of g is defined by $\int_a^b g(t)\, \nabla t = G(b) - G(a)$. For more details on time scales one can see [1; 12; 13].

Now we describe the diamond-α derivative and integral, referring the reader to [21; 22; 24; 25; 26; 27] for more on this calculus.

Let $\mathbb{T}$ be a time scale and f differentiable on $\mathbb{T}$ in the Δ and ∇ senses. For $t \in \mathbb{T}_k^k$ we define the diamond-α dynamic derivative $f^{\diamond^\alpha}(t)$ by

$$f^{\diamond^\alpha}(t) = \alpha f^\Delta(t) + (1 - \alpha) f^\nabla(t), \quad 0 \le \alpha \le 1. \tag{8.10}$$

Thus, f is diamond-α differentiable if and only if f is Δ and ∇ differentiable. The diamond-α derivative reduces to the standard Δ derivative for $\alpha = 1$, or the standard ∇ derivative for $\alpha = 0$. Also, it gives a "weighted derivative" for $\alpha \in (0, 1)$. Diamond-α derivatives have shown in computational experiments to provided efficient and balanced approximation formulae, leading to the design of more reliable numerical methods [24; 25].

Let $f, g : \mathbb{T} \to \mathbb{R}$ be diamond-α differentiable at $t \in \mathbb{T}_k^k$. Then,

(i) $f \pm g : \mathbb{T} \to \mathbb{R}$ is diamond-α differentiable at $t \in \mathbb{T}_k^k$ with

$$(f \pm g)^{\diamond^\alpha}(t) = (f)^{\diamond^\alpha}(t) \pm (g)^{\diamond^\alpha}(t). \tag{8.11}$$

(ii) For any constant c, $cf : \mathbb{T} \to \mathbb{R}$ is diamond-α differentiable at $t \in \mathbb{T}_k^k$ with

$$(cf)^{\diamond^\alpha}(t) = c(f)^{\diamond^\alpha}(t). \tag{8.12}$$

(iii) $fg : \mathbb{T} \to \mathbb{R}$ is diamond-α differentiable at $t \in \mathbb{T}_k^k$ with

$$(fg)^{\diamond^\alpha}(t) = (f)^{\diamond^\alpha}(t) g(t) + \alpha f^\sigma(t) g^\Delta(t) + (1 - \alpha) f^\rho(t) g^\nabla(t). \tag{8.13}$$

Let $a, t \in \mathbb{T}$, and $h : \mathbb{T} \to \mathbb{R}$. Then, the diamond-$\alpha$ integral from a to t of h is defined by

$$\int_a^t h(\tau)\, \diamond_\alpha \tau = \alpha \int_a^t h(\tau)\, \Delta\tau + (1 - \alpha) \int_a^t h(\tau)\, \nabla\tau, \quad 0 \le \alpha \le 1. \tag{8.14}$$

We may notice the absence of an anti-derivative for the $\diamond_\alpha$ combined derivative. For $t \in \mathbb{T}_k^k$, in general

$$\left(\int_a^t h(\tau)\, \diamond_\alpha \tau \right)^{\diamond^\alpha} \ne h(t). \tag{8.15}$$

Although the fundamental theorem of calculus does not hold for the $\diamond_\alpha$-integral, other properties hold true. Let $a, b, t \in \mathbb{T}$, $c \in \mathbb{R}$. Then,

(i) $\displaystyle\int_a^t \{f(\tau) \pm g(\tau)\} \Diamond_\alpha \tau = \int_a^t f(\tau) \Diamond_\alpha \tau \pm \int_a^t g(\tau) \Diamond_\alpha \tau;$ (8.16)

(ii) $\displaystyle\int_a^t cf(\tau) \Diamond_\alpha \tau = c \int_a^t f(t) \Diamond_\alpha \tau;$

(iii) $\displaystyle\int_a^t f(\tau) \Diamond_\alpha \tau = \int_a^b f(\tau) \Diamond_\alpha \tau + \int_b^t f(\tau) \Diamond_\alpha \tau;$

(iv) If $f(t) \geq 0$ for all t, then $\displaystyle\int_a^b f(t) \Diamond_\alpha t \geq 0;$

(v) If $f(t) \leq g(t)$ for all t, then $\displaystyle\int_a^b f(t) \Diamond_\alpha t \leq \int_a^b g(t) \Diamond_\alpha t;$

(vi) If $f(t) \geq 0$ for all t, then $f(t) \equiv 0$ if and only if $\displaystyle\int_a^b f(t) \Diamond_\alpha t = 0;$

(vii) $\displaystyle\int_a^b c \Diamond_\alpha t = c(b-a);$

(viii) $\displaystyle\left| \int_a^b f(t) \Diamond_\alpha t \right| \leq \int_a^b |f(t)| \Diamond_\alpha t.$

We would use Jensen's diamond-α integral inequalities.

Theorem 8.1. *(Jensen's inequality, see [26]). Tet $\mathbb{T}$ be a time scale, $a, b \in \mathbb{T}$ with $a < b$, and $c, d \in \mathbb{R}$. If $g \in C([a,b]_{\mathbb{T}}, (c,d))$ and $f \in C((c,d), \mathbb{R})$ is convex, then*

$$f\left(\frac{\int_a^b g(s) \Diamond_\alpha s}{b-a} \right) \leq \frac{\int_a^b f(g(s)) \Diamond_\alpha s}{b-a}.$$ (8.17)

Also we need the extended Jensen's inequality on time scales via diamond-α integral.

Theorem 8.2. *(Generalized Jensen's inequality, see [26]) Let $\mathbb{T}$ be a time scale, $a, b \in \mathbb{T}$ with $a < b$, $c, d \in \mathbb{R}$, $g \in C([a,b]_{\mathbb{T}}, (c,d))$, and $h \in C([a,b]_{\mathbb{T}}, \mathbb{R})$ with*

$$\int_a^b |h(s)| \Diamond_\alpha s > 0.$$ (8.18)

If $f \in C((c,d), \mathbb{R})$ is convex, then

$$f\left(\frac{\int_a^b |h(s)| g(s) \Diamond_\alpha s}{\int_a^b |h(s)| \Diamond_\alpha s} \right) \leq \frac{\int_a^b |h(s)| f(g(s)) \Diamond_\alpha s}{\int_a^b |h(s)| \Diamond_\alpha s}.$$ (8.19)

We further need

Theorem 8.3. *(Hölder's Inequality, see [14]) For continuous functions $f, g : [a,b]_{\mathbb{T}} \to \mathbb{R}$, we have:*

$$\int_a^b |f(t) g(t)| \Diamond_\alpha t \leq \left[\int_a^b |f(t)|^p \Diamond_\alpha t \right]^{\frac{1}{p}} \left[\int_a^b |g(t)|^q \Diamond_\alpha t \right]^{\frac{1}{q}},$$ (8.20)

where $p > 1$, and $q = \frac{p}{p-1}$.

We obtain

Theorem 8.4. *(Generalization of Hölder's inequality)* Let $f_i \in C\left([a,b]_{\mathbb{T}}, \mathbb{R}\right)$, $i = 1, \ldots, n$, and $p_i > 1$ such that $\sum_{i=1}^{n} \frac{1}{p_i} = 1$. Then

$$\int_a^b \prod_{i=1}^n |f_i(t)| \, \Diamond_\alpha t \leq \prod_{i=1}^n \left(\int_a^b |f_i(t)|^{p_i} \, \Diamond_\alpha t \right)^{\frac{1}{p_i}}. \tag{8.21}$$

Proof. Using (8.20) and induction hypothesis, exactly as in [28]. $\qquad\square$

Comment. Bt Tietze's extension theorem of General Topology we easily derive that a continuous function f of $\prod_{i=1}^n \left([a_i, b_i] \cap \mathbb{T}_i\right)$ (where $\mathbb{T}_i$, $i = 1, \ldots, n \in \mathbb{N}$ are time scales) is bounded, since its continuous extension F on $\prod_{i=1}^n [a_i, b_i] \subseteq \mathbb{R}^n$ is bounded, $n \in \mathbb{N}$.

Comment. It is regarding the univariate functions. Based on [15] we see that the Cauchy Time scales delta Δ and nabla ∇ integrals are equal to definite Riemann Time Scales Δ, ∇ integrals, respectively. Thus, the diamond-α-Cauchy integral (8.14) is a diamond α-Riemann integral over continuous functions. Of course the last integral exists, since continuous functions are Riemann Δ and ∇-integrable, and it is equal to the corresponding α-Lebesgue integral, by [15].

In particular the dominated and bounded convergence theorems hold true with respect to the Lebesgue-Δ, ∇ measures.

Comment. Let $\mathbb{T}_1$, $\mathbb{T}_2$ be time scales and $f : [a,b]_{\mathbb{T}_1} \times [c,d]_{\mathbb{T}_2} \to \mathbb{R}$ be continuous. By [9] and [10] we get that f is Riemann Δ and ∇-integrable over $[a,b)_{\mathbb{T}_1} \times [c,d)_{\mathbb{T}_2}$ and $(a,b]_{\mathbb{T}_1} \times (c,d]_{\mathbb{T}_2}$, respectively. Hence by [9], [10], f is Lebesgue Δ and ∇-integrable there.

Thus by Fubini's theorem we get

$$\int_a^b \left(\int_c^d f(x,y) \, \Delta y \right) \Delta x = \int_c^d \left(\int_a^b f(x,y) \, \Delta x \right) \Delta y, \tag{8.22}$$

and

$$\int_a^b \left(\int_c^d f(x,y) \, \nabla y \right) \nabla x = \int_c^d \left(\int_a^b f(x,y) \, \nabla x \right) \nabla y. \tag{8.23}$$

We define $(\alpha \in [0,1])$

$$\int_a^b \left(\int_c^d f(x,y) \, \Diamond_\alpha y \right) \Diamond_\alpha x$$

$$= \alpha \int_a^b \left(\int_c^d f(x,y) \, \Delta y \right) \Delta x + (1-\alpha) \int_a^b \left(\int_c^d f(x,y) \, \nabla y \right) \nabla x. \tag{8.24}$$

One can generalize (8.24) for multiple integrals.

So for f continuous we get the $\Diamond_\alpha$-Fubini's theorem main property:

$$\int_a^b \left(\int_c^d f(x,y) \Diamond_\alpha y \right) \Diamond_\alpha x = \int_c^d \left(\int_a^b f(x,y) \Diamond_\alpha x \right) \Diamond_\alpha y. \qquad (8.25)$$

We make

Remark 8.1. Let $\mathbb{T}_1$, $\mathbb{T}_2$ be time scales and $f : [a,b]_{\mathbb{T}_1} \times [c,d]_{\mathbb{T}_2} \to \mathbb{R}$ be continuous. Consider

$$g(x) = \int_c^d f(x,y) \Diamond_\alpha y$$

$$= \alpha \int_c^d f(x,y) \Delta y + (1-\alpha) \int_c^d f(x,y) \nabla y, \qquad (8.26)$$

$\alpha \in [0,1]$, $\forall\, x \in [a,b]_{\mathbb{T}_1}$.

We prove that g is continuous on $[a,b]_{\mathbb{T}_1}$. Let $x_n \to x$, where $\{x_n\}_{n \in \mathbb{N}}$, $x \in [a,b]_{\mathbb{T}_1}$ then $f(x_n,y) \to f(x,y)$, as $n \to \infty$, $\forall\, y \in [c,d]_{\mathbb{T}_2}$. Furthermore there exists $M > 0$ such that $|f(x_n,y)|$, $|f(x,y)| \le M$, $\forall\, y \in [c,d]_{\mathbb{T}_2}$. Hence by Lebesgue's bounded convergence theorem ([9]) we get that

$$\lim_{n \to \infty} \int_c^d f(x_n,y) \Delta y = \int_c^d f(x,y) \Delta y, \qquad (8.27)$$

and

$$\lim_{n \to \infty} \int_c^d f(x_n,y) \nabla y = \int_c^d f(x,y) \nabla y. \qquad (8.28)$$

Combining (8.27) and (8.28) we obtain $g(x_n) \to g(x)$, as $n \to \infty$, proving the continuity of g.

Comment. In [7] we proved that if $\Phi : \mathbb{R}_+ := [0,\infty) \to \mathbb{R}$ is convex and increasing, then Φ is continuous on $\mathbb{R}_+$. Furthermore for $x < 0$ we extend $\Phi(x) := \Phi(-x)$, to the symmetric branch of Φ. Both branches of Φ make a convex function on $(-\infty,\infty)$.

So now we can apply Jensen's inequality also on $\mathbb{R}$. Plus, it is well known that if $\Phi : (A,B) \subseteq \mathbb{R} \to \mathbb{R}$ is convex, then Φ is continuous on (A,B).

8.2   Main Results

We present inequalities on $\Diamond_\alpha$- integral operators.

Theorem 8.5. Let $\mathbb{T}_1$, $\mathbb{T}_2$ be time scales, $a,b \in \mathbb{T}_1$; $c,d \in \mathbb{T}_2$; $k(x,y)$ is a kernel function with $x \in [a,b]_{\mathbb{T}_1}$, $y \in [c,d]_{\mathbb{T}_2}$; k is continuous function from $[a,b]_{\mathbb{T}_1} \times [c,d]_{\mathbb{T}_2}$ into $\mathbb{R}_+$. Consider

$$K(x) := \int_c^d k(x,y) \Diamond_\alpha y, \quad \forall\, x \in [a,b]_{\mathbb{T}_1}. \qquad (8.29)$$

We assume that $K(x) > 0$, $\forall\, x \in [a,b]_{\mathrm{T}_1}$. Consider $f : [c,d]_{\mathrm{T}_2} \to \mathbb{R}$ continuous, and the $\Diamond_\alpha$-integral operator function

$$g(x) := \int_c^d k(x,y) f(y) \Diamond_\alpha y, \tag{8.30}$$

$\forall\, x \in [a,b]_{\mathrm{T}_1}$.

Consider also the weight function

$$u : [a,b]_{\mathrm{T}_1} \to \mathbb{R}_+, \tag{8.31}$$

which is continuous.

Define further the function

$$v(y) := \int_a^b \frac{u(x) k(x,y)}{K(x)} \Diamond_\alpha x, \tag{8.32}$$

$\forall\, y \in [c,d]_{\mathrm{T}_2}$.

Let I denote any of $(0,\infty)$ or $[0,\infty)$, and $\Phi : I \to \mathbb{R}$ be a convex and increasing function. In particular we assume that

$$|f| \left([c,d]_{\mathrm{T}_2}\right) \subseteq I. \tag{8.33}$$

Then

$$\int_a^b u(x) \Phi\left(\frac{|g(x)|}{K(x)}\right) \Diamond_\alpha x \leq \int_c^d v(y) \Phi\left(|f(y)|\right) \Diamond_\alpha y. \tag{8.34}$$

Proof. We see that

$$\int_a^b u(x) \Phi\left(\frac{|g(x)|}{K(x)}\right) \Diamond_\alpha x = \int_a^b u(x) \Phi\left(\frac{1}{K(x)} \left| \int_c^d k(x,y) f(y) \Diamond_\alpha y \right| \right) \Diamond_\alpha x \tag{8.35}$$

$$\leq \int_a^b u(x) \Phi\left(\frac{1}{K(x)} \int_c^d k(x,y) |f(y)| \Diamond_\alpha y\right) \Diamond_\alpha x$$

(by generalized Jensen's inequality, see Theorem 8.2 and Comment 8.1)

$$\leq \int_a^b \frac{u(x)}{K(x)} \left(\int_c^d k(x,y) \Phi\left(|f(y)|\right) \Diamond_\alpha y\right) \Diamond_\alpha x$$

$$= \int_a^b \left(\int_c^d \frac{u(x) k(x,y)}{K(x)} \Phi\left(|f(y)|\right) \Diamond_\alpha y\right) \Diamond_\alpha x \tag{8.36}$$

(by (8.25))

$$= \int_c^d \left(\int_a^b \frac{u(x) k(x,y)}{K(x)} \Phi\left(|f(y)|\right) \Diamond_\alpha x\right) \Diamond_\alpha y$$

$$= \int_c^d \Phi\left(|f(y)|\right) \left(\int_a^b \frac{u(x) k(x,y)}{K(x)} \Diamond_\alpha x\right) \Diamond_\alpha y = \int_c^d v(y) \Phi\left(|f(y)|\right) \Diamond_\alpha y, \tag{8.37}$$

proving the claim. $\qquad\square$

We continue with

Theorem 8.6. *All as in Theorem 8.5, however now Φ is not necessarily increasing and only from $(0, \infty)$ into $\mathbb{R}$. Additionally we assume that f is of fixed strict sign. Then*

$$\int_a^b u(x) \Phi \left(\frac{|g(x)|}{K(x)} \right) \Diamond_\alpha x \le \int_c^d v(y) \Phi(|f(y)|) \Diamond_\alpha y. \tag{8.38}$$

Proof. We notice that

$$|g(x)| = \left| \int_c^d k(x,y) f(y) \Diamond_\alpha y \right| = \int_c^d k(x,y) |f(y)| \Diamond_\alpha y. \tag{8.39}$$

Therefore we have

$$\int_a^b u(x) \Phi \left(\frac{|g(x)|}{K(x)} \right) \Diamond_\alpha x = \int_a^b u(x) \Phi \left(\frac{1}{K(x)} \left| \int_c^d k(x,y) f(y) \Diamond_\alpha y \right| \right) \Diamond_\alpha x$$

$$= \int_a^b u(x) \Phi \left(\frac{1}{K(x)} \int_c^d k(x,y) |f(y)| \Diamond_\alpha y \right) \Diamond_\alpha x. \tag{8.40}$$

The rest follows as in the proof of Theorem 8.5.   □

Corollary 8.1. *(to Theorem 8.6)* *It holds*

$$\int_a^b u(x) \ln \left(\frac{|g(x)|}{K(x)} \right) \Diamond_\alpha x \ge \int_c^d v(y) \ln(|f(y)|) \Diamond_\alpha y. \tag{8.41}$$

Proof. Apply (8.38) for $\Phi(x) = -\ln x$, which a convex function with domain $(0, \infty)$.   □

Corollary 8.2. *(to Theorem 8.5)* *It holds*

$$\int_a^b u(x) e^{\frac{|g(x)|}{K(x)}} \Diamond_\alpha x \le \int_c^d v(y) e^{|f(y)|} \Diamond_\alpha y. \tag{8.42}$$

Proof. Apply (8.34) for $\Phi(x) = e^x$, $x \ge 0$.   □

Notation 8.1. Let $\mathbb{T}_1$, $\mathbb{T}_2$ be time scales, $a, b \in \mathbb{T}_1$; $c, d \in \mathbb{T}_2$; $k_i(x, y)$ is a kernel function with $x \in [a, b]_{\mathbb{T}_1}$, $y \in [c, d]_{\mathbb{T}_2}$; k_i is continuous function from $[a, b]_{\mathbb{T}_1} \times [c, d]_{\mathbb{T}_2}$ into $\mathbb{R}_+$ for $i = 1, \ldots, m \in \mathbb{N}$. Consider

$$K_i(x) := \int_c^d k_i(x, y) \Diamond_\alpha y, \quad \forall x \in [a, b]_{\mathbb{T}_1}, \tag{8.43}$$

$i = 1, \ldots, m$.

We assume that $K_i(x) > 0$, $\forall x \in [a, b]_{\mathbb{T}_1}$, $i = 1, \ldots, m$. Consider $f_i : [c, d]_{\mathbb{T}_2} \to \mathbb{R}$ continuous, $i = 1, \ldots, m$, and the $\Diamond_\alpha$-integral operator function

$$g_i(x) := \int_c^d k_i(x, y) f_i(y) \Diamond_\alpha y, \tag{8.44}$$

$\forall\, x \in [a,b]_{\mathbb{T}_1}$, $i = 1, \ldots, m$.

Consider also the weight function

$$u : [a,b]_{\mathbb{T}_1} \to \mathbb{R}_+, \tag{8.45}$$

which is continuous.

Define further the function

$$\lambda_m(y) := \int_a^b \frac{u(x) \prod_{i=1}^m k_i(x,y)}{\prod_{i=1}^m K_i(x)} \Diamond_\alpha x, \tag{8.46}$$

$\forall\, y \in [c,d]_{\mathbb{T}_2}$.

Here $\Phi_i : \mathbb{R}_+ \to \mathbb{R}_+$, $i = 1, \ldots, m$, are convex and increasing functions.

We give

Theorem 8.7. *All as in Notation 8.1. Let $j \in \{1, \ldots, m\}$ be fixed. Then*

$$\int_a^b u(x) \prod_{i=1}^m \Phi_i\left(\frac{|g_i(x)|}{K_i(x)}\right) \Diamond_\alpha x \tag{8.47}$$

$$\leq \left(\prod_{\substack{i=1 \\ i\neq j}}^m \int_c^d \Phi_i(|f_i(y)|) \Diamond_\alpha y\right) \left(\int_c^d \Phi_j(|f_j(y)|) \lambda_m(y) \Diamond_\alpha y\right).$$

Proof. We demonstrate the proof for $m = 3$. For general m it follows the same way. Here we use the extended Jensen's inequality, see Theorem 8.2 and Comment 8.1, $\Diamond_\alpha$-Fubini's theorem, see (8.25), and that Φ_i are increasing.

We have

$$\int_a^b u(x) \prod_{i=1}^3 \Phi_i\left(\frac{|g_i(x)|}{K_i(x)}\right) \Diamond_\alpha x$$

$$= \int_a^b u(x) \prod_{i=1}^3 \Phi_i\left(\left|\frac{1}{K_i(x)} \int_c^d k_i(x,y) f_i(y) \Diamond_\alpha y\right|\right) \Diamond_\alpha x \tag{8.48}$$

$$\leq \int_a^b u(x) \prod_{i=1}^3 \Phi_i\left(\frac{1}{K_i(x)} \int_c^d k_i(x,y) |f_i(y)| \Diamond_\alpha y\right) \Diamond_\alpha x$$

$$\leq \int_a^b u(x) \prod_{i=1}^3 \left(\frac{1}{K_i(x)} \int_c^d k_i(x,y) \Phi_i(|f_i(y)|) \Diamond_\alpha y\right) \Diamond_\alpha x$$

$$= \int_a^b \left(\frac{u(x)}{\prod_{i=1}^3 K_i(x)}\right) \left(\prod_{i=1}^3 \int_c^d k_i(x,y) \Phi_i(|f_i(y)|) \Diamond_\alpha y\right) \Diamond_\alpha x$$

(calling $\theta(x) := \dfrac{u(x)}{\prod_{i=1}^{3} K_i(x)}$)

$$= \int_a^b \theta(x) \left(\prod_{i=1}^{3} \int_c^d k_i(x,y) \, \Phi_i\left(|f_i(y)|\right) \Diamond_\alpha y \right) \Diamond_\alpha x \qquad (8.49)$$

$$= \int_a^b \theta(x) \left[\int_c^d \left(\prod_{i=1}^{2} \int_c^d k_i(x,y) \, \Phi_i\left(|f_i(y)|\right) \Diamond_\alpha y \right) \right.$$

$$\left. \cdot k_3(x,y) \, \Phi_3\left(|f_3(y)|\right) \Diamond_\alpha y \right] \Diamond_\alpha x$$

$$= \int_a^b \left(\int_c^d \theta(x) \left(\prod_{i=1}^{2} \int_c^d k_i(x,y) \, \Phi_i\left(|f_i(y)|\right) \Diamond_\alpha y \right) \right.$$

$$\left. \cdot k_3(x,y) \, \Phi_3\left(|f_3(y)|\right) \Diamond_\alpha y \right) \Diamond_\alpha x$$

$$= \int_c^d \left(\int_a^b \theta(x) \left(\prod_{i=1}^{2} \int_c^d k_i(x,y) \, \Phi_i\left(|f_i(y)|\right) \Diamond_\alpha y \right) \right.$$

$$\left. \cdot k_3(x,y) \, \Phi_3\left(|f_3(y)|\right) \Diamond_\alpha x \right) \Diamond_\alpha y$$

$$= \int_c^d \Phi_3\left(|f_3(y)|\right) \left(\int_a^b \theta(x) \, k_3(x,y) \left(\prod_{i=1}^{2} \int_c^d k_i(x,y) \, \Phi_i\left(|f_i(y)|\right) \Diamond_\alpha y \right) \Diamond_\alpha x \right) \Diamond_\alpha y$$

$$\qquad (8.50)$$

$$= \int_c^d \Phi_3\left(|f_3(y)|\right) \left[\int_a^b \theta(x) \, k_3(x,y) \left(\int_c^d \left\{ \int_c^d k_1(x,y) \, \Phi_1\left(|f_1(y)|\right) \Diamond_\alpha y \right\} \right. \right.$$

$$\left. \left. \cdot k_2(x,y) \, \Phi_2\left(|f_2(y)|\right) \Diamond_\alpha y \right) \Diamond_\alpha x \right] \Diamond_\alpha y$$

$$= \int_c^d \Phi_3\left(|f_3(y)|\right) \left[\int_a^b \left(\int_c^d \theta(x) \, k_2(x,y) \, k_3(x,y) \, \Phi_2\left(|f_2(y)|\right) \right. \right. \qquad (8.51)$$

$$\left. \left. \cdot \left\{ \int_c^d k_1(x,y) \, \Phi_1\left(|f_1(y)|\right) \Diamond_\alpha y \right\} \Diamond_\alpha y \right) \Diamond_\alpha x \right] \Diamond_\alpha y$$

$$= \left(\int_c^d \Phi_3\left(|f_3(y)|\right) \Diamond_\alpha y \right) \left[\int_a^b \left(\int_c^d \theta(x) \, k_2(x,y) \, k_3(x,y) \, \Phi_2\left(|f_2(y)|\right) \right. \right.$$

$$\left. \left. \cdot \left\{ \int_c^d k_1(x,y) \, \Phi_1\left(|f_1(y)|\right) \Diamond_\alpha y \right\} \Diamond_\alpha y \right) \Diamond_\alpha x \right]$$

$$= \left(\int_c^d \Phi_3 \left(|f_3 \left(y \right)| \right) \Diamond_\alpha y \right) \left[\int_c^d \left(\int_a^b \theta \left(x \right) k_2 \left(x, y \right) k_3 \left(x, y \right) \Phi_2 \left(|f_2 \left(y \right)| \right) \right. \right.$$

$$\left. \left. \cdot \left\{ \int_c^d k_1 \left(x, y \right) \Phi_1 \left(|f_1 \left(y \right)| \right) \Diamond_\alpha y \right\} \Diamond_\alpha x \right) \Diamond_\alpha y \right] \qquad (8.52)$$

$$= \left(\int_c^d \Phi_3 \left(|f_3 \left(y \right)| \right) \Diamond_\alpha y \right) \left[\int_c^d \Phi_2 \left(|f_2 \left(y \right)| \right) \left(\int_a^b \theta \left(x \right) k_2 \left(x, y \right) k_3 \left(x, y \right) \right. \right.$$

$$\left. \left. \cdot \left(\int_c^d k_1 \left(x, y \right) \Phi_1 \left(|f_1 \left(y \right)| \right) \Diamond_\alpha y \right) \Diamond_\alpha x \right) \Diamond_\alpha y \right]$$

$$= \left(\int_c^d \Phi_3 \left(|f_3 \left(y \right)| \right) \Diamond_\alpha y \right) \left[\int_c^d \Phi_2 \left(|f_2 \left(y \right)| \right) \left\{ \int_a^b \left(\int_c^d \theta \left(x \right) \prod_{i=1}^3 k_i \left(x, y \right) \right. \right. \right.$$

$$\left. \left. \left. \cdot \Phi_1 \left(|f_1 \left(y \right)| \right) \Diamond_\alpha y \right) \Diamond_\alpha x \right\} \Diamond_\alpha y \right]$$

$$= \left(\int_c^d \Phi_3 \left(|f_3 \left(y \right)| \right) \Diamond_\alpha y \right) \left(\int_c^d \Phi_2 \left(|f_2 \left(y \right)| \right) \Diamond_\alpha y \right)$$

$$\cdot \left(\int_a^b \left(\int_c^d \theta \left(x \right) \prod_{i=1}^3 k_i \left(x, y \right) \Phi_1 \left(|f_1 \left(y \right)| \right) \Diamond_\alpha y \right) \Diamond_\alpha x \right) \qquad (8.53)$$

$$= \left(\prod_{i=2}^3 \int_c^d \Phi_i \left(|f_i \left(y \right)| \right) \Diamond_\alpha y \right)$$

$$\cdot \left(\int_c^d \left(\int_a^b \theta \left(x \right) \prod_{i=1}^3 k_i \left(x, y \right) \Phi_1 \left(|f_1 \left(y \right)| \right) \Diamond_\alpha x \right) \Diamond_\alpha y \right)$$

$$= \left(\prod_{i=2}^3 \int_c^d \Phi_i \left(|f_i \left(y \right)| \right) \Diamond_\alpha y \right)$$

$$\cdot \left(\int_c^d \Phi_1 \left(|f_1 \left(y \right)| \right) \left(\int_a^b \theta \left(x \right) \prod_{i=1}^3 k_i \left(x, y \right) \Diamond_\alpha x \right) \Diamond_\alpha y \right)$$

$$= \left(\prod_{i=2}^3 \int_c^d \Phi_i \left(|f_i \left(y \right)| \right) \Diamond_\alpha y \right) \left(\int_c^d \Phi_1 \left(|f_1 \left(y \right)| \right) \lambda_3 \left(y \right) \Diamond_\alpha y \right), \qquad (8.54)$$

proving the claim. $\qquad \square$

Corollary 8.3. *(to Theorem 8.7) It holds*

$$\int_a^b u(x) e^{\sum_{i=1}^m \frac{|g_i(x)|}{K_i(x)}} \Diamond_\alpha x \tag{8.55}$$

$$\leq \left(\prod_{\substack{i=1 \\ i \neq j}}^m \int_c^d e^{|f_i(y)|} \Diamond_\alpha y \right) \left(\int_c^d e^{|f_j(y)|} \lambda_m(y) \Diamond_\alpha y \right).$$

Proof.   Apply $\Phi_i(x) = e^x$, $x \geq 0$, for all $i = 1, \ldots, m$.   □

We continue with

Theorem 8.8. *All as in Theorem 8.7, but now* $\Phi_i : (0, \infty) \to \mathbb{R}_+$, $i = 1, \ldots, m$, *are convex and not necessarily increasing. Furthermore all f_i, $i = 1, \ldots, m$, are of fixed strict sign. Then (8.47) is valid.*

Proof.   Similar to Theorem 8.6, and Theorem 8.7.   □

We give the following application:

Corollary 8.4. *All as in Theorem 8.8, with* $\Phi_i(x) = -\ln x$, $i = 1, \ldots, m \in \mathbb{N}$. *It holds*

$$(-1)^m \int_a^b u(x) \prod_{i=1}^m \ln\left(\frac{|g_i(x)|}{K_i(x)} \right) \Diamond_\alpha x \tag{8.56}$$

$$\leq (-1)^m \left(\prod_{\substack{i=1 \\ i \neq j}}^m \int_c^d \ln(|f_i(y)|) \Diamond_\alpha y \right) \left(\int_c^d \ln(|f_j(y)|) \lambda_m(y) \Diamond_\alpha y \right).$$

Proof.   By (8.47).   □

We continue with

Theorem 8.9. *All as in Notation 8.1. Define*

$$u_i(y) := \int_a^b u(x) \frac{k_i(x, y)}{K_i(x)} \Diamond_\alpha x, \tag{8.57}$$

$\forall \, y \in [c, d]_{\mathbb{T}_2}$, $i = 1, \ldots, m \in \mathbb{N}$.
   Let $p_i > 1 : \sum_{i=1}^m \frac{1}{p_i} = 1$. Then

$$\int_a^b u(x) \prod_{i=1}^m \Phi_i\left(\frac{|g_i(x)|}{K_i(x)} \right) \Diamond_\alpha x \tag{8.58}$$

$$\leq \prod_{i=1}^m \left(\int_c^d u_i(y) \, \Phi_i(|f_i(y)|)^{p_i} \Diamond_\alpha y \right)^{\frac{1}{p_i}}.$$

Proof. Notice that Φ_i, $i = 1, \ldots, m$, are continuous functions. Here we use the generalized Hölder's inequality, see Theorem 8.4. We have

$$\int_a^b u(x) \prod_{i=1}^m \Phi_i \left(\frac{|g_i(x)|}{K_i(x)} \right) \Diamond_\alpha x$$

$$= \int_a^b \prod_{i=1}^m \left(u(x)^{\frac{1}{p_i}} \Phi_i \left(\frac{|g_i(x)|}{K_i(x)} \right) \right) \Diamond_\alpha x \tag{8.59}$$

$$\leq \prod_{i=1}^m \left(\int_a^b u(x) \Phi_i \left(\frac{|g_i(x)|}{K_i(x)} \right)^{p_i} \Diamond_\alpha x \right)^{\frac{1}{p_i}}$$

(notice here that $\Phi_i^{p_i}$, $i = 1, \ldots, m$, are convex, increasing and continuous, non-negative functions, and by Theorem 8.5 we get)

$$\leq \prod_{i=1}^m \left(\int_c^d u_i(y) \Phi_i (|f_i(y)|)^{p_i} \Diamond_\alpha y \right)^{\frac{1}{p_i}}, \tag{8.60}$$

proving the claim. $\qquad\qquad\square$

We also give

Theorem 8.10. *All as in Theorem 8.9, but now* $\Phi_i : (0, \infty) \to \mathbb{R}_+$, $i = 1, \ldots, m$, *are convex and not necessarily increasing. Furthermore all* f_i, $i = 1, \ldots, m$, *are of fixed strict sign. Then (8.58) is valid.*

Proof. Similar to Theorem 8.9, and by using Theorem 8.6. $\qquad\qquad\square$

Corollary 8.5. *(to Theorem 8.9) Let* $\alpha_i \geq 1$, $i = 1, \ldots, m$. *Then*

$$\int_a^b u(x) \prod_{i=1}^m \left(\frac{|g_i(x)|}{K_i(x)} \right)^{\alpha_i} \Diamond_\alpha x \tag{8.61}$$

$$\leq \prod_{i=1}^m \left(\int_c^d u_i(y) (|f_i(y)|)^{\alpha_i p_i} \Diamond_\alpha y \right)^{\frac{1}{p_i}}.$$

Proof. Apply (8.58) for $\Phi_i(x) = x^{\alpha_i}$, $x \geq 0$, $i = 1, \ldots, m$. $\qquad\qquad\square$

Corollary 8.6. *(to Theorem 8.9) It holds*

$$\int_a^b u(x) e^{\sum_{i=1}^m \frac{|g_i(x)|}{K_i(x)}} \Diamond_\alpha x \tag{8.62}$$

$$\leq \prod_{i=1}^m \left(\int_c^d u_i(y) e^{p_i |f_i(y)|} \Diamond_\alpha y \right)^{\frac{1}{p_i}}.$$

Proof. Apply (8.58) for $\Phi_i(x) = e^x$, $x \geq 0$, for all $i = 1, \ldots, m$. $\qquad\qquad\square$

We need

Definition 8.1. ([3]) Let $\mathbb{T}$ be a time scale. Consider the coordinate wise rd-continuous functions $h_\alpha : \mathbb{T} \times \mathbb{T} \to \mathbb{R}$, $\alpha \geq 0$, such that $h_0(t, s) = 1$,

$$h_{\alpha+1}(t, s) = \int_s^t h_\alpha(\tau, s) \, \Delta\tau, \tag{8.63}$$

$\forall \, s, t \in \mathbb{T}$.

When $\mathbb{T} = \mathbb{R}$, then $\sigma(t) = t$, we define

$$h_\alpha(t, s) := \frac{(t - s)^\alpha}{\Gamma(\alpha + 1)}, \quad \alpha \geq 0. \tag{8.64}$$

When $\mathbb{T} = \mathbb{Z}$, then $\sigma(t) = t + 1$, $t \in \mathbb{Z}$, and

$$h_k(t, s) = \frac{(t - s)^{(k)}}{k!}, \quad \forall \, k \in \mathbb{N}_0 = \mathbb{N} \cup \{0\}, \tag{8.65}$$

$\forall \, t, s \in \mathbb{Z}$, where $t^{(0)} = 1$, $t^{(k)} = \prod_{i=0}^{k-1}(t - i)$ for $k \in \mathbb{N}$. Also it holds

$$\int_a^b f(t) \, \Delta t = \sum_{t=a}^{b-1} f(t), \quad a < b; \ a, b \in \mathbb{Z}. \tag{8.66}$$

We need

Definition 8.2. ([3]) For $\alpha \geq 1$ we define the time scale Δ-Riemann-Liouville type fractional integral $(a, b \in \mathbb{T})$

$$K_a^\alpha f(t) = \int_a^t h_{\alpha-1}(t, \sigma(\tau)) f(\tau) \, \Delta\tau, \tag{8.67}$$

(by [11] is an integral on $[a, t) \cap \mathbb{T}$)

$$K_a^0 f = f,$$

where $f \in L_1([a, b) \cap \mathbb{T})$ (Lebesgue Δ-integrable functions on $[a, b) \cap \mathbb{T}$, see [15], [9], [10]), $t \in [a, b] \cap \mathbb{T}$.

Notice $K_a^1 f(t) = \int_a^t f(\tau) \, \Delta\tau$ is absolutely continuous in $t \in [a, b] \cap \mathbb{T}$, see [11].

Lemma 8.1. ([3]) Let $\alpha > 1$, $f \in L_1([a, b) \cap \mathbb{T})$. If additionally $h_{\alpha-1}(s, \sigma(t))$ is Lebesgue Δ-measurable on $([a, b) \cap \mathbb{T})^2$, then $K_a^\alpha f \in L_1([a, b) \cap \mathbb{T})$.

We need

Definition 8.3. ([3]) Assume $\mathbb{T}$ time scale such that $\mathbb{T}^k = \mathbb{T}$.

Let $\mu > 2 : m - 1 < \mu < m \in \mathbb{N}$, i.e. $m = \lceil \mu \rceil$ (ceiling of the number), $\tilde{\nu} = m - \mu$ $(0 < \tilde{\nu} < 1)$.

Here we take $f \in C_{rd}^m([a, b] \cap \mathbb{T})$. Clearly here ([15]) f^{Δ^m} is a Lebesgue Δ-integrable function. Assume $h_{\tilde{\nu}}(s, \sigma(t))$ is continuous on $([a, b] \cap \mathbb{T})^2$.

We define the delta fractional derivative on time scale $\mathbb{T}$ of order $\mu-1$ as follows:

$$\Delta_{a*}^{\mu-1} f(t) = \left(K_a^{\tilde{\nu}+1} f^{\Delta^m}\right)(t) = \int_a^t h_{\tilde{\nu}}(t, \sigma(\tau)) f^{\Delta^m}(\tau) \Delta\tau, \qquad (8.68)$$

$\forall t \in [a, b] \cap \mathbb{T}$.

Notice here that $\Delta_{a*}^{\mu-1} f \in C([a, b] \cap \mathbb{T})$ by a simple argument using dominated convergence theorem in Lebesgue Δ-sense.

If $\mu = m$, then $\tilde{\nu} = 0$ and by (8.68) we get

$$\Delta_{a*}^{m-1} f(t) = K_a^1 f^{\Delta^m}(t) = f^{\Delta^{m-1}}(t). \qquad (8.69)$$

More generally, by [11], given that $f^{\Delta^{m-1}}$ is everywhere finite and absolutely continuous on $[a, b] \cap \mathbb{T}$, then f^{Δ^m} exists Δ-a.e. and is Lebesgue Δ-integrable on $[a, t) \cap \mathbb{T}$, $\forall t \in [a, b] \cap \mathbb{T}$ and one can plug it into (8.68).

We need

Definition 8.4. ([4]) Consider the coordinate wise ld-continuous functions $\widehat{h}_\alpha : \mathbb{T} \times \mathbb{T} \to \mathbb{R}$, $\alpha \geq 0$, such that $\widehat{h}_0(t, s) = 1$,

$$\widehat{h}_{\alpha+1}(t, s) = \int_s^t \widehat{h}_\alpha(\tau, s) \nabla\tau, \qquad (8.70)$$

$\forall s, t \in \mathbb{T}$.

In the case of $\mathbb{T} = \mathbb{R}$; then $\rho(t) = t$, and $\widehat{h}_k(t, s) = \frac{(t-s)^k}{k!}$, $k \in \mathbb{N}_0$, and define

$$\widehat{h}_\alpha(t, s) := \frac{(t-s)^\alpha}{\Gamma(\alpha+1)}, \qquad \alpha \geq 0. \qquad (8.71)$$

Let $T = \mathbb{Z}$, then $\rho(t) = t - 1$, $t \in \mathbb{Z}$. Define $t^{\overline{0}} := 1$, $t^{\overline{k}} := t(t+1) \dots (t+k-1)$, $k \in \mathbb{N}$, and by (8.70) we have $\widehat{h}_k(t, s) = \frac{(t-s)^{\overline{k}}}{k!}$, $s, t \in \mathbb{Z}$, $k \in \mathbb{N}_0$.

Here $\int_{t_0}^t \nabla\tau = \sum_{t_0+1}^t$.

We need

Definition 8.5. ([4]) For $\alpha \geq 1$ we define the time scale ∇-Riemann-Liouville type fractional integral $(a, b \in \mathbb{T})$

$$J_a^\alpha f(t) = \int_a^t \widehat{h}_{\alpha-1}(t, \rho(\tau)) f(\tau) \nabla\tau, \qquad (8.72)$$

(by [11] the last integral is on $(a, t] \cap \mathbb{T}$)

$$J_a^0 f(t) = f(t),$$

where $f \in L_1((a, b] \cap \mathbb{T})$ (Lebesgue ∇-integrable functions on $(a, b] \cap \mathbb{T}$, see [9], [10], [15]), $t \in [a, b] \cap \mathbb{T}$.

Notice $J_a^1 f(t) = \int_a^t f(\tau) \nabla\tau$ is absolutely continuous in $t \in [a, b] \cap \mathbb{T}$, see [11].

Lemma 8.2. ([4]) Let $\alpha > 1$, $f \in L_1((a, b] \cap \mathbb{T})$. If additionally $\widehat{h}_{\alpha-1}(s, \rho(t))$ is Lebesgue ∇-measurable on $((a, b] \cap \mathbb{T})^2$, then $J_a^\alpha f \in L_1((a, b] \cap \mathbb{T})$.

We also need

Definition 8.6. ([4]) Assume $\mathbb{T}_k = \mathbb{T}$.

Let $\mu > 2$ such that $m - 1 < \mu < m \in \mathbb{N}$, i.e. $m = \lceil \mu \rceil$, $\tilde{\nu} = m - \mu$ ($0 < \tilde{\nu} < 1$).

Let $f \in C_{ld}^m ([a, b] \cap \mathbb{T})$. Clearly here ([15]) f^{∇^m} is a Lebesgue ∇-integrable function. Assume $\widehat{h}_{\tilde{\nu}} (s, \rho(t))$ is continuous on $([a, b] \cap \mathbb{T})^2$.

We define the nabla fractional derivative on time scale $\mathbb{T}$ of order $\mu - 1$ as follows:

$$\nabla_{a*}^{\mu-1} f(t) = \left(J_a^{\tilde{\nu}+1} f^{\nabla^m} \right)(t) = \int_a^t \widehat{h}_{\tilde{\nu}} (t, \rho(\tau)) f^{\nabla^m} (\tau) \nabla \tau, \qquad (8.73)$$

$\forall \, t \in [a, b] \cap \mathbb{T}$.

Notice here that $\nabla_{a*}^{\mu-1} f \in C([a, b] \cap \mathbb{T})$ by a simple argument using dominated convergence theorem in Lebesgue ∇-sense.

If $\mu = m$, then $\tilde{\nu} = 0$ and by (8.73) we get

$$\nabla_{a*}^{m-1} f(t) = J_a^1 f^{\nabla^m} (t) = f^{\nabla^{m-1}} (t). \qquad (8.74)$$

More generally, by [11], given that $f^{\nabla^{m-1}}$ is everywhere finite and absolutely continuous on $[a, b] \cap \mathbb{T}$, then f^{∇^m} exists ∇-a.e. and is Lebesgue ∇-integrable on $(a, t] \cap \mathbb{T}$, $\forall \, t \in [a, b] \cap \mathbb{T}$, and one can plug it into (8.73).

We present

Theorem 8.11. *Let $\mathbb{T}$ be a time scale, and $a, b \in \mathbb{T}$, $a < b$, with $\sigma(a) = a$. Let $\alpha \geq 1$, h_α as in (8.63), and $K_a^\alpha f$ as in (8.67), where $f \in L_1([a, b) \cap \mathbb{T})$. Assume further that $h_{\alpha-1} (s, \sigma(t))$ is Lebesgue Δ-measurable on $([a, b) \cap \mathbb{T})^2$.*
Call

$$K^* (x) := \int_a^b \chi_{[a,x)} (y) |h_{\alpha-1} (x, \sigma(y))| \Delta y \qquad (8.75)$$

$$= \int_a^x |h_{\alpha-1} (x, \sigma(y))| \Delta y,$$

$\forall \, x \in ([a, b] \cap \mathbb{T})$, where $\chi_{[a,x)} (y)$ is the characteristic function on $[a, x) \cap \mathbb{T}$.
Assume that $K^* (x) > 0$ (delta) Lebesgue measure Δ-a.e. in $x \in ([a, b] \cap \mathbb{T})$.
Consider also the weight function

$$u : ([a, b) \cap \mathbb{T}) \to \mathbb{R}_+, \qquad (8.76)$$

which is (delta) Lebesgue Δ-measurable. Assume that the function

$$x \to \frac{u(x)}{K^* (x)} \chi_{[a,x)} (y) |h_{\alpha-1} (x, \sigma(y))|$$

is Δ-integrable on $([a, b) \cap \mathbb{T})$ for each fixed $y \in ([a, b) \cap \mathbb{T})$.
Define v^* on $([a, b) \cap \mathbb{T})$ by

$$v^* (y) := \int_a^b \frac{u(x)}{K^* (x)} \chi_{[a,x)} (y) |h_{\alpha-1} (x, \sigma(y))| \Delta x < \infty. \qquad (8.77)$$

Let $\Phi : \mathbb{R}_+ \to \mathbb{R}$ *be a convex and increasing function. Then*

$$\int_a^b u(x) \Phi \left(\frac{|K_a^\alpha f(x)|}{K^*(x)} \right) \Delta x \leq \int_a^b v^*(x) \Phi(|f(x)|) \Delta x, \qquad (8.78)$$

under the further assumptions:

(i) $f, \Phi(|f|)$ *are both* $\chi_{[a,x)}(y)|h_{\alpha-1}(x, \sigma(y))| \Delta y$-*integrable,* Δ-*a.e. in* $x \in ([a,b) \cap \mathbb{T})$,

(ii) $v^* \Phi(|f|)$ *is* Δ-*Lebesgue integrable.*

Proof. By [7], Φ is continuous on $\mathbb{R}_+$. Here we use Jensen's inequality, Tonelli's theorem and Fubini's theorem, all on Time Scales setting. Also we use that Φ is convex and increasing. We extend $\Phi(x) := \Phi(-x)$, $x < 0$, then both branches of Φ make a convex function on $\mathbb{R}$ and we can apply Jensen's inequality.

Next we have

$$K_a^\alpha f(x) = \int_a^b \chi_{[a,x)}(y) h_{\alpha-1}(x, \sigma(y)) f(y) \Delta y, \quad \forall \, x \in [a,b] \cap \mathbb{T},$$

and

$$|K_a^\alpha f(x)| \leq \int_a^b \chi_{[a,x)}(y) |h_{\alpha-1}(x, \sigma(y))| |f(y)| \Delta y.$$

We see that

$$\int_a^b u(x) \Phi \left(\frac{|K_a^\alpha f(x)|}{K^*(x)} \right) \Delta x$$

$$\leq \int_a^b u(x) \Phi \left(\frac{1}{K_*(x)} \int_a^b \chi_{[a,x)}(y) |h_{\alpha-1}(x, \sigma(y))| |f(y)| \Delta y \right) \Delta x$$

(by Jensen's inequality)

$$\leq \int_a^b \frac{u(x)}{K^*(x)} \left(\int_a^b \chi_{[a,x)}(y) |h_{\alpha-1}(x, \sigma(y))| \Phi(|f(y)|) \Delta y \right) \Delta x$$

$$= \int_a^b \left(\int_a^b \frac{u(x)}{K^*(x)} \chi_{[a,x)}(y) |h_{\alpha-1}(x, \sigma(y))| \Phi(|f(y)|) \Delta y \right) \Delta x$$

(by Fubini's theorem)

$$= \int_a^b \left(\int_a^b \frac{u(x)}{K^*(x)} \chi_{[a,x)}(y) |h_{\alpha-1}(x, \sigma(y))| \Phi(|f(y)|) \Delta x \right) \Delta y$$

$$= \int_a^b \Phi(|f(y)|) \left(\int_a^b \frac{u(x)}{K^*(x)} \chi_{[a,x)}(y) |h_{\alpha-1}(x, \sigma(y))| \Delta x \right) \Delta y$$

$$= \int_a^b \Phi(|f(y)|) v^*(y) \Delta y,$$

completing the proof of the theorem. $\qquad \qquad \square$

The counterpart of last result follows:

Theorem 8.12. *Let* $\mathbb{T}$ *be a time scale, and* $a, b \in \mathbb{T} - \{\min \mathbb{T}\}$, $a < b$, *with* $\rho(a) = a$. *Let* $\alpha \geq 1$, $\widehat{h}_\alpha$ *as in (8.70), and* $J_a^\alpha f$ *as in (8.72), where* $f \in L_1((a, b] \cap \mathbb{T})$. *Assume further that* $\widehat{h}_{\alpha-1}(s, \rho(t))$ *is Lebesgue* ∇-*measurable on* $((a, b] \cap \mathbb{T})^2$.
   Call

$$K_*(x) := \int_a^b \chi_{[a,x)}(y) \left| \widehat{h}_{\alpha-1}(x, \rho(y)) \right| \nabla y \tag{8.79}$$

$$= \int_a^x \left| \widehat{h}_{\alpha-1}(x, \rho(y)) \right| \nabla y,$$

$\forall\ x \in ([a, b] \cap \mathbb{T})$. *Assume that* $K_*(x) > 0$ *(nabla) Lebesgue measure* ∇-*a.e. in* $x \in ([a, b] \cap \mathbb{T})$.
   Consider also the weight function

$$w : ((a, b] \cap \mathbb{T}) \to \mathbb{R}_+, \tag{8.80}$$

which is (nabla) Lebesgue ∇-*measurable.*
   Assume that the function

$$x \to \frac{w(x)}{K_*(x)} \chi_{[a,x)}(y) \left| \widehat{h}_{\alpha-1}(x, \rho(y)) \right|$$

is ∇-*integrable on* $((a, b] \cap \mathbb{T})$ *for each fixed* $y \in ((a, b] \cap \mathbb{T})$.
   Define v_* *on* $((a, b] \cap \mathbb{T})$ *by*

$$v_*(y) := \int_a^b \frac{w(x)}{K_*(x)} \chi_{[a,x)}(y) \left| \widehat{h}_{\alpha-1}(x, \rho(y)) \right| \nabla x < \infty. \tag{8.81}$$

Let $\Phi : \mathbb{R}_+ \to \mathbb{R}$ *be a convex and increasing function. Then*

$$\int_a^b w(x) \Phi \left(\frac{|J_a^\alpha f(x)|}{K_*(x)} \right) \nabla x \leq \int_a^b v_*(x) \Phi(|f(x)|) \nabla x, \tag{8.82}$$

under the further assumptions:
   (i) $f, \Phi(|f|)$ *are both* $\chi_{[a,x)}(y) \left| \widehat{h}_{\alpha-1}(x, \rho(y)) \right| \nabla y$-*integrable,* ∇-*a.e. in* $x \in ((a, b] \cap \mathbb{T})$,
   (ii) $v_* \Phi(|f|)$ *is* ∇-*Lebesgue integrable.*

Proof. Similar to the proof of Theorem 8.11.   □

   We continue with

Theorem 8.13. *Let* $\mathbb{T}$ *be a time scale, and* $a, b \in \mathbb{T}$, $a < b$, *with* $\sigma(a) = a$. *Let* $\alpha \geq 1$, h_α *as in (8.63), and* $K_a^\alpha f_i$ *as in (8.67), where* $f_i \in L_1([a, b) \cap \mathbb{T})$, $i = 1, \ldots, m \in \mathbb{N}$. *Assume further that* $h_{\alpha-1}(s, \sigma(t))$ *is Lebesgue* Δ-*measurable on* $([a, b) \cap \mathbb{T})^2$. *Call*

$$K^*(x) := \int_a^b \chi_{[a,x)}(y) |h_{\alpha-1}(x, \sigma(y))| \Delta y \tag{8.83}$$

$\forall\, x \in ([a,b] \cap \mathbb{T})$. *Assume that* $K^{*}(x) > 0$, Δ-*a.e. in* $x \in ([a,b] \cap \mathbb{T})$.
Here the weight function

$$u : ([a,b) \cap \mathbb{T}) \to \mathbb{R}_{+},$$

is Lebesgue Δ-*measurable. Assume that the function*

$$x \to u(x)\,\chi_{[a,x)}(y)\left(\frac{|h_{\alpha-1}(x,\sigma(y))|}{K^{*}(x)}\right)^{m}$$

is Lebesgue Δ-*integrable on* $([a,b) \cap \mathbb{T})$ *for each fixed* $y \in ([a,b) \cap \mathbb{T})$.
Define v_{m}^{*} *on* $([a,b) \cap \mathbb{T})$ *by*

$$v_{m}^{*}(y) := \int_{a}^{b} u(x)\,\chi_{[a,x)}(y)\left(\frac{|h_{\alpha-1}(x,\sigma(y))|}{K^{*}(x)}\right)^{m}\Delta x < \infty. \tag{8.84}$$

Let $\Phi_{i} : \mathbb{R}_{+} \to \mathbb{R}_{+}$, $i = 1,\dots,m$, *are convex and increasing functions.*
Then

$$\int_{a}^{b} u(x)\prod_{i=1}^{m}\Phi_{i}\left(\frac{|K_{a}^{\alpha}f_{i}(x)|}{K^{*}(x)}\right)\Delta x \tag{8.85}$$

$$\leq \left(\prod_{i=2}^{m}\int_{a}^{b}\Phi_{i}\left(|f_{i}(y)|\right)\Delta y\right)\left(\int_{a}^{b}\Phi_{1}\left(|f_{1}(y)|\right)v_{m}^{*}(y)\,\Delta y\right),$$

under the further assumptions:
   (i) $f_{i}, \Phi_{i}(|f_{i}|)$ *are both* $\chi_{[a,x)}(y)\,|h_{\alpha-1}(x,\sigma(y))|\,\Delta y$-*integrable,* Δ-*a.e. in* $x \in ([a,b) \cap \mathbb{T})$, $i = 1,\dots,m$, *and*
   (ii) $v_{m}^{*}\Phi_{1}(|f_{1}|)$, $\Phi_{2}(|f_{2}|)$, $\Phi_{3}(|f_{3}|),\dots,\Phi_{m}(|f_{m}|)$, *are all* Δ-*Lebesgue integrable.*

Proof. As in [8], and Theorem 8.7 here. □

The counterpart of last result follows

Theorem 8.14. *Let* $\mathbb{T}$ *be a time scale, and* $a,b \in \mathbb{T} - \{\min \mathbb{T}\}$, $a < b$, *with* $\rho(a) = a$.
Let $\alpha \geq 1$, $\widehat{h}_{\alpha}$ *as in (8.70), and* $J_{a}^{\alpha}f_{i}$ *as in (8.72), where* $f_{i} \in L_{1}((a,b] \cap \mathbb{T})$,
$i = 1,\dots,m \in \mathbb{N}$. *Assume further that* $\widehat{h}_{\alpha-1}(s,\rho(t))$ *is Lebesgue* ∇-*measurable on* $((a,b] \cap \mathbb{T})^{2}$. *Call*

$$K_{*}(x) := \int_{a}^{b}\chi_{[a,x)}(y)\left|\widehat{h}_{\alpha-1}(x,\rho(y))\right|\nabla y \tag{8.86}$$

$\forall\, x \in ([a,b] \cap \mathbb{T})$. *Assume that* $K_{*}(x) > 0$, ∇-*a.e. in* $x \in ([a,b] \cap \mathbb{T})$.
Here the weight function

$$w : ((a,b] \cap \mathbb{T}) \to \mathbb{R}_{+},$$

is Lebesgue ∇-*measurable. Assume that the function*

$$x \to w(x)\,\chi_{[a,x)}(y)\left(\frac{\left|\widehat{h}_{\alpha-1}(x,\rho(y))\right|}{K_{*}(x)}\right)^{m}$$

is Lebesgue ∇-integrable on $((a, b] \cap \mathbb{T})$ for each fixed $y \in ((a, b] \cap \mathbb{T})$.
   Define v_{m*} on $((a, b] \cap \mathbb{T})$ by

$$v_{m*}(y) := \int_a^b w(x) \chi_{[a,x)}(y) \left(\frac{\left|\widehat{h}_{\alpha-1}(x, \rho(y))\right|}{K_*(x)} \right)^m \nabla x < \infty. \tag{8.87}$$

Let $\Phi_i : \mathbb{R}_+ \to \mathbb{R}_+$, $i = 1, \ldots, m$, are convex and increasing functions.
   Then

$$\int_a^b w(x) \prod_{i=1}^m \Phi_i \left(\frac{|J_a^\alpha f_i(x)|}{K_*(x)} \right) \nabla x \tag{8.88}$$

$$\leq \left(\prod_{i=2}^m \int_a^b \Phi_i(|f_i(y)|) \nabla y \right) \left(\int_a^b \Phi_1(|f_1(y)|) v_{m*}(y) \nabla y \right),$$

under the further assumptions:
   (i) $f_i, \Phi_i(|f_i|)$ are both $\chi_{[a,x)}(y) \left|\widehat{h}_{\alpha-1}(x, \rho(y))\right| \nabla y$-integrable, ∇-a.e. in $x \in ((a, b] \cap \mathbb{T})$, $i = 1, \ldots, m$, and
   (ii) $v_{m*}\Phi_1(|f_1|), \Phi_2(|f_2|), \Phi_3(|f_3|), \ldots, \Phi_m(|f_m|)$, are all ∇-Lebesgue integrable.

Proof.   As in [8], and Theorem 8.7 here. $\square$

   We continue with

Theorem 8.15. *Let $\mathbb{T}$ be a time scale, and $a, b \in \mathbb{T}$, $a < b$, with $\sigma(a) = a$. Let $\alpha \geq 1$, h_α as in (8.63), and $K_a^\alpha f_i$ as in (8.67), where $f_i \in L_1([a, b) \cap \mathbb{T})$, $i = 1, \ldots, m \in \mathbb{N}$. Assume further that $h_{\alpha-1}(s, \sigma(t))$ is Lebesgue Δ-measurable on $([a, b) \cap \mathbb{T})^2$. Call*

$$K^*(x) := \int_a^b \chi_{[a,x)}(y) |h_{\alpha-1}(x, \sigma(y))| \Delta y \tag{8.89}$$

$\forall x \in ([a, b] \cap \mathbb{T})$. *Assume that $K^*(x) > 0$, Δ-a.e. in $x \in ([a, b] \cap \mathbb{T})$.*
   Here the weight function

$$u : ([a, b) \cap \mathbb{T}) \to \mathbb{R}_+,$$

is Lebesgue Δ-measurable. Assume that the function

$$x \to u(x) \chi_{[a,x)}(y) \left(\frac{|h_{\alpha-1}(x, \sigma(y))|}{K^*(x)} \right)$$

is Lebesgue Δ-integrable on $([a, b) \cap \mathbb{T})$ for each fixed $y \in ([a, b) \cap \mathbb{T})$.
   Define v^* on $([a, b) \cap \mathbb{T})$ by

$$v^*(y) := \int_a^b \frac{u(x)}{K^*(x)} \chi_{[a,x)}(y) |h_{\alpha-1}(x, \sigma(y))| \Delta x < \infty. \tag{8.90}$$

Let $p_i > 1 : \sum_{i=1}^m \frac{1}{p_i} = 1$. Let the functions $\Phi_i : \mathbb{R}_+ \to \mathbb{R}_+$, $i = 1, \ldots, m$, be convex and increasing.

Then

$$\int_a^b u\,(x) \prod_{i=1}^m \Phi_i \left(\frac{|K_a^\alpha f_i\,(x)|}{K^*\,(x)} \right) \Delta x \tag{8.91}$$

$$\leq \left(\prod_{i=1}^m \int_a^b v^*\,(y)\, \Phi_i\,(|f_i\,(y)|)^{p_i}\, \Delta y \right)^{\frac{1}{p_i}},$$

under the further assumptions:

(i) f_i, $\Phi_i\,(|f_i|)^{p_i}$ *are both* $\chi_{[a,x)}\,(y)\,|h_{\alpha-1}\,(x,\sigma\,(y))|\,\Delta y$-*integrable,* Δ-*a.e. in* $x \in ([a,b) \cap \mathbb{T})$, *for all* $i = 1, \ldots, m$, *and*

(ii) $v^* \Phi_i\,(|f_i|)^{p_i}$ *is* Δ-*Lebesgue integrable,* $i = 1, \ldots, m$.

Proof. As in [5], and Theorem 8.9 here. $\qquad\square$

The counterpart of last result follows.

Theorem 8.16. *Let* $\mathbb{T}$ *be a time scale, and* $a, b \in \mathbb{T} - \{\min \mathbb{T}\}$, $a < b$, *with* $\rho\,(a) = a$. *Let* $\alpha \geq 1$, $\widehat{h}_\alpha$ *as in (8.70), and* $J_a^\alpha f_i$ *as in (8.72), where* $f_i \in L_1\,((a,b] \cap \mathbb{T})$, $i = 1, \ldots, m \in \mathbb{N}$. *Assume further that* $\widehat{h}_{\alpha-1}\,(s, \rho\,(t))$ *is Lebesgue* ∇-*measurable on* $((a,b] \cap \mathbb{T})^2$. *Call*

$$K_*\,(x) := \int_a^b \chi_{[a,x)}\,(y)\,\left| \widehat{h}_{\alpha-1}\,(x, \rho\,(y)) \right| \nabla y, \tag{8.92}$$

$\forall\, x \in ([a,b] \cap \mathbb{T})$. *Assume that* $K_*\,(x) > 0$, ∇-*a.e. in* $x \in ([a,b] \cap \mathbb{T})$.

Here the weight function

$$w : ((a,b] \cap \mathbb{T}) \to \mathbb{R}_+,$$

is Lebesgue ∇-*measurable. Assume that the function*

$$x \to w\,(x)\, \chi_{[a,x)}\,(y) \left(\frac{\left| \widehat{h}_{\alpha-1}\,(x, \rho\,(y)) \right|}{K_*\,(x)} \right)$$

is Lebesgue ∇-*integrable on* $((a,b] \cap \mathbb{T})$ *for each fixed* $y \in ((a,b] \cap \mathbb{T})$.

Define v_* *on* $((a,b] \cap \mathbb{T})$ *by*

$$v_*\,(y) := \int_a^b \frac{w\,(x)}{K_*\,(x)} \chi_{[a,x)}\,(y) \left| \widehat{h}_{\alpha-1}\,(x, \rho\,(y)) \right| \nabla x < \infty. \tag{8.93}$$

Let $p_i > 1 : \sum_{i=1}^m \frac{1}{p_i} = 1$. *Let the functions* $\Phi_i : \mathbb{R}_+ \to \mathbb{R}_+$, $i = 1, \ldots, m$, *be convex and increasing.*

Then

$$\int_a^b w\,(x) \prod_{i=1}^m \Phi_i \left(\frac{|J_a^\alpha f_i\,(x)|}{K_*\,(x)} \right) \nabla x \tag{8.94}$$

$$\leq \left(\prod_{i=1}^m \int_a^b v_*\,(y)\, \Phi_i\,(|f_i\,(y)|)^{p_i}\, \nabla y \right)^{\frac{1}{p_i}},$$

under the further assumptions:

(i) $f_i, \Phi_i \left(|f_i| \right)^{p_i}$ *are both* $\chi_{[a,x)} \left(y \right) \left| \widehat{h}_{\alpha-1} \left(x, \rho \left(y \right) \right) \right| \nabla y$-*integrable,* ∇-*a.e. in* $x \in ((a,b] \cap \mathbb{T})$, *for all* $i = 1, \ldots, m$, *and*

(ii) $v_* \Phi_i \left(|f_i| \right)^{p_i}$ *is* ∇-*Lebesgue integrable,* $i = 1, \ldots, m$.

Proof.   As in [5], and Theorem 8.9 here. $\square$

We give

Corollary 8.7. *(to Theorem 8.13) It holds*

$$\int_a^b u \left(x \right) e^{\sum_{i=1}^m \left(\frac{\left| K_a^\alpha f_i \left(x \right) \right|}{K^* \left(x \right)} \right)} \Delta x \tag{8.95}$$

$$\leq \left(\prod_{i=2}^m \int_a^b e^{|f_i(y)|} \Delta y \right) \left(\int_a^b e^{|f_1(y)|} v_m^* \left(y \right) \Delta y \right),$$

under the assumptions:

(i) $f_i, e^{|f_i|}$ *are* $\chi_{[a,x)} \left(y \right) |h_{\alpha-1} \left(x, \sigma \left(y \right) \right)| \Delta y$-*integrable,* Δ-*a.e.*   *in* $x \in ([a,b) \cap \mathbb{T})$, $i = 1, \ldots, m$, *and*

(ii) $v_m^* e^{|f_1|}, e^{|f_2|}, e^{|f_3|}, \ldots, e^{|f_m|}$, *are all* Δ-*integrable.*

Corollary 8.8. *(to Theorem 8.16) It holds*

$$\int_a^b w \left(x \right) e^{\sum_{i=1}^m \left(\frac{\left| J_a^\alpha f_i \left(x \right) \right|}{K_* \left(x \right)} \right)} \nabla x \tag{8.96}$$

$$\leq \left(\prod_{i=1}^m \int_a^b v_* \left(y \right) e^{p_i |f_i(y)|} \nabla y \right)^{\frac{1}{p_i}},$$

under the assumptions:

(i) $f_i, e^{p_i |f_i|}$ *are both* $\chi_{[a,x)} \left(y \right) \left| \widehat{h}_{\alpha-1} \left(x, \rho \left(y \right) \right) \right| \nabla y$-*integrable,* ∇-*a.e. in* $x \in ((a,b] \cap \mathbb{T})$, *for all* $i = 1, \ldots, m$, *and*

(ii) $v_* e^{p_i |f_i|}$ *is* ∇-*Lebesgue integrable,* $i = 1, \ldots, m$.

We continue with

Theorem 8.17. *Let* $\mathbb{T}$ *be a time scale, and* $a, b \in \mathbb{T}$, $a < b$, *with* $\sigma \left(a \right) = a$. *Let all as in Definition 8.3, with* $f_i \in C_{rd}^m \left([a,b] \cap \mathbb{T} \right)$, $i = 1, \ldots, m_* \in \mathbb{N}$; $h_{\widetilde{v}}$ *as in (8.63). Call*

$$K_1 \left(x \right) := \int_a^b \chi_{[a,x)} \left(y \right) |h_{\widetilde{v}} \left(x, \sigma \left(y \right) \right)| \Delta y \tag{8.97}$$

$\forall \, x \in ([a,b] \cap \mathbb{T})$. *Assume that* $K_1 \left(x \right) > 0$, Δ-*a.e. in* $x \in ([a,b] \cap \mathbb{T})$.

Here the weight function

$$u : ([a,b) \cap \mathbb{T}) \rightarrow \mathbb{R}_+,$$

is Lebesgue Δ-measurable. Assume that the function

$$x \to u(x) \chi_{[a,x)}(y) \left(\frac{|h_{\widetilde{\nu}}(x, \sigma(y))|}{K_1(x)} \right)^{m_*}$$

is Lebesgue Δ-integrable on $([a, b) \cap \mathbb{T})$ for each fixed $y \in ([a, b) \cap \mathbb{T})$.
  Define φ_{m_*} on $([a, b) \cap \mathbb{T})$ by

$$\varphi_{m_*}(y) := \int_a^b u(x) \chi_{[a,x)}(y) \left(\frac{|h_{\widetilde{\nu}}(x, \sigma(y))|}{K_1(x)} \right)^{m_*} \Delta x < \infty. \tag{8.98}$$

Here $\Phi_i : \mathbb{R}_+ \to \mathbb{R}_+$, $i = 1, \ldots, m_*$, are convex and increasing functions.
  Then

$$\int_a^b u(x) \prod_{i=1}^{m_*} \Phi_i \left(\frac{\left| \Delta_{a*}^{\mu-1} f_i(x) \right|}{K_1(x)} \right) \Delta x \tag{8.99}$$

$$\leq \left(\prod_{i=2}^{m_*} \int_a^b \Phi_i \left(\left| f_i^{\Delta^m}(y) \right| \right) \Delta y \right) \left(\int_a^b \Phi_1 \left(\left| f_1^{\Delta^m}(y) \right| \right) \varphi_{m_*}(y) \Delta y \right),$$

under the assumption that φ_{m_*} is Δ-Lebesgue integrable on $([a, b) \cap \mathbb{T})$.

Proof. By Theorem 8.13. □

  We also derive

Theorem 8.18. *Let* $\mathbb{T}$ *be a time scale, and* $a, b \in \mathbb{T} - \{\min \mathbb{T}\}$, $a < b$, *with* $\rho(a) = a$. *Let all as in Definition 8.6, with* $f_i \in C_{ld}^m([a, b] \cap \mathbb{T})$, $i = 1, \ldots, m_* \in \mathbb{N}$; $h_{\widetilde{\nu}}$ *as in* (8.70). *Call*

$$K_2(x) := \int_a^b \chi_{[a,x)}(y) \left| \widehat{h}_{\widetilde{\nu}}(x, \rho(y)) \right| \nabla y, \tag{8.100}$$

$\forall x \in ([a, b] \cap \mathbb{T})$. *Assume that* $K_2(x) > 0$, ∇-*a.e. in* $x \in ([a, b] \cap \mathbb{T})$.
  Here the weight function

$$w : ((a, b] \cap \mathbb{T}) \to \mathbb{R}_+,$$

is Lebesgue ∇-*measurable. Assume that the function*

$$x \to w(x) \chi_{[a,x)}(y) \left(\frac{\left| \widehat{h}_{\widetilde{\nu}}(x, \rho(y)) \right|}{K_2(x)} \right)$$

is Lebesgue ∇-*integrable on* $((a, b] \cap \mathbb{T})$ *for each fixed* $y \in ((a, b] \cap \mathbb{T})$.
  Define ψ *on* $((a, b] \cap \mathbb{T})$ *by*

$$\psi(y) := \int_a^b \frac{w(x)}{K_2(x)} \chi_{[a,x)}(y) \left| \widehat{h}_{\widetilde{\nu}}(x, \rho(y)) \right| \nabla x < \infty. \tag{8.101}$$

Let $p_i > 1 : \sum_{i=1}^{m_*} \frac{1}{p_i} = 1$. *Let the functions* $\Phi_i : \mathbb{R}_+ \to \mathbb{R}_+$, $i = 1, \ldots, m_*$, *be convex and increasing.*

Then

$$\int_a^b w(x) \prod_{i=1}^m \Phi_i \left(\frac{\left| \nabla_a^{\mu-1} f_i(x) \right|}{K_2(x)} \right) \nabla x \tag{8.102}$$

$$\leq \left(\prod_{i=1}^m \int_a^b \psi(y) \, \Phi_i \left(\left| f_i^{\nabla^m}(y) \right| \right)^{p_i} \nabla y \right)^{\frac{1}{p_i}},$$

under the assumption that ψ is ∇-Lebesgue integrable on $((a,b] \cap \mathbb{T})$.

Proof.   By Theorem 8.16. $\square$

We finish with

Remark 8.2. (to Theorem 8.5)
(i) Let $\mathbb{T}_1 = \mathbb{R}$, $\mathbb{T}_2 = \mathbb{Z}$; $0 \leq \alpha \leq 1$. Then

$$\int_a^b \cdot \Diamond_\alpha x = \int_a^b \cdot dx, \tag{8.103}$$

and

$$\int_c^d \cdot \Diamond_\alpha y = \alpha \sum_{y=c}^{d-1} \cdot + (1-\alpha) \sum_{y=c+1}^d \cdot. \tag{8.104}$$

Assume $k : [a,b] \times [c,d]_\mathbb{Z} \to \mathbb{R}_+$, a continuous function. So here

$$K(x) = \alpha \sum_{y=c}^{d-1} k(x,y) + (1-\alpha) \sum_{y=c+1}^d k(x,y) > 0, \tag{8.105}$$

$\forall \, x \in [a,b]$, and
$f : [c,d]_\mathbb{Z} \to \mathbb{R}$, with

$$g(x) = \alpha \sum_{y=c}^{d-1} k(x,y) f(y) + (1-\alpha) \sum_{y=c+1}^d k(x,y) f(y), \tag{8.106}$$

$\forall \, x \in [a,b]$.
Here $u : [a,b] \to \mathbb{R}_+$ continuous, and

$$v(y) = \int_a^b \frac{u(x) \, k(x,y)}{K(x)} dx, \tag{8.107}$$

$\forall \, y \in [c,d]_\mathbb{Z}$.
Let $\Phi : \mathbb{R}_+ \to \mathbb{R}$ convex and increasing function. Then, by (8.34), we obtain

$$\int_a^b u(x) \, \Phi \left(\frac{|g(x)|}{K(x)} \right) dx \tag{8.108}$$

$$\leq \alpha \sum_{y=c}^{d-1} v(y) \, \Phi \left(|f(y)| \right) + (1-\alpha) \sum_{y=c+1}^d v(y) \, \Phi \left(|f(y)| \right).$$

(ii) Let $\mathbb{T}_1 = \mathbb{Z}$, $\mathbb{T}_2 = \mathbb{R}$. Then

$$\int_a^b \cdot \Diamond_\alpha x = \alpha \sum_{x=a}^{b-1} \cdot + (1-\alpha) \sum_{x=a+1}^{b} \cdot , \tag{8.109}$$

and

$$\int_c^d \cdot \Diamond_\alpha y = \int_c^d \cdot dy. \tag{8.110}$$

Assume $k : [a,b]_\mathbb{Z} \times [c,d] \to \mathbb{R}_+$, a continuous function. So here

$$K(x) = \int_c^d k(x,y)\, dy > 0, \quad \forall x \in [a,b]_\mathbb{Z}. \tag{8.111}$$

Consider $f : [c,d] \to \mathbb{R}$ continuous and

$$g(x) = \int_c^d k(x,y) f(y)\, dy, \quad \forall x \in [a,b]_\mathbb{Z}. \tag{8.112}$$

Let $u : [a,b]_\mathbb{Z} \to \mathbb{R}_+$. Here it is

$$v(y) = \alpha \sum_{x=a}^{b-1} \frac{u(x)\, k(x,y)}{K(x)} + (1-\alpha) \sum_{x=a+1}^{b} \frac{u(x)\, k(x,y)}{K(x)}, \tag{8.113}$$

$\forall\, y \in [c,d]$.

Let $\Phi : \mathbb{R}_+ \to \mathbb{R}$ convex and increasing function. Then, by (8.34), we derive

$$\alpha \sum_{x=a}^{b-1} u(x)\, \Phi\left(\frac{|g(x)|}{K(x)}\right) + (1-\alpha) \sum_{x=a+1}^{b} u(x)\, \Phi\left(\frac{|g(x)|}{K(x)}\right) \tag{8.114}$$

$$\leq \int_c^d v(y)\, \Phi\left(|f(y)|\right) dy.$$

Bibliography

1. R. Agarwal, M. Bohner, D. O'Regan, A. Peterson, *Dynamic equations on time scales: a survey*, J. Comput. Appl. Math. 141 (1-2) (2002), 1-26.
2. R. Agarwal, M. Bohner, A. Peterson, *Inequalities on time scales: a survey*, Math. Inequal. Appl. 4 (4) (2001), 535-557.
3. G.A. Anastassiou, *Principles of delta fractional calculus on time scales and inequalities*, Math. Comput. Modelling 52 (3-4) (2010), 556-566.
4. G.A. Anastassiou, *Foundations of nabla fractional calculus on time scales and inequalities*, Comput. Math. Appl. 59 (12) (2010), 3750-3762.
5. G.A. Anastassiou, *Fractional integral inequalities involving convexity*, Sarajevo J. Math. Special Issue Honoring 60th Birthday of M. Kulenovich 8 (21) (2012), 203-233.
6. G.A. Anastassiou, *Integral operator inequalities on time scales*, Internat. J. Difference Equations 7 (2) (2012), 111-137.
7. G.A. Anastassiou, *Rational inequalities for integral operators under convexity*, Commun. Appl. Anal. 16 (2) (2012), 179-210.
8. G.A. Anastassiou, *Univariate Hardy type fractional inequalities*, in G. Anastassiou, O. Duman (eds.), Proceedings of International Conference in Applied Mathematics and Approximation Theory 2012, Ankara, Turkey, May 17-20, 2012, Tobb Univ. of Economics and Technology, Springer, New York, 2013.
9. M. Bohner, G.S. Guseinov, *Multiple Lebesgue integration on time scales*, Adv. Difference Equations 2006 (2006), Article ID 26391, 1-12.
10. M. Bohner, G.S. Guseinov, *Double integral calculus of variations on time scales*, Comput. Math. Appl. 54 (2007), 45-57.
11. M. Bohner, H. Luo, *Singular second-order multipoint dynamic boundary value problems with mixed derivatives*, Adv. Difference Equations 2006 (2006), Article ID 54989, 1-15.
12. M. Bohner, A. Peterson, *Dynamic Equations on Time Scales: An Introduction with Applications*, Birkhäuser, Boston, 2001.
13. M. Bohner, A. Peterson, *Advances in Dynamic Equations on Time Scales*, Birkhäuser, Boston, 2003.
14. R.A.C. Ferreira, M.R. Sidi Ammi, D.F.M. Torres, *Diamond-alpha integral inequalities on time scales*, Int. J. Math. Stat. 5 (A09) (2009), 52-59.
15. G.S. Guseinov, *Integration on time scales*, J. Math. Anal. Appl. 285 (2003), 107-127.
16. H.G. Hardy, *Notes on some points in the integral calculus*, Messenger Math. 47 (10) (1918), 145-150.
17. S. Hilger, *Analysis on measure chains — a unified approach to continuous and discrete calculus*, Results Math. 18 (1-2) (1990), 18-56.

18. S. Hilger, *Differential and difference calculus — unified!*, Nonlinear Anal. 30 (5) (1997), 2683-2694.

19. S. Iqbal, K. Krulic, J. Pecaric, *On an inequality of H.G. Hardy*, J. Inequalities Appl. 2010 (2010), Article ID 264347, 23 pages.

20. A.A. Kilbas, H.M. Srivastava, J.J. Trujillo, *Theory and Applications of Fractional Differential Equations*, North-Holland Mathematics Studies, Vol. 204, Elsevier, New York, 2006.

21. D. Mozyrska, D.F.M. Torres, *Diamond-alpha polynomial series on time scales*, Int. J. Math. Stat. 5 (A09) (2009), 92-101.

22. J.W. Rogers, Jr., Q. Sheng, *Notes on the diamond-α dynamic derivative on time scales*, J. Math. Anal. Appl. 326 (1) (2007), 228-241.

23. S.G. Samko, A.A. Kilbas, O.I. Marichev, *Fractional Integral and Derivatives: Theory and Applications*, Gordon and Breach Science Publishers, Yverdon, Switzerland, 1993.

24. Q. Sheng, *Hybrid approximations via second order combined dynamic derivatives on time scales*, Electron. J. Qual. Theory Differ. Equ. 2007 (17), (2007), 13 pp. (electronic).

25. Q. Sheng, M. Fadag, J. Henderson, J.M. Davis, *An exploration of combined dynamic derivatives on time scales and their applications*, Nonlinear Anal. Real World Appl. 7 (3) (2006), 395-413.

26. M.R. Sidi Ammi, R.A.C. Ferreira, D.F.M. Torres, *Diamond-α Jensen's inequality on time scales*, J. Inequalities Appl. 2008 (2008), Article ID 576876, 13 pages.

27. M.R. Sidi Ammi, D.F.M. Torres, *Combined dynamic Grüss inequalities on time scales*, J. Math. Sci. 161 (6), (2012), 792-802.

28. W.-S. Cheung, *Generalizations of Hölder's inequality*, Internat. J. Math. Math. Sci. 26 (1) (2001), 7-10.

Chapter 9

About Vectorial Integral Operator Inequalities Using Convexity over Time Scales

Here we obtain a wide range of vectorial integral operator general inequalities on time scales using convexity. Our treatment is combined by using the diamond-alpha integral. When that fails in the fractional setting we employ the delta and nabla integrals. We give many interesting applications. It follows [9].

9.1 Background

We start with the definition of the Riemann-Liouville fractional integrals, see [23]. Let $[a, b]$, $(-\infty < a < b < \infty)$ be a finite interval on the real axis $\mathbb{R}$. The Riemann-Liouville fractional integrals $I_{a+}^{\alpha} f$ and $I_{b-}^{\alpha} f$ of order $\alpha > 0$ are defined by

$$\left(I_{a+}^{\alpha} f\right)(x) = \frac{1}{\Gamma(\alpha)} \int_{a}^{x} f(t)(x - t)^{\alpha-1} dt, \quad (x > a), \tag{9.1}$$

$$\left(I_{b-}^{\alpha} f\right)(x) = \frac{1}{\Gamma(\alpha)} \int_{x}^{b} f(t)(t - x)^{\alpha-1} dt, \quad (x < b), \tag{9.2}$$

respectively. Here $\Gamma(\alpha)$ is the Gamma function. These integrals are called the left-sided and the right-sided fractional integrals. We mention a basic property of the operators $I_{a+}^{\alpha} f$ and $I_{b-}^{\alpha} f$ of order $\alpha > 0$, see also [27]. The result says that the fractional integral operators $I_{a+}^{\alpha} f$ and $I_{b-}^{\alpha} f$ are bounded in $L_p(a, b)$, $1 \le p \le \infty$, that is

$$\left\|I_{a+}^{\alpha} f\right\|_p \le K \left\|f\right\|_p, \quad \left\|I_{b-}^{\alpha} f\right\|_p \le K \left\|f\right\|_p \tag{9.3}$$

where

$$K = \frac{(b - a)^{\alpha}}{\alpha \Gamma(\alpha)}. \tag{9.4}$$

Inequality (9.3), that is the result involving the left-sided fractional integral, was proved by H. G. Hardy in one of his first papers, see [19]. He did not write down the constant, but the calculation of the constant was hidden inside his proof.

So we are motivated by (9.3), and also [5], [8], [6], [22], [2], and we will prove analogous properties on Time Scales. But first we need some background on Time Scales, see also [13].

A time scale $\mathbb{T}$ is an arbitrary nonempty closed subset of the real numbers. The time scales calculus was initiated by S. Hilger in his PhD thesis in order to unify discrete and continuous analysis [20; 21]. Let $\mathbb{T}$ be a time scale with the topology that it inherits from the real numbers. For $t \in \mathbb{T}$, we define the forward jump operator $\sigma : \mathbb{T} \to \mathbb{T}$ by

$$\sigma(t) = \inf\{s \in \mathbb{T} : s > t\}, \qquad (9.5)$$

and the backward jump operator $\rho : \mathbb{T} \to \mathbb{T}$ by

$$\rho(t) = \sup\{s \in \mathbb{T} : s < t\}. \qquad (9.6)$$

If $\sigma(t) > t$ we say that t is right-scattered, while if $\rho(t) < t$ we say that t is left-scattered. Points that are simultaneously right-scattered and left-scattered are said to be isolated. If $\sigma(t) = t$, then t is called right-dense; if $\rho(t) = t$, then t is called left-dense. The mappings $\mu, \nu : \mathbb{T} \to [0, +\infty)$ defined by $\mu(t) := \sigma(t) - t$ and $\nu(t) := t - \rho(t)$ are called, respectively, the forward and backward graininess function.

Given a time scale $\mathbb{T}$, we introduce the sets $\mathbb{T}^k$, $\mathbb{T}_k$, and $\mathbb{T}_k^k$ as follows. If $\mathbb{T}$ has a left-scattered maximum t_1, then $\mathbb{T}^k = \mathbb{T} - \{t_1\}$, otherwise $\mathbb{T}^k = \mathbb{T}$. If $\mathbb{T}$ has a right-scattered minimum t_2, then $\mathbb{T}_k = \mathbb{T} - \{t_2\}$, otherwise $\mathbb{T}_k = \mathbb{T}$. Finally, $\mathbb{T}_k^k = \mathbb{T}^k \cap \mathbb{T}_k$.

Let $f : \mathbb{T} \to \mathbb{R}$ be a real valued function on a time scale $\mathbb{T}$. Then, for $t \in \mathbb{T}^k$, we define $f^\Delta(t)$ to be the number, if one exists, such that for all $\epsilon > 0$, there is a neighborhood U of t such that for all $s \in U$,

$$\left| f(\sigma(t)) - f(s) - f^\Delta(t)(\sigma(t) - s) \right| \le \epsilon |\sigma(t) - s|. \qquad (9.7)$$

We say that f is delta differentiable on $\mathbb{T}^k$ provided $f^\Delta(t)$ exists for all $t \in \mathbb{T}^k$. Similarly, for $t \in \mathbb{T}_k$ we define $f^\nabla(t)$ to be the number, if one exists, such that for all $\epsilon > 0$, there is a neighborhood V of t such that for all $s \in V$

$$\left| f(\rho(t)) - f(s) - f^\nabla(t)(\rho(t) - s) \right| \le \epsilon |\rho(t) - s|. \qquad (9.8)$$

We say that f is nabla differentiable on $\mathbb{T}_k$, provided that $f^\nabla(t)$ exists for all $t \in \mathbb{T}_k$.

For $f : \mathbb{T} \to \mathbb{R}$ we define the function $f^\sigma : \mathbb{T} \to \mathbb{R}$ by $f^\sigma(t) = f(\sigma(t))$ for all $t \in \mathbb{T}$, that is $f^\sigma = f \circ \sigma$. Similarly, we define the function $f^\rho : \mathbb{T} \to \mathbb{R}$ by $f^\rho(t) = f(\rho(t))$ for all $t \in \mathbb{T}$, that is, $f^\rho = f \circ \rho$.

A function $f : \mathbb{T} \to \mathbb{R}$ is called rd-continuous, provided it is continuous at all right-dense points in $\mathbb{T}$ and its left-sided limits finite at all left-dense points in $\mathbb{T}$. A function $f : \mathbb{T} \to \mathbb{R}$ is called ld-continuous, provided it is continuous at all left-dense points in $\mathbb{T}$ and its right-sided limits finite at all right-dense points in $\mathbb{T}$.

A function $F : \mathbb{T} \to \mathbb{R}$ is called a delta antiderivative of $f : \mathbb{T} \to \mathbb{R}$ provided that $F^\Delta(t) = f(t)$ holds for all $t \in \mathbb{T}^k$. Then the delta integral of f is defined by

$$\int_a^b f(t)\,\Delta t = F(b) - F(a). \qquad (9.9)$$

A function $G : \mathbb{T} \to \mathbb{R}$ is called a nabla antiderivative of $g : \mathbb{T} \to \mathbb{R}$, provided $G^{\nabla}(t) = g(t)$ holds for all $t \in \mathbb{T}_k$. Then the nabla integral of g is defined by $\int_a^b g(t)\, \nabla t = G(b) - G(a)$. For more details on time scales one can see [1; 13; 14].

Now we describe the diamond-α derivative and integral, referring the reader to [24; 26; 28; 29; 30; 31] for more on this calculus.

Let $\mathbb{T}$ be a time scale and f differentiable on $\mathbb{T}$ in the Δ and ∇ senses. For $t \in \mathbb{T}_k^k$ we define the diamond-α dynamic derivative $f^{\diamond^{\alpha}}(t)$ by

$$f^{\diamond^{\alpha}}(t) = \alpha f^{\Delta}(t) + (1 - \alpha) f^{\nabla}(t), \quad 0 \leq \alpha \leq 1. \tag{9.10}$$

Thus, f is diamond-α differentiable if and only if f is Δ and ∇ differentiable. The diamond-α derivative reduces to the standard Δ derivative for $\alpha = 1$, or the standard ∇ derivative for $\alpha = 0$. Also, it gives a "weighted derivative" for $\alpha \in (0, 1)$. Diamond-α derivatives have shown in computational experiments to provided efficient and balanced approximation formulae, leading to the design of more reliable numerical methods [28; 29].

Let $f, g : \mathbb{T} \to \mathbb{R}$ be diamond-α differentiable at $t \in \mathbb{T}_k^k$. Then,

(i) $f \pm g : \mathbb{T} \to \mathbb{R}$ is diamond-α differentiable at $t \in \mathbb{T}_k^k$ with

$$(f \pm g)^{\diamond^{\alpha}}(t) = (f)^{\diamond^{\alpha}}(t) \pm (g)^{\diamond^{\alpha}}(t). \tag{9.11}$$

(ii) For any constant c, $cf : \mathbb{T} \to \mathbb{R}$ is diamond-α differentiable at $t \in \mathbb{T}_k^k$ with

$$(cf)^{\diamond^{\alpha}}(t) = c(f)^{\diamond^{\alpha}}(t). \tag{9.12}$$

(iii) $fg : \mathbb{T} \to \mathbb{R}$ is diamond-α differentiable at $t \in \mathbb{T}_k^k$ with

$$(fg)^{\diamond^{\alpha}}(t) = (f)^{\diamond^{\alpha}}(t)\, g(t) + \alpha f^{\sigma}(t)\, g^{\Delta}(t) + (1 - \alpha) f^{\rho}(t)\, g^{\nabla}(t). \tag{9.13}$$

Let $a, t \in \mathbb{T}$, and $h : \mathbb{T} \to \mathbb{R}$. Then, the diamond-$\alpha$ integral from a to t of h is defined by

$$\int_a^t h(\tau)\, \diamond_{\alpha}\tau = \alpha \int_a^t h(\tau)\, \Delta\tau + (1 - \alpha) \int_a^t h(\tau)\, \nabla\tau, \quad 0 \leq \alpha \leq 1. \tag{9.14}$$

We may notice the absence of an anti-derivative for the $\diamond_{\alpha}$ combined derivative. For $t \in \mathbb{T}_k^k$, in general

$$\left(\int_a^t h(\tau)\, \diamond_{\alpha}\tau \right)^{\diamond^{\alpha}} \neq h(t). \tag{9.15}$$

Although the fundamental theorem of calculus does not hold for the $\diamond_{\alpha}$-integral, other properties hold true. Let $a, b, t \in \mathbb{T}$, $c \in \mathbb{R}$. Then,

(i) $\displaystyle\int_a^t \{f(\tau) \pm g(\tau)\} \Diamond_\alpha \tau = \int_a^t f(\tau) \Diamond_\alpha \tau \pm \int_a^t g(\tau) \Diamond_\alpha \tau;$ (9.16)

(ii) $\displaystyle\int_a^t cf(\tau) \Diamond_\alpha \tau = c \int_a^t f(t) \Diamond_\alpha \tau;$

(iii) $\displaystyle\int_a^t f(\tau) \Diamond_\alpha \tau = \int_a^b f(\tau) \Diamond_\alpha \tau + \int_b^t f(\tau) \Diamond_\alpha \tau;$

(iv) If $f(t) \geq 0$ for all t, then $\displaystyle\int_a^b f(t) \Diamond_\alpha t \geq 0;$

(v) If $f(t) \leq g(t)$ for all t, then $\displaystyle\int_a^b f(t) \Diamond_\alpha t \leq \int_a^b g(t) \Diamond_\alpha t;$

(vi) If $f(t) \geq 0$ for all t, then $f(t) \equiv 0$ if and only if $\displaystyle\int_a^b f(t) \Diamond_\alpha t = 0;$

(vii) $\displaystyle\int_a^b c \Diamond_\alpha t = c(b-a);$

(viii) $\left| \displaystyle\int_a^b f(t) \Diamond_\alpha t \right| \leq \int_a^b |f(t)| \Diamond_\alpha t.$

We mention

Theorem 9.1. *(multivariate Jensen inequality, see also [16, p. 76], [25]) Let f be a convex function defined on an open and convex subset $C \subseteq \mathbb{R}^n$, and let $X = (X_1, \ldots, X_n)$ be a random vector such that Probability $(X \in C) = 1$. Assume also $E(|X|)$, $E(|f(X)|) < \infty$, E stands for the expectation. Then $EX \in C$, and*

$$f(EX) \leq Ef(X). \tag{9.17}$$

We would use Jensen's diamond-α inequality

Theorem 9.2. *([15], Jensen's inequality with multiple variables). Tet $\mathbb{T}$ a time scale, $S \subseteq \mathbb{R}^n$ open and convex, $a, b \in \mathbb{T}$ and $f : S \to \mathbb{R}$ a continuous convex function. Let $h, g_1, \ldots, g_n \in C(\mathbb{T}, \mathbb{R})$ such that $\int_a^b |h(t)| \Diamond_\alpha t > 0$ and $g_1([a,b]) \times \cdots \times g_n([a,b]) \subset S$. Then*

$$f\left(\frac{\int_a^b |h(t)| g_1(t) \Diamond_\alpha t}{\int_a^b |h(t)| \Diamond_\alpha t}, \ldots, \frac{\int_a^b |h(t)| g_n(t) \Diamond_\alpha t}{\int_a^b |h(t)| \Diamond_\alpha t} \right) \leq \frac{\int_a^b |h(t)| f(g_1(t), \ldots, g_n(t)) \Diamond_\alpha t}{\int_a^b |h(t)| \Diamond_\alpha t} \tag{9.18}$$

Remark 9.1. i) By [10] and [18] we conclude that the multivariate Jensen's inequality is valid for the delta-Δ and nabla-∇ Lebesgue integrable functions and measures, respectively.

ii) Let $\Phi : \mathbb{R}_+^n \to \mathbb{R}$, we call it increasing per coordinate, iff whenever $x_i \leq y_i$, $i = 1, \ldots, n$; $x_i, y_i \in \mathbb{R}_+$, then $\Phi(x_1, \ldots, x_n) \leq \Phi(y_1, \ldots, y_n)$.

In [5], we proved that if $\Phi : \mathbb{R}_+^n \to \mathbb{R}$ is convex and increasing per coordinate, then Φ is continuous.

We can extend Φ to $\Phi : \mathbb{R}^n \to \mathbb{R}$ and still be convex, by defining

$$\Phi(x_1, \ldots, x_j, \ldots, x_n) := \Phi(x_1, \ldots, -x_j, \ldots, x_n), \tag{9.19}$$

for any $x_j < 0$, $j \in \{1, \ldots, n\}$, see Lemma 9.1.

Hence we can apply Jensen's inequality using Φ on $\mathbb{R}^n$. It is well known, that a convex function on an open and convex subset of $\mathbb{R}^n$ is continuous.

We further need

Theorem 9.3. *(Hölder's Inequality, see [17]) For continuous functions f, g : $[a, b]_{\mathbb{T}} \to \mathbb{R}$, we have:*

$$\int_a^b |f(t) g(t)| \, \Diamond_\alpha t \leq \left[\int_a^b |f(t)|^p \, \Diamond_\alpha t \right]^{\frac{1}{p}} \left[\int_a^b |g(t)|^q \, \Diamond_\alpha t \right]^{\frac{1}{q}}, \tag{9.20}$$

where $p > 1$, and $q = \frac{p}{p-1}$.

We obtain

Theorem 9.4. *(Generalization of Hölder's inequality) Let $f_i \in C([a, b]_{\mathbb{T}}, \mathbb{R})$, $i = 1, \ldots, n$, and $p_i > 1$ such that $\sum_{i=1}^n \frac{1}{p_i} = 1$. Then*

$$\int_a^b \prod_{i=1}^n |f_i(t)| \, \Diamond_\alpha t \leq \prod_{i=1}^n \left(\int_a^b |f_i(t)|^{p_i} \, \Diamond_\alpha t \right)^{\frac{1}{p_i}}. \tag{9.21}$$

Proof. Using (9.20) and induction hypothesis, exactly as in [32]. $\qquad \square$

Comment. By Tietze's extension theorem of General Topology we easily derive that a continuous function f of $\prod_{i=1}^n ([a_i, b_i] \cap \mathbb{T}_i)$ (where $\mathbb{T}_i$, $i = 1, \ldots, n \in \mathbb{N}$ are time scales) is bounded, since its continuous extension F on $\prod_{i=1}^n [a_i, b_i] \subseteq \mathbb{R}^n$ is bounded, $n \in \mathbb{N}$.

Comment. It is regarding the univariate functions. Based on [18] we see that the Cauchy Time scales delta Δ and nabla ∇ integrals are equal to definite Riemann Time Scales Δ, ∇ integrals, respectively. Thus, the diamond-α-Cauchy integral (9.14) is a diamond α-Riemann integral over continuous functions. Of course the last integral exists, since continuous functions are Riemann Δ and ∇-integrable, and it is equal to the corresponding α-Lebesgue integral, by [18].

In particular the dominated and bounded convergence theorems hold true with respect to the Lebesgue-Δ, ∇ measures.

Comment. Let $\mathbb{T}_1$, $\mathbb{T}_2$ be time scales and $f : [a, b]_{\mathbb{T}_1} \times [c, d]_{\mathbb{T}_2} \to \mathbb{R}$ be continuous. By [10] and [11] we get that f is Riemann Δ and ∇-integrable over $[a, b)_{\mathbb{T}_1} \times [c, d)_{\mathbb{T}_2}$

and $(a, b]_{\mathbb{T}_1} \times (c, d]_{\mathbb{T}_2}$, respectively. Hence by [10], [11], f is Lebesgue Δ and ∇-integrable there.

Thus by Fubini's theorem we get

$$\int_a^b \left(\int_c^d f(x, y) \,\Delta y \right) \Delta x = \int_c^d \left(\int_a^b f(x, y) \,\Delta x \right) \Delta y, \qquad (9.22)$$

and

$$\int_a^b \left(\int_c^d f(x, y) \,\nabla y \right) \nabla x = \int_c^d \left(\int_a^b f(x, y) \,\nabla x \right) \nabla y. \qquad (9.23)$$

We define $(\alpha \in [0, 1])$

$$\int_a^b \left(\int_c^d f(x, y) \,\Diamond_\alpha y \right) \Diamond_\alpha x$$

$$= \alpha \int_a^b \left(\int_c^d f(x, y) \,\Delta y \right) \Delta x + (1 - \alpha) \int_a^b \left(\int_c^d f(x, y) \,\nabla y \right) \nabla x. \qquad (9.24)$$

One can generalize (9.24) for multiple integrals.

So for f continuous we get the $\Diamond_\alpha$-Fubini's theorem main property:

$$\int_a^b \left(\int_c^d f(x, y) \,\Diamond_\alpha y \right) \Diamond_\alpha x = \int_c^d \left(\int_a^b f(x, y) \,\Diamond_\alpha x \right) \Diamond_\alpha y. \qquad (9.25)$$

We make

Remark 9.2. Let $\mathbb{T}_1$, $\mathbb{T}_2$ be time scales and $f : [a, b]_{\mathbb{T}_1} \times [c, d]_{\mathbb{T}_2} \to \mathbb{R}$ be continuous. Consider

$$g(x) = \int_c^d f(x, y) \,\Diamond_\alpha y$$

$$= \alpha \int_c^d f(x, y) \,\Delta y + (1 - \alpha) \int_c^d f(x, y) \,\nabla y, \qquad (9.26)$$

$\alpha \in [0, 1]$, $\forall\, x \in [a, b]_{\mathbb{T}_1}$.

We prove that g is continuous on $[a, b]_{\mathbb{T}_1}$. Let $x_n \to x$, where $\{x_n\}_{n \in \mathbb{N}}$, $x \in [a, b]_{\mathbb{T}_1}$ then $f(x_n, y) \to f(x, y)$, as $n \to \infty$, $\forall\, y \in [c, d]_{\mathbb{T}_2}$. Furthermore there exists $M > 0$ such that $|f(x_n, y)|$, $|f(x, y)| \le M$, $\forall\, y \in [c, d]_{\mathbb{T}_2}$. Hence by Lebesgue's bounded convergence theorem ([10]) we get that

$$\lim_{n \to \infty} \int_c^d f(x_n, y) \,\Delta y = \int_c^d f(x, y) \,\Delta y, \qquad (9.27)$$

and

$$\lim_{n \to \infty} \int_c^d f(x_n, y) \,\nabla y = \int_c^d f(x, y) \,\nabla y. \qquad (9.28)$$

Combining (9.27) and (9.28) we obtain $g(x_n) \to g(x)$, as $n \to \infty$, proving the continuity of g.

For completeness we give

Lemma 9.1. *Let* $\Phi : \mathbb{R}^n_+ \to \mathbb{R}$ *increasing per coordinate and convex. We extend* Φ *on* $\mathbb{R}^n$ *by defining* $\Phi(x_1, \ldots, x_j, \ldots, x_n) := \Phi(x_1, \ldots, -x_j, \ldots, x_n)$, *for any* $x_j < 0$, $j \in \{1, \ldots, n\}$. *Then* Φ *is convex on* $\mathbb{R}^n$.

Proof. Let $0 \le \lambda \le 1$, $(x_1, \ldots, x_n), (y_1, \ldots, y_n) \in \mathbb{R}^n$. Then we observe that

$$\Phi(\lambda(x_1, \ldots, x_n) + (1 - \lambda)(y_1, \ldots, y_n))$$

$$= \Phi(\lambda x_1 + (1 - \lambda) y_1, \ldots, \lambda x_n + (1 - \lambda) y_n)$$

$$= \Phi(|\lambda x_1 + (1 - \lambda) y_1|, \ldots, |\lambda x_n + (1 - \lambda) y_n|)$$

$$\le \Phi(\lambda |x_1| + (1 - \lambda) |y_1|, \ldots, \lambda |x_n| + (1 - \lambda) |y_n|)$$

$$= \Phi(\lambda(|x_1|, \ldots, |x_n|) + (1 - \lambda)(|y_1|, \ldots, |y_n|))$$

$$\le \lambda \Phi(|x_1|, \ldots, |x_n|) + (1 - \lambda) \Phi(|y_1|, \ldots, |y_n|)$$

$$= \lambda \Phi(x_1, \ldots, x_n) + (1 - \lambda) \Phi(y_1, \ldots, y_n),$$

proving the claim.   □

Notation 9.1. Let $(\Omega_1, \Sigma_1, \mu_1)$ and $(\Omega_2, \Sigma_2, \mu_2)$ be measure spaces with positive σ-finite measures, and let $k : \Omega_1 \times \Omega_2 \to \mathbb{R}$ be a nonnegative measurable function, $k(x, \cdot)$ measurable on Ω_2 and

$$K(x) = \int_{\Omega_2} k(x, y) \, d\mu_2(y), \quad x \in \Omega_1. \tag{9.29}$$

We suppose that $K(x) > 0$ a.e. on Ω_1, and by a weight function u (shortly: a weight), we mean a nonnegative measurable function on the actual set. Let the measurable functions $g_i : \Omega_1 \to \mathbb{R}$, $i = 1, \ldots, n$, with the representation

$$g_i(x) = \int_{\Omega_2} k(x, y) f_i(y) \, d\mu_2(y), \tag{9.30}$$

where $f_i : \Omega_2 \to \mathbb{R}$ are measurable functions, $i = 1, \ldots, n$.

Denote by $\overrightarrow{x} = x := (x_1, \ldots, x_n) \in \mathbb{R}^n$, $\overrightarrow{g} := (g_1, \ldots, g_n)$ and $\overrightarrow{f} := (f_1, \ldots, f_n)$.

We consider here $\Phi : \mathbb{R}^n_+ \to \mathbb{R}$ a convex function, which is increasing per coordinate.

Example 9.1. (for Φ).

1) Given g_i is convex and increasing on $\mathbb{R}_+$, then $\Phi(x_1, \ldots, x_n) := \sum_{i=1}^n g_i(x_i)$ is convex on $\mathbb{R}^n_+$, and increasing per coordinate; the same properties hold for:

2) $\|x\|_p = \left(\sum_{i=1}^n x_i^p \right)^{\frac{1}{p}}$, $p \ge 1$,

3) $\|x\|_\infty = \max_{i \in \{1, \ldots, n\}} x_i$,

4) $\sum_{i=1}^{n} x_i^2$,

5) $\sum_{i=1}^{n} \left(i \cdot x_i^2 \right)$,

6) $\sum_{i=1}^{n} \sum_{j=1}^{i} x_j^2$,

7) $\ln \left(\sum_{i=1}^{n} e^{x_i} \right)$,

8) let g_j are convex and increasing per coordinate on $\mathbb{R}_+^n$, then so is $\sum_{j=1}^{m} e^{g_j(x)}$,

and so is $\ln \left(\sum_{j=1}^{m} e^{g_j(x)} \right)$, $x \in \mathbb{R}_+^n$.

Notation 9.2. From now on we may write

$$\overrightarrow{g}(x) = \int_{\Omega_2} k(x,y) \overrightarrow{f}(y) \, d\mu_2(y), \tag{9.31}$$

which means

$$(g_1(x), \ldots, g_n(x)) = \left(\int_{\Omega_2} k(x,y) f_1(y) \, d\mu_2(y), \ldots, \int_{\Omega_2} k(x,y) f_n(y) \, d\mu_2(y) \right). \tag{9.32}$$

Similarly, we may write

$$|\overrightarrow{g}(x)| = \left| \int_{\Omega_2} k(x,y) \overrightarrow{f}(y) \, d\mu_2(y) \right|, \tag{9.33}$$

and we would mean

$$(|g_1(x)|, \ldots, |g_n(x)|)$$

$$= \left(\left| \int_{\Omega_2} k(x,y) f_1(y) \, d\mu_2(y) \right|, \ldots, \left| \int_{\Omega_2} k(x,y) f_n(y) \, d\mu_2(y) \right| \right). \tag{9.34}$$

We also can write that

$$|\overrightarrow{g}(x)| \leq \int_{\Omega_2} k(x,y) \left| \overrightarrow{f}(y) \right| d\mu_2(y), \tag{9.35}$$

and we mean the fact that

$$|g_i(x)| \leq \int_{\Omega_2} k(x,y) |f_i(y)| \, d\mu_2(y), \tag{9.36}$$

for all $i = 1, \ldots, n$. Similarly for other properties.

Notation 9.3. Next let $(\Omega_1, \Sigma_1, \mu_1)$ and $(\Omega_2, \Sigma_2, \mu_2)$ be measure spaces with positive σ-finite measures, and let $k_j : \Omega_1 \times \Omega_2 \to \mathbb{R}$ be a nonnegative measurable function, $k_j(x, \cdot)$ measurable on Ω_2 and

$$K_j(x) = \int_{\Omega_2} k_j(x,y) \, d\mu_2(y), \quad x \in \Omega_1, j = 1, \ldots, m. \tag{9.37}$$

We suppose that $K_j(x) > 0$ a.e. on Ω_1.

Let the measurable functions $g_{ji} : \Omega_1 \to \mathbb{R}$ with the representation

$$g_{ji}(x) = \int_{\Omega_2} k_j(x,y) f_{ji}(y) \, d\mu_2(y), \tag{9.38}$$

where $f_{ji} : \Omega_2 \to \mathbb{R}$ are measurable functions, $i = 1, \ldots, n$ and $j = 1, \ldots, m$.

Denote the function vectors $\overrightarrow{g_j} := (g_{j1}, g_{j2}, \ldots, g_{jn})$ and $\overrightarrow{f_j} := (f_{j1}, \ldots, f_{jn})$, $j = 1, \ldots, m$.

We say $\overrightarrow{f_j}$ is integrable with respect to measure μ, iff all f_{ji} are integrable with respect to μ.

We consider here $\Phi_j : \mathbb{R}_+^n \to \mathbb{R}_+$, $j = 1, \ldots, m$, convex functions that are increasing per coordinate.

Again u is a weight function on Ω_1.

9.2   Main Results

We present vectorial inequalities on $\diamondsuit_\alpha$-integral operators.

Theorem 9.5. *Let $\mathbb{T}_1$, $\mathbb{T}_2$ be time scales, $a, b \in \mathbb{T}_1$; $c, d \in \mathbb{T}_2$; $k(x,y)$ is a kernel function with $x \in [a,b]_{\mathbb{T}_1}$, $y \in [c,d]_{\mathbb{T}_2}$; k is continuous function from $[a,b]_{\mathbb{T}_1} \times [c,d]_{\mathbb{T}_2}$ into $\mathbb{R}_+$. Consider*

$$K(x) := \int_c^d k(x,y) \diamondsuit_\alpha y, \quad \forall \, x \in [a,b]_{\mathbb{T}_1}. \tag{9.39}$$

We assume that $K(x) > 0$, $\forall \, x \in [a,b]_{\mathbb{T}_1}$. Consider $f_i : [c,d]_{\mathbb{T}_2} \to \mathbb{R}$ continuous, and the $\diamondsuit_\alpha$-integral operator function

$$g_i(x) := \int_c^d k(x,y) f_i(y) \diamondsuit_\alpha y, \tag{9.40}$$

$\forall \, x \in [a,b]_{\mathbb{T}_1}$, $i = 1, \ldots, n$. Let $\overrightarrow{g} := (g_1, \ldots, g_n)$, $\overrightarrow{f} := (f_1, \ldots, f_n)$.
Consider also the weight function

$$u : [a,b]_{\mathbb{T}_1} \to \mathbb{R}_+, \tag{9.41}$$

which is continuous.
Define further the function

$$v(y) := \int_a^b \frac{u(x) k(x,y)}{K(x)} \diamondsuit_\alpha x, \tag{9.42}$$

$\forall \, y \in [c,d]_{\mathbb{T}_2}$.
Let I denote any of $(0, \infty)^n$ or $[0, \infty)^n$, and $\Phi : I \to \mathbb{R}$ be a convex and increasing per coordinate function. In particular we assume that

$$|f_i| ([c,d]_{\mathbb{T}_2}) \subseteq I, \; i = 1, \ldots, n. \tag{9.43}$$

Then

$$\int_a^b u(x) \Phi\left(\frac{\left| \overrightarrow{g(x)} \right|}{K(x)} \right) \diamondsuit_\alpha x \leq \int_c^d v(y) \Phi\left(\left| \overrightarrow{f(y)} \right| \right) \diamondsuit_\alpha y. \tag{9.44}$$

Proof. We see that

$$\int_a^b u(x) \, \Phi\left(\frac{\left|\overrightarrow{g(x)}\right|}{K(x)}\right) \Diamond_\alpha x = \int_a^b u(x) \, \Phi\left(\frac{1}{K(x)} \left|\int_c^d k(x,y) \, \overrightarrow{f(y)} \Diamond_\alpha y\right|\right) \Diamond_\alpha x \tag{9.45}$$

$$\leq \int_a^b u(x) \, \Phi\left(\frac{1}{K(x)} \int_c^d k(x,y) \left|\overrightarrow{f(y)}\right| \Diamond_\alpha y\right) \Diamond_\alpha x$$

(by generalized Jensen's inequality, see Theorem 9.2 and Comment 9.1)

$$\leq \int_a^b \frac{u(x)}{K(x)} \left(\int_c^d k(x,y) \, \Phi\left(\left|\overrightarrow{f(y)}\right|\right) \Diamond_\alpha y\right) \Diamond_\alpha x$$

$$= \int_a^b \left(\int_c^d \frac{u(x) \, k(x,y)}{K(x)} \Phi\left(\left|\overrightarrow{f(y)}\right|\right) \Diamond_\alpha y\right) \Diamond_\alpha x \tag{9.46}$$

(by (9.25))

$$= \int_c^d \left(\int_a^b \frac{u(x) \, k(x,y)}{K(x)} \Phi\left(\left|\overrightarrow{f(y)}\right|\right) \Diamond_\alpha x\right) \Diamond_\alpha y$$

$$= \int_c^d \Phi\left(\left|\overrightarrow{f(y)}\right|\right) \left(\int_a^b \frac{u(x) \, k(x,y)}{K(x)} \Diamond_\alpha x\right) \Diamond_\alpha y = \int_c^d v(y) \, \Phi\left(\left|\overrightarrow{f(y)}\right|\right) \Diamond_\alpha y, \tag{9.47}$$

proving the claim.   □

We continue with

Theorem 9.6. *All as in Theorem 9.5, however now Φ is not necessarily increasing per coordinate and only from $(0,\infty)^n$ into $\mathbb{R}$. Additionally we assume that each f_i is of fixed strict sign, $i = 1, \ldots, n$. Then*

$$\int_a^b u(x) \, \Phi\left(\frac{\left|\overrightarrow{g(x)}\right|}{K(x)}\right) \Diamond_\alpha x \leq \int_c^d v(y) \, \Phi\left(\left|\overrightarrow{f(y)}\right|\right) \Diamond_\alpha y. \tag{9.48}$$

Proof. We notice that

$$\left|\overrightarrow{g(x)}\right| = \left|\int_c^d k(x,y) \, \overrightarrow{f(y)} \Diamond_\alpha y\right| = \int_c^d k(x,y) \left|\overrightarrow{f(y)}\right| \Diamond_\alpha y. \tag{9.49}$$

Therefore we have

$$\int_a^b u(x) \, \Phi\left(\frac{\left|\overrightarrow{g(x)}\right|}{K(x)}\right) \Diamond_\alpha x = \int_a^b u(x) \, \Phi\left(\frac{1}{K(x)} \left|\int_c^d k(x,y) \, \overrightarrow{f(y)} \Diamond_\alpha y\right|\right) \Diamond_\alpha x$$

$$= \int_a^b u(x) \, \Phi\left(\frac{1}{K(x)} \int_c^d k(x,y) \left|\overrightarrow{f(y)}\right| \Diamond_\alpha y\right) \Diamond_\alpha x. \tag{9.50}$$

The rest follows as in the proof of Theorem 9.5.   □

Corollary 9.1. *(to Theorem 9.6) It holds*

$$\int_a^b u(x) \sum_{i=1}^n \ln\left(\frac{|g_i(x)|}{K(x)}\right) \Diamond_\alpha x \geq \int_c^d v(y) \sum_{i=1}^n \ln\left(|f_i(y)|\right) \Diamond_\alpha y. \tag{9.51}$$

Proof. Apply (9.48) for $\Phi(x) = -\sum_{i=1}^n \ln x_i$, which a convex function with domain $(0, \infty)^n$. $\square$

Corollary 9.2. *(to Theorem 9.5) It holds*

$$\int_a^b u(x) \sum_{i=1}^n e^{\frac{|g_i(x)|}{K(x)}} \Diamond_\alpha x \leq \int_c^d v(y) \sum_{i=1}^n e^{|f_i(y)|} \Diamond_\alpha y. \tag{9.52}$$

Proof. Apply (9.44) for $\Phi(x) = \sum_{i=1}^n e^{x_i}$, $x_i \geq 0$. $\square$

Notation 9.4. Let $\mathbb{T}_1$, $\mathbb{T}_2$ be time scales, $a, b \in \mathbb{T}_1$; $c, d \in \mathbb{T}_2$; $k_j(x, y)$ is a kernel function with $x \in [a, b]_{\mathbb{T}_1}$, $y \in [c, d]_{\mathbb{T}_2}$; k_j is continuous function from $[a, b]_{\mathbb{T}_1} \times [c, d]_{\mathbb{T}_2}$ into $\mathbb{R}_+$ for $j = 1, \ldots, m \in \mathbb{N}$. Consider

$$K_j(x) := \int_c^d k_j(x, y) \Diamond_\alpha y, \quad \forall\, x \in [a, b]_{\mathbb{T}_1}, \tag{9.53}$$

$j = 1, \ldots, m$.

We assume that $K_j(x) > 0$, $\forall\, x \in [a, b]_{\mathbb{T}_1}$, $j = 1, \ldots, m$. Consider $f_{ji} : [c, d]_{\mathbb{T}_2} \to \mathbb{R}$ continuous, $i = 1, \ldots, n$, and $j = 1, \ldots, m$, and the $\Diamond_\alpha$-integral operator function

$$g_{ji}(x) := \int_c^d k_j(x, y) f_{ji}(y) \Diamond_\alpha y, \tag{9.54}$$

$\forall\, x \in [a, b]_{\mathbb{T}_1}$, $i = 1, \ldots, n$, $j = 1, \ldots, m$.

Denote the function vectors $\overrightarrow{g_j} := (g_{j1}, \ldots, g_{jn})$ and $\overrightarrow{f_j} := (f_{j1}, \ldots, f_{jn})$, $j = 1, \ldots, m$.

Consider also the weight function

$$u : [a, b]_{\mathbb{T}_1} \to \mathbb{R}_+, \tag{9.55}$$

which is continuous.

Define further the function

$$\lambda_m(y) := \int_a^b \frac{u(x) \prod_{j=1}^m k_j(x, y)}{\prod_{j=1}^m K_j(x)} \Diamond_\alpha x, \tag{9.56}$$

$\forall\, y \in [c, d]_{\mathbb{T}_2}$.

Here $\Phi_j : \mathbb{R}_+^n \to \mathbb{R}_+$, $j = 1, \ldots, m$, are convex and increasing per coordinate functions.

We give

Theorem 9.7. *All as in Notation 9.4. Let $\rho \in \{1, \ldots, m\}$ be fixed. Then*

$$\int_a^b u(x) \prod_{j=1}^m \Phi_j\left(\frac{|\overrightarrow{g_j(x)}|}{K_j(x)}\right) \Diamond_\alpha x \tag{9.57}$$

$$\leq \left(\prod_{\substack{j=1 \\ j\neq\rho}}^{m} \int_c^d \Phi_j \left(\left| \overrightarrow{f_j(y)} \right| \right) \Diamond_\alpha y \right) \left(\int_c^d \Phi_\rho \left(\left| \overrightarrow{f_\rho(y)} \right| \right) \lambda_m(y) \Diamond_\alpha y \right).$$

Proof. We demonstrate the proof for $m = 3$. For general m it follows the same way. Here we use the multivariate Jensen's inequality, see Theorem 9.2 and Comment 9.1, $\Diamond_\alpha$-Fubini's theorem, see (9.25), and that Φ_j are increasing per coordinate.

We have

$$\int_a^b u(x) \prod_{j=1}^3 \Phi_j \left(\frac{\left| \overrightarrow{g_j(x)} \right|}{K_j(x)} \right) \Diamond_\alpha x$$

$$= \int_a^b u(x) \prod_{j=1}^3 \Phi_j \left(\left| \frac{1}{K_j(x)} \int_c^d k_j(x,y) \overrightarrow{f_j(y)} \Diamond_\alpha y \right| \right) \Diamond_\alpha x \qquad (9.58)$$

$$\leq \int_a^b u(x) \prod_{j=1}^3 \Phi_j \left(\frac{1}{K_j(x)} \int_c^d k_j(x,y) \left| \overrightarrow{f_j(y)} \right| \Diamond_\alpha y \right) \Diamond_\alpha x$$

$$\leq \int_a^b u(x) \prod_{j=1}^3 \left(\frac{1}{K_j(x)} \int_c^d k_j(x,y) \Phi_j \left(\left| \overrightarrow{f_j(y)} \right| \right) \Diamond_\alpha y \right) \Diamond_\alpha x$$

$$= \int_a^b \left(\frac{u(x)}{\prod\limits_{j=1}^3 K_j(x)} \right) \left(\prod_{j=1}^3 \int_c^d k_j(x,y) \Phi_j \left(\left| \overrightarrow{f_j(y)} \right| \right) \Diamond_\alpha y \right) \Diamond_\alpha x$$

(calling $\theta(x) := \dfrac{u(x)}{\prod\limits_{j=1}^3 K_j(x)}$)

$$= \int_a^b \theta(x) \left(\prod_{j=1}^3 \int_c^d k_j(x,y) \Phi_j \left(\left| \overrightarrow{f_j(y)} \right| \right) \Diamond_\alpha y \right) \Diamond_\alpha x \qquad (9.59)$$

$$= \int_a^b \theta(x) \left[\int_c^d \left(\prod_{j=1}^2 \int_c^d k_j(x,y) \Phi_j \left(\left| \overrightarrow{f_j(y)} \right| \right) \Diamond_\alpha y \right) \right.$$

$$\left. \cdot k_3(x,y) \Phi_3 \left(\left| \overrightarrow{f_3(y)} \right| \right) \Diamond_\alpha y \right] \Diamond_\alpha x$$

$$= \int_a^b \left(\int_c^d \theta(x) \left(\prod_{j=1}^2 \int_c^d k_j(x,y) \Phi_j \left(\left| \overrightarrow{f_j(y)} \right| \right) \Diamond_\alpha y \right) \right.$$

$$k_3\left(x,y\right)\Phi_3\left(\left|\overrightarrow{f_3\left(y\right)}\right|\right)\Diamond_\alpha y\right)\Diamond_\alpha x$$

$$=\int_c^d\left(\int_a^b\theta\left(x\right)\left(\prod_{j=1}^2\int_c^d k_j\left(x,y\right)\Phi_j\left(\left|\overrightarrow{f_j\left(y\right)}\right|\right)\Diamond_\alpha y\right)\right.$$

$$\left.\cdot\,k_3\left(x,y\right)\Phi_3\left(\left|\overrightarrow{f_3\left(y\right)}\right|\right)\Diamond_\alpha x\right)\Diamond_\alpha y$$

$$=\int_c^d\Phi_3\left(\left|\overrightarrow{f_3\left(y\right)}\right|\right)\left(\int_a^b\theta\left(x\right)k_3\left(x,y\right)\left(\prod_{j=1}^2\int_c^d k_j\left(x,y\right)\Phi_j\left(\left|\overrightarrow{f_j\left(y\right)}\right|\right)\Diamond_\alpha y\right)\Diamond_\alpha x\right)\Diamond_\alpha y$$

$$(9.60)$$

$$=\int_c^d\Phi_3\left(\left|\overrightarrow{f_3\left(y\right)}\right|\right)\left[\int_a^b\theta\left(x\right)k_3\left(x,y\right)\left(\int_c^d\left\{\int_c^d k_1\left(x,y\right)\Phi_1\left(\left|\overrightarrow{f_1\left(y\right)}\right|\right)\Diamond_\alpha y\right\}\right.\right.$$

$$\left.\left.\cdot\,k_2\left(x,y\right)\Phi_2\left(\left|\overrightarrow{f_2\left(y\right)}\right|\right)\Diamond_\alpha y\right)\Diamond_\alpha x\right]\Diamond_\alpha y$$

$$=\int_c^d\Phi_3\left(\left|\overrightarrow{f_3\left(y\right)}\right|\right)\left[\int_a^b\left(\int_c^d\theta\left(x\right)k_2\left(x,y\right)k_3\left(x,y\right)\Phi_2\left(\left|\overrightarrow{f_2\left(y\right)}\right|\right)\right.\right.\qquad(9.61)$$

$$\left.\left.\cdot\left\{\int_c^d k_1\left(x,y\right)\Phi_1\left(\left|\overrightarrow{f_1\left(y\right)}\right|\right)\Diamond_\alpha y\right\}\Diamond_\alpha y\right)\Diamond_\alpha x\right]\Diamond_\alpha y$$

$$=\left(\int_c^d\Phi_3\left(\left|\overrightarrow{f_3\left(y\right)}\right|\right)\Diamond_\alpha y\right)\left[\int_a^b\left(\int_c^d\theta\left(x\right)k_2\left(x,y\right)k_3\left(x,y\right)\Phi_2\left(\left|\overrightarrow{f_2\left(y\right)}\right|\right)\right.\right.$$

$$\left.\left.\cdot\left\{\int_c^d k_1\left(x,y\right)\Phi_1\left(\left|\overrightarrow{f_1\left(y\right)}\right|\right)\Diamond_\alpha y\right\}\Diamond_\alpha y\right)\Diamond_\alpha x\right]$$

$$=\left(\int_c^d\Phi_3\left(\left|\overrightarrow{f_3\left(y\right)}\right|\right)\Diamond_\alpha y\right)\left[\int_c^d\left(\int_a^b\theta\left(x\right)k_2\left(x,y\right)k_3\left(x,y\right)\Phi_2\left(\left|\overrightarrow{f_2\left(y\right)}\right|\right)\right.\right.$$

$$\left.\left.\cdot\left\{\int_c^d k_1\left(x,y\right)\Phi_1\left(\left|\overrightarrow{f_1\left(y\right)}\right|\right)\Diamond_\alpha y\right\}\Diamond_\alpha x\right)\Diamond_\alpha y\right]\qquad(9.62)$$

$$=\left(\int_c^d\Phi_3\left(\left|\overrightarrow{f_3\left(y\right)}\right|\right)\Diamond_\alpha y\right)\left[\int_c^d\Phi_2\left(\left|\overrightarrow{f_2\left(y\right)}\right|\right)\left(\int_a^b\theta\left(x\right)k_2\left(x,y\right)k_3\left(x,y\right)\right.\right.$$

$$\left.\left.\cdot\left(\int_c^d k_1\left(x,y\right)\Phi_1\left(\left|\overrightarrow{f_1\left(y\right)}\right|\right)\Diamond_\alpha y\right)\Diamond_\alpha x\right)\Diamond_\alpha y\right]$$

$$= \left(\int_c^d \Phi_3 \left(\left| \overrightarrow{f_3(y)} \right| \right) \Diamond_\alpha y \right) \left[\int_c^d \Phi_2 \left(\left| \overrightarrow{f_2(y)} \right| \right) \left\{ \int_a^b \left(\int_c^d \theta(x) \prod_{j=1}^3 k_j(x,y) \right. \right. \right.$$

$$\left. \left. \left. \cdot \Phi_1 \left(\left| \overrightarrow{f_1(y)} \right| \right) \Diamond_\alpha y \right) \Diamond_\alpha x \right\} \Diamond_\alpha y \right]$$

$$= \left(\int_c^d \Phi_3 \left(\left| \overrightarrow{f_3(y)} \right| \right) \Diamond_\alpha y \right) \left(\int_c^d \Phi_2 \left(\left| \overrightarrow{f_2(y)} \right| \right) \Diamond_\alpha y \right)$$

$$\cdot \left(\int_a^b \left(\int_c^d \theta(x) \prod_{j=1}^3 k_j(x,y) \Phi_1 \left(\left| \overrightarrow{f_1(y)} \right| \right) \Diamond_\alpha y \right) \Diamond_\alpha x \right) \qquad (9.63)$$

$$= \left(\prod_{j=2}^3 \int_c^d \Phi_j \left(\left| \overrightarrow{f_j(y)} \right| \right) \Diamond_\alpha y \right)$$

$$\cdot \left(\int_c^d \left(\int_a^b \theta(x) \prod_{j=1}^3 k_j(x,y) \Phi_1 \left(\left| \overrightarrow{f_1(y)} \right| \right) \Diamond_\alpha x \right) \Diamond_\alpha y \right)$$

$$= \left(\prod_{j=2}^3 \int_c^d \Phi_j \left(\left| \overrightarrow{f_j(y)} \right| \right) \Diamond_\alpha y \right)$$

$$\cdot \left(\int_c^d \Phi_1 \left(\left| \overrightarrow{f_1(y)} \right| \right) \left(\int_a^b \theta(x) \prod_{j=1}^3 k_j(x,y) \Diamond_\alpha x \right) \Diamond_\alpha y \right)$$

$$= \left(\prod_{j=2}^3 \int_c^d \Phi_j \left(\left| \overrightarrow{f_j(y)} \right| \right) \Diamond_\alpha y \right) \left(\int_c^d \Phi_1 \left(\left| \overrightarrow{f_1(y)} \right| \right) \lambda_3(y) \Diamond_\alpha y \right), \qquad (9.64)$$

proving the claim.   □

Corollary 9.3. *(to Theorem 9.7)* It holds

$$\int_a^b u(x) \prod_{j=1}^m \left(\sum_{i=1}^n e^{\frac{|g_{ji}(x)|}{K_j(x)}} \right) \Diamond_\alpha x \qquad (9.65)$$

$$\leq \left(\prod_{\substack{j=1 \\ j \neq \rho}}^m \int_c^d \sum_{i=1}^n e^{|f_{ji}(y)|} \Diamond_\alpha y \right) \left(\int_c^d \left(\sum_{i=1}^n e^{|f_{\rho i}(y)|} \right) \lambda_m(y) \Diamond_\alpha y \right).$$

Proof.   Apply $\Phi_j(x) = \sum_{i=1}^n e^{x_i}$, $x_i \geq 0$, for all $j = 1, \ldots, m$.   □

We continue with

Theorem 9.8. *All as in Theorem 9.7, but now $\Phi_j : (0, \infty)^n \to \mathbb{R}_+$, $j = 1, \ldots, m$, are convex and not necessarily increasing per coordinate. Furthermore all f_{ji}, $i = 1, \ldots, n$, $j = 1, \ldots, m$, are of fixed strict sign. Then (9.57) is valid.*

Proof. Similar to Theorem 9.6, and Theorem 9.7. $\qquad\qquad\qquad\square$

We give the following application:

Corollary 9.4. *All as in Theorem 9.8, with $\Phi_j(x) = -\sum_{i=1}^n \ln x_i$, $j = 1, \ldots, m \in \mathbb{N}$. It holds*

$$(-1)^m \int_a^b u(x) \prod_{j=1}^m \left(\sum_{i=1}^n \ln \left(\frac{|g_{ji}(x)|}{K_j(x)} \right) \right) \diamondsuit_\alpha x \qquad (9.66)$$

$$\leq (-1)^m \left(\prod_{\substack{j=1 \\ j \neq \rho}}^m \int_c^d \sum_{i=1}^n \ln \left(|f_{ji}(y)| \right) \diamondsuit_\alpha y \right) \left(\int_c^d \left(\sum_{i=1}^n \ln \left(|f_{\rho i}(y)| \right) \right) \lambda_m(y) \diamondsuit_\alpha y \right).$$

Proof. By (9.57). $\qquad\qquad\qquad\square$

We continue with

Theorem 9.9. *All as in Notation 9.4. Define*

$$u_j(y) := \int_a^b u(x) \frac{k_j(x, y)}{K_j(x)} \diamondsuit_\alpha x, \qquad (9.67)$$

$\forall\, y \in [c, d]_{\mathbb{T}_2}$, $j = 1, \ldots, m \in \mathbb{N}$.
   Let $p_j > 1 : \sum_{j=1}^m \frac{1}{p_j} = 1$. Then

$$\int_a^b u(x) \prod_{j=1}^m \Phi_j \left(\frac{\left| \overrightarrow{g_j(x)} \right|}{K_j(x)} \right) \diamondsuit_\alpha x \qquad (9.68)$$

$$\leq \prod_{j=1}^m \left(\int_c^d u_j(y) \Phi_j \left(\left| \overrightarrow{f_j(y)} \right| \right)^{p_j} \diamondsuit_\alpha y \right)^{\frac{1}{p_j}}.$$

Proof. Notice that Φ_j, $j = 1, \ldots, m$, are continuous functions. Here we use the generalized Hölder's inequality, see Theorem 9.4. We have

$$\int_a^b u(x) \prod_{j=1}^m \Phi_j \left(\frac{\left| \overrightarrow{g_j(x)} \right|}{K_j(x)} \right) \diamondsuit_\alpha x$$

$$= \int_a^b \prod_{j=1}^m \left(u(x)^{\frac{1}{p_j}} \Phi_j \left(\frac{\left| \overrightarrow{g_j(x)} \right|}{K_j(x)} \right) \right) \diamondsuit_\alpha x \qquad (9.69)$$

$$\leq \prod_{j=1}^{m} \left(\int_a^b u(x) \, \Phi_j \left(\frac{\left| \overrightarrow{g_j(x)} \right|}{K_j(x)} \right)^{p_j} \Diamond_\alpha x \right)^{\frac{1}{p_j}}$$

(notice here that $\Phi_j^{p_j}$, $j = 1, \ldots, m$, are convex, increasing per coordinate and continuous, non-negative functions, and by Theorem 9.5 we get)

$$\leq \prod_{j=1}^{m} \left(\int_c^d u_j(y) \, \Phi_j \left(|f_j(y)| \right)^{p_j} \Diamond_\alpha y \right)^{\frac{1}{p_j}}, \tag{9.70}$$

proving the claim. $\qquad\qquad\qquad\qquad\qquad\qquad\qquad\qquad\qquad\qquad\qquad \square$

We also give

Theorem 9.10. *All as in Theorem 9.9, but now* $\Phi_j : (0, \infty)^n \to \mathbb{R}_+$, $j = 1, \ldots, m$, *are convex and not necessarily increasing per coordinate. Furthermore all* f_{ji}, $j = 1, \ldots, m$, $i = 1, \ldots, n$, *are each of fixed strict sign. Then (9.68) is valid.*

Proof. Similar to Theorem 9.9, and by using Theorem 9.6. $\qquad\qquad\qquad \square$

Corollary 9.5. *(to Theorem 9.9) It holds*

$$\int_a^b u(x) \prod_{j=1}^{m} \left\| \frac{\left| \overrightarrow{g_j(x)} \right|}{K_j(x)} \right\|_{p_j} \Diamond_\alpha x \tag{9.71}$$

$$\leq \prod_{j=1}^{m} \left(\int_c^d u_j(y) \left\| \overrightarrow{f_j(y)} \right\|^{p_j} \Diamond_\alpha y \right)^{\frac{1}{p_j}}.$$

Proof. Apply (9.68) for $\Phi_j(x) = \|x\|_{p_j}$, $x_i \geq 0$, $i = 1, \ldots, n$, $j = 1, \ldots, m$. $\quad \square$

Corollary 9.6. *(to Theorem 9.9) It holds*

$$\int_a^b u(x) \prod_{j=1}^{m} \left(\sum_{i=1}^{n} e^{\frac{|g_{ji}(x)|}{K_j(x)}} \right) \Diamond_\alpha x \tag{9.72}$$

$$\leq \prod_{j=1}^{m} \left(\int_c^d u_j(y) \left(\sum_{i=1}^{n} e^{|f_{ji}(y)|} \right)^{p_j} \Diamond_\alpha y \right)^{\frac{1}{p_j}}.$$

Proof. Apply (9.68) for $\Phi_j(x) = \sum_{i=1}^{n} e^{x_i}$, $x_i \geq 0$, $i = 1, \ldots, n$, for all $j = 1, \ldots, m$. $\qquad\qquad\qquad\qquad\qquad\qquad\qquad\qquad\qquad\qquad\qquad\qquad \square$

We need

Definition 9.1. ([3]) Let $\mathbb{T}$ be a time scale. Consider the coordinate wise rd-continuous functions $h_\alpha : \mathbb{T} \times \mathbb{T} \to \mathbb{R}$, $\alpha \geq 0$, such that $h_0(t, s) = 1$,

$$h_{\alpha+1}(t, s) = \int_s^t h_\alpha(\tau, s) \, \Delta\tau, \tag{9.73}$$

$\forall \, s, t \in \mathbb{T}$.

When $\mathbb{T} = \mathbb{R}$, then $\sigma(t) = t$, we define

$$h_\alpha(t, s) := \frac{(t - s)^\alpha}{\Gamma(\alpha + 1)}, \quad \alpha \geq 0. \tag{9.74}$$

When $\mathbb{T} = \mathbb{Z}$, then $\sigma(t) = t + 1$, $t \in \mathbb{Z}$, and

$$h_k(t, s) = \frac{(t - s)^{(k)}}{k!}, \quad \forall \, k \in \mathbb{N}_0 = \mathbb{N} \cup \{0\}, \tag{9.75}$$

$\forall \, t, s \in \mathbb{Z}$, where $t^{(0)} = 1$, $t^{(k)} = \prod_{i=0}^{k-1} (t - i)$ for $k \in \mathbb{N}$. Also it holds

$$\int_a^b f(t) \, \Delta t = \sum_{t=a}^{b-1} f(t), \quad a < b; \ a, b \in \mathbb{Z}. \tag{9.76}$$

We need

Definition 9.2. ([3]) For $\alpha \geq 1$ we define the time scale Δ-Riemann-Liouville type fractional integral $(a, b \in \mathbb{T})$

$$K_a^\alpha f(t) = \int_a^t h_{\alpha-1}(t, \sigma(\tau)) \, f(\tau) \, \Delta \tau, \tag{9.77}$$

(by [12] is an integral on $[a, t] \cap \mathbb{T}$)

$$K_a^0 f = f,$$

where $f \in L_1([a, b] \cap \mathbb{T})$ (Lebesgue Δ-integrable functions on $[a, b] \cap \mathbb{T}$, see [18], [10], [11]), $t \in [a, b] \cap \mathbb{T}$.

Notice $K_a^1 f(t) = \int_a^t f(\tau) \, \Delta \tau$ is absolutely continuous in $t \in [a, b] \cap \mathbb{T}$, see [12].

Lemma 9.2. ([3]) *Let* $\alpha > 1$, $f \in L_1([a, b] \cap \mathbb{T})$. *If additionally* $h_{\alpha-1}(s, \sigma(t))$ *is Lebesgue* Δ-*measurable on* $([a, b] \cap \mathbb{T})^2$, *then* $K_a^\alpha f \in L_1([a, b] \cap \mathbb{T})$.

We need

Definition 9.3. ([3]) Assume $\mathbb{T}$ time scale such that $\mathbb{T}^k = \mathbb{T}$.

Let $\mu > 2 : m - 1 < \mu < m \in \mathbb{N}$, i.e. $m = \lceil \mu \rceil$ (ceiling of the number), $\widetilde{\nu} = m - \mu$ $(0 < \widetilde{\nu} < 1)$.

Here we take $f \in C_{rd}^m([a, b] \cap \mathbb{T})$. Clearly here ([18]) f^{Δ^m} is a Lebesgue Δ-integrable function. Assume $h_{\widetilde{\nu}}(s, \sigma(t))$ is continuous on $([a, b] \cap \mathbb{T})^2$.

We define the delta fractional derivative on time scale $\mathbb{T}$ of order $\mu - 1$ as follows:

$$\Delta_{a*}^{\mu-1} f(t) = \left(K_a^{\widetilde{\nu}+1} f^{\Delta^m} \right)(t) = \int_a^t h_{\widetilde{\nu}}(t, \sigma(\tau)) \, f^{\Delta^m}(\tau) \, \Delta \tau, \tag{9.78}$$

$\forall \, t \in [a, b] \cap \mathbb{T}$.

Notice here that $\Delta_{a*}^{\mu-1} f \in C([a, b] \cap \mathbb{T})$ by a simple argument using dominated convergence theorem in Lebesgue Δ-sense.

If $\mu = m$, then $\widetilde{\nu} = 0$ and by (9.78) we get

$$\Delta_{a*}^{m-1} f(t) = K_a^1 f^{\Delta^m}(t) = f^{\Delta^{m-1}}(t). \tag{9.79}$$

More generally, by [12], given that $f^{\Delta^{m-1}}$ is everywhere finite and absolutely continuous on $[a, b] \cap \mathbb{T}$, then f^{Δ^m} exists Δ-a.e. and is Lebesgue Δ-integrable on $[a, t] \cap \mathbb{T}$, $\forall \, t \in [a, b] \cap \mathbb{T}$ and one can plug it into (9.78).

We need

Definition 9.4. ([4]) Consider the coordinate wise ld-continuous functions $\widehat{h}_\alpha :$ $\mathbb{T} \times \mathbb{T} \to \mathbb{R}$, $\alpha \geq 0$, such that $\widehat{h}_0(t,s) = 1$,

$$\widehat{h}_{\alpha+1}(t,s) = \int_s^t \widehat{h}_\alpha(\tau,s)\,\nabla\tau, \tag{9.80}$$

$\forall\, s,t \in \mathbb{T}$.

In the case of $\mathbb{T} = \mathbb{R}$; then $\rho(t) = t$, and $\widehat{h}_k(t,s) = \frac{(t-s)^k}{k!}$, $k \in \mathbb{N}_0$, and define

$$\widehat{h}_\alpha(t,s) := \frac{(t-s)^\alpha}{\Gamma(\alpha+1)}, \qquad \alpha \geq 0. \tag{9.81}$$

Let $T = \mathbb{Z}$, then $\rho(t) = t-1$, $t \in \mathbb{Z}$. Define $t^{\overline{0}} := 1$, $t^{\overline{k}} := t(t+1)..(t+k-1)$, $k \in \mathbb{N}$, and by (9.80) we have $\widehat{h}_k(t,s) = \frac{(t-s)^{\overline{k}}}{k!}$, $s,t \in \mathbb{Z}$, $k \in \mathbb{N}_0$.

Here $\int_{t_0}^t \nabla\tau = \sum_{t_0+1}^t$.

We need

Definition 9.5. ([4]) For $\alpha \geq 1$ we define the time scale ∇-Riemann-Liouville type fractional integral $(a,b \in \mathbb{T})$

$$J_a^\alpha f(t) = \int_a^t \widehat{h}_{\alpha-1}(t,\rho(\tau))\,f(\tau)\,\nabla\tau, \tag{9.82}$$

(by [12] the last integral is on $(a,t] \cap \mathbb{T}$)

$$J_a^0 f(t) = f(t),$$

where $f \in L_1((a,b] \cap \mathbb{T})$ (Lebesgue ∇-integrable functions on $(a,b] \cap \mathbb{T}$, see [10], [11], [18]), $t \in [a,b] \cap \mathbb{T}$.

Notice $J_a^1 f(t) = \int_a^t f(\tau)\,\nabla\tau$ is absolutely continuous in $t \in [a,b] \cap \mathbb{T}$, see [12].

Lemma 9.3. ([4]) Let $\alpha > 1$, $f \in L_1((a,b] \cap \mathbb{T})$. If additionally $\widehat{h}_{\alpha-1}(s,\rho(t))$ is Lebesgue ∇-measurable on $((a,b] \cap \mathbb{T})^2$, then $J_a^\alpha f \in L_1((a,b] \cap \mathbb{T})$.

We also need

Definition 9.6. ([4]) Assume $\mathbb{T}_k = \mathbb{T}$.

Let $\mu > 2$ such that $m-1 < \mu < m \in \mathbb{N}$, i.e. $m = \lceil \mu \rceil$, $\widetilde{\nu} = m - \mu$ $(0 < \widetilde{\nu} < 1)$.

Let $f \in C_{ld}^m([a,b] \cap \mathbb{T})$. Clearly here ([18]) f^{∇^m} is a Lebesgue ∇-integrable function. Assume $\widehat{h}_{\widetilde{\nu}}(s,\rho(t))$ is continuous on $([a,b] \cap \mathbb{T})^2$.

We define the nabla fractional derivative on time scale $\mathbb{T}$ of order $\mu-1$ as follows:

$$\nabla_{a*}^{\mu-1} f(t) = \left(J_a^{\widetilde{\nu}+1} f^{\nabla^m} \right)(t) = \int_a^t \widehat{h}_{\widetilde{\nu}}(t,\rho(\tau))\,f^{\nabla^m}(\tau)\,\nabla\tau, \tag{9.83}$$

$\forall\, t \in [a,b] \cap \mathbb{T}$.

Notice here that $\nabla_{a*}^{\mu-1} f \in C([a,b] \cap \mathbb{T})$ by a simple argument using dominated convergence theorem in Lebesgue ∇-sense.

If $\mu = m$, then $\widetilde{\nu} = 0$ and by (9.83) we get

$$\nabla_{a*}^{m-1} f(t) = J_a^1 f^{\nabla^m}(t) = f^{\nabla^{m-1}}(t). \tag{9.84}$$

More generally, by [12], given that $f^{\nabla^{m-1}}$ is everywhere finite and absolutely continuous on $[a,b] \cap \mathbb{T}$, then f^{∇^m} exists ∇-a.e. and is Lebesgue ∇-integrable on $(a,t] \cap \mathbb{T}$, $\forall\, t \in [a,b] \cap \mathbb{T}$, and one can plug it into (9.83).

We present

Theorem 9.11. *Let $\mathbb{T}$ be a time scale, and $a,b \in \mathbb{T}$, $a < b$, with $\sigma(a) = a$. Let $\alpha \geq 1$, h_α as in (9.73), and $K_a^\alpha f_i$ as in (9.77), where $f_i \in L_1([a,b] \cap \mathbb{T})$, $i = 1,\ldots,n$. Assume further that $h_{\alpha-1}(s,\sigma(t))$ is Lebesgue Δ-measurable on $([a,b] \cap \mathbb{T})^2$.*
Let $\overrightarrow{f} := (f_1,\ldots,f_n)$, $\overrightarrow{K_a^\alpha f} := (K_a^\alpha f_1,\ldots,K_a^\alpha f_n)$.
Call

$$K^*(x) := \int_a^b \chi_{[a,x)}(y) \, |h_{\alpha-1}(x,\sigma(y))| \, \Delta y \tag{9.85}$$

$$= \int_a^x |h_{\alpha-1}(x,\sigma(y))| \, \Delta y,$$

$\forall\, x \in ([a,b] \cap \mathbb{T})$, where $\chi_{[a,x)}(y)$ is the characteristic function on $[a,x) \cap \mathbb{T}$.
Assume that $K^*(x) > 0$ (delta) Lebesgue measure Δ-a.e. in $x \in ([a,b] \cap \mathbb{T})$.
Consider also the weight function

$$u : ([a,b) \cap \mathbb{T}) \to \mathbb{R}_+, \tag{9.86}$$

which is (delta) Lebesgue Δ-measurable. Assume that the function

$$x \to \frac{u(x)}{K^*(x)} \chi_{[a,x)}(y) \, |h_{\alpha-1}(x,\sigma(y))|$$

is Δ-integrable on $([a,b) \cap \mathbb{T})$ for each fixed $y \in ([a,b) \cap \mathbb{T})$.
Define v^* on $([a,b) \cap \mathbb{T})$ by

$$v^*(y) := \int_a^b \frac{u(x)}{K^*(x)} \chi_{[a,x)}(y) \, |h_{\alpha-1}(x,\sigma(y))| \, \Delta x < \infty. \tag{9.87}$$

Let $\Phi : \mathbb{R}_+^n \to \mathbb{R}$ be a convex and increasing per coordinate function. Then

$$\int_a^b u(x) \, \Phi\left(\frac{\left|\overrightarrow{K_a^\alpha f(x)}\right|}{K^*(x)}\right) \Delta x \leq \int_a^b v^*(x) \, \Phi\left(\left|\overrightarrow{f(x)}\right|\right) \Delta x, \tag{9.88}$$

under the further assumptions:
(i) $\overrightarrow{f}, \Phi\left(\left|\overrightarrow{f}\right|\right)$ are both $\chi_{[a,x)}(y) \, |h_{\alpha-1}(x,\sigma(y))| \, \Delta y$-integrable, Δ-a.e. in $x \in ([a,b) \cap \mathbb{T})$,
(ii) $v^* \Phi\left(\left|\overrightarrow{f}\right|\right)$ is Δ-Lebesgue integrable.

Proof. By [5], Φ is continuous on $\mathbb{R}_+$. Here we use the multivariate Jensen's inequality, Tonelli's theorem and Fubini's theorem, all on Time Scales setting. Also we use that Φ is convex and increasing per coordinate. We extend $\Phi(x)$ on $\mathbb{R}^n$, see Lemma 9.1, so we can apply multivariate Jensen's inequality.

Next we have

$$K_a^\alpha f_i(x) = \int_a^b \chi_{[a,x)}(y) h_{\alpha-1}(x, \sigma(y)) f_i(y) \Delta y, \quad \forall \, x \in [a, b] \cap \mathbb{T},$$

and

$$|K_a^\alpha f_i(x)| \le \int_a^b \chi_{[a,x)}(y) |h_{\alpha-1}(x, \sigma(y))| |f_i(y)| \Delta y, \quad i = 1, \dots, n.$$

We see that

$$\int_a^b u(x) \Phi\left(\frac{\left| \overrightarrow{K_a^\alpha f(x)} \right|}{K^*(x)} \right) \Delta x$$

$$\le \int_a^b u(x) \Phi\left(\frac{1}{K^*(x)} \int_a^b \chi_{[a,x)}(y) |h_{\alpha-1}(x, \sigma(y))| \left| \overrightarrow{f(y)} \right| \Delta y \right) \Delta x$$

(by multivariate Jensen's inequality)

$$\le \int_a^b \frac{u(x)}{K^*(x)} \left(\int_a^b \chi_{[a,x)}(y) |h_{\alpha-1}(x, \sigma(y))| \, \Phi\left(\left| \overrightarrow{f(y)} \right| \right) \Delta y \right) \Delta x$$

$$= \int_a^b \left(\int_a^b \frac{u(x)}{K^*(x)} \chi_{[a,x)}(y) |h_{\alpha-1}(x, \sigma(y))| \, \Phi\left(\left| \overrightarrow{f(y)} \right| \right) \Delta y \right) \Delta x$$

(by Fubini's theorem)

$$= \int_a^b \left(\int_a^b \frac{u(x)}{K^*(x)} \chi_{[a,x)}(y) |h_{\alpha-1}(x, \sigma(y))| \, \Phi\left(\left| \overrightarrow{f(y)} \right| \right) \Delta x \right) \Delta y$$

$$= \int_a^b \Phi\left(\left| \overrightarrow{f(y)} \right| \right) \left(\int_a^b \frac{u(x)}{K^*(x)} \chi_{[a,x)}(y) |h_{\alpha-1}(x, \sigma(y))| \Delta x \right) \Delta y$$

$$= \int_a^b \Phi\left(\left| \overrightarrow{f(y)} \right| \right) v^*(y) \Delta y,$$

completing the proof of the theorem. $\qquad\square$

The counterpart of last result follows:

Theorem 9.12. *Let* $\mathbb{T}$ *be a time scale, and* $a, b \in \mathbb{T} - \{\min \mathbb{T}\}$, $a < b$, *with* $\rho(a) = a$. *Let* $\alpha \ge 1$, $\widehat{h}_\alpha$ *as in (9.80), and* $J_a^\alpha f_i$ *as in (9.82), where* $f_i \in L_1((a, b] \cap \mathbb{T})$, $i = 1, \dots, n$. *Let* $\overrightarrow{f} := (f_1, \dots, f_n)$, $\overrightarrow{J_a^\alpha f} := (J_a^\alpha f_1, \dots, J_a^\alpha f_n)$. *Assume further that* $\widehat{h}_{\alpha-1}(s, \rho(t))$ *is Lebesgue* ∇-*measurable on* $((a, b] \cap \mathbb{T})^2$.

Call

$$K_* (x) := \int_a^b \chi_{[a,x)} (y) \left| \widehat{h}_{\alpha-1} (x, \rho (y)) \right| \nabla y \qquad (9.89)$$

$$= \int_a^x \left| \widehat{h}_{\alpha-1} (x, \rho (y)) \right| \nabla y,$$

$\forall \, x \in ([a, b] \cap \mathbb{T})$. *Assume that $K_* (x) > 0$ (nabla) Lebesgue measure ∇-a.e. in $x \in ([a, b] \cap \mathbb{T})$.*

Consider also the weight function

$$w : ((a, b] \cap \mathbb{T}) \to \mathbb{R}_+, \qquad (9.90)$$

which is (nabla) Lebesgue ∇-measurable.

Assume that the function

$$x \to \frac{w (x)}{K_* (x)} \chi_{[a,x)} (y) \left| \widehat{h}_{\alpha-1} (x, \rho (y)) \right|$$

is ∇-integrable on $((a, b] \cap \mathbb{T})$ for each fixed $y \in ((a, b] \cap \mathbb{T})$.

Define v_* on $((a, b] \cap \mathbb{T})$ by

$$v_* (y) := \int_a^b \frac{w (x)}{K_* (x)} \chi_{[a,x)} (y) \left| \widehat{h}_{\alpha-1} (x, \rho (y)) \right| \nabla x < \infty. \qquad (9.91)$$

Let $\Phi : \mathbb{R}_+^n \to \mathbb{R}$ be a convex and increasing per coordinate function. Then

$$\int_a^b w (x) \Phi \left(\frac{\left| \overrightarrow{J_a^\alpha f (x)} \right|}{K_* (x)} \right) \nabla x \le \int_a^b v_* (x) \Phi \left(\left| \overrightarrow{f (x)} \right| \right) \nabla x, \qquad (9.92)$$

under the further assumptions:

(i) $\overrightarrow{f}, \Phi \left(\left| \overrightarrow{f} \right| \right)$ are both $\chi_{[a,x)} (y) \left| \widehat{h}_{\alpha-1} (x, \rho (y)) \right| \nabla y$-integrable, ∇-a.e. in $x \in ((a, b] \cap \mathbb{T})$,

(ii) $v_* \Phi \left(\left| \overrightarrow{f} \right| \right)$ is ∇-Lebesgue integrable.

Proof. Similar to the proof of Theorem 9.11. $\qquad\qquad\qquad \square$

We continue with

Theorem 9.13. *Let $\mathbb{T}$ be a time scale, and $a, b \in \mathbb{T}$, $a < b$, with $\sigma (a) = a$. Let $\alpha \ge 1$, h_α as in (9.73), and $K_a^\alpha f_{ji}$ as in (9.77), where $f_{ji} \in L_1 ([a, b] \cap \mathbb{T})$, $j = 1, \dots, m \in \mathbb{N}$, $i = 1, \dots, n \in \mathbb{N}$. Let $\overrightarrow{f_j} := (f_{j1}, \dots, j_{jn})$, $\overrightarrow{K_a^\alpha f_j} := (K_a^\alpha f_{j1}, \dots, K_a^\alpha f_{jn})$, $j = 1, \dots, m$. Assume further that $h_{\alpha-1} (s, \sigma (t))$ is Lebesgue Δ-measurable on $([a, b) \cap \mathbb{T})^2$ Call*

$$K^* (x) := \int_a^b \chi_{[a,x)} (y) \left| h_{\alpha-1} (x, \sigma (y)) \right| \Delta y \qquad (9.93)$$

$\forall \, x \in ([a, b] \cap \mathbb{T})$. *Assume that $K^* (x) > 0$, Δ-a.e. in $x \in ([a, b] \cap \mathbb{T})$.*

Here the weight function

$$u : ([a, b) \cap \mathbb{T}) \to \mathbb{R}_+,$$

is Lebesgue Δ-measurable. Assume that the function

$$x \to u(x) \chi_{[a,x)}(y) \left(\frac{|h_{\alpha-1}(x, \sigma(y))|}{K^*(x)} \right)^m$$

is Lebesgue Δ-integrable on $([a, b) \cap \mathbb{T})$ for each fixed $y \in ([a, b) \cap \mathbb{T})$.
    Define v_m^* on $([a, b) \cap \mathbb{T})$ by

$$v_m^*(y) := \int_a^b u(x) \chi_{[a,x)}(y) \left(\frac{|h_{\alpha-1}(x, \sigma(y))|}{K^*(x)} \right)^m \Delta x < \infty. \tag{9.94}$$

Let $\Phi_j : \mathbb{R}_+^n \to \mathbb{R}_+$, $j = 1, \dots, m$, are convex and increasing per coordinate functions.
    Then

$$\int_a^b u(x) \prod_{j=1}^m \Phi_j \left(\frac{\left| \overrightarrow{K_a^\alpha f_j(x)} \right|}{K^*(x)} \right) \Delta x \tag{9.95}$$

$$\leq \left(\prod_{j=2}^m \int_a^b \Phi_j \left(\left| \overrightarrow{f_j(y)} \right| \right) \Delta y \right) \left(\int_a^b \Phi_1 \left(\left| \overrightarrow{f_1(y)} \right| \right) v_m^*(y) \Delta y \right),$$

under the further assumptions:
    (i) $\overrightarrow{f_j}, \Phi_j \left(\left| \overrightarrow{f_j} \right| \right)$ are both $\chi_{[a,x)}(y) |h_{\alpha-1}(x, \sigma(y))| \Delta y$-integrable, Δ-a.e. in $x \in ([a, b) \cap \mathbb{T})$, $j = 1, \dots, m$, and
    (ii) $v_m^* \Phi_1 \left(\left| \overrightarrow{f_1} \right| \right)$, $\Phi_2 \left(\left| \overrightarrow{f_2} \right| \right)$, $\Phi_3 \left(\left| \overrightarrow{f_3} \right| \right), \dots, \Phi_m \left(\left| \overrightarrow{f_m} \right| \right)$, are all Δ-Lebesgue integrable.

Proof.   As in [5], and Theorem 9.7 here. $\square$

The counterpart of last result follows

Theorem 9.14. *Let $\mathbb{T}$ be a time scale, and $a, b \in \mathbb{T} - \{\min \mathbb{T}\}$, $a < b$, with $\rho(a) = a$. Let $\alpha \geq 1$, $\widehat{h}_\alpha$ as in (9.80), and $J_a^\alpha f_{ji}$ as in (9.82), where $f_{ji} \in L_1((a, b] \cap \mathbb{T})$, for $j = 1, \dots, m \in \mathbb{N}$, $i = 1, \dots, n \in \mathbb{N}$. Let $\overrightarrow{f_j} := (f_{j1}, \dots, f_{jn})$, $\overrightarrow{J_a^\alpha f_j} := (J_a^\alpha f_{j1}, \dots, J_a^\alpha f_{jn})$, $j = 1, \dots, m$. Assume further that $\widehat{h}_{\alpha-1}(s, \rho(t))$ is Lebesgue ∇-measurable on $((a, b] \cap \mathbb{T})^2$. Call*

$$K_*(x) := \int_a^b \chi_{[a,x)}(y) \left| \widehat{h}_{\alpha-1}(x, \rho(y)) \right| \nabla y \tag{9.96}$$

$\forall x \in ([a, b] \cap \mathbb{T})$. Assume that $K_*(x) > 0$, ∇-a.e. in $x \in ([a, b] \cap \mathbb{T})$.
    Here the weight function

$$w : ((a, b] \cap \mathbb{T}) \to \mathbb{R}_+,$$

is Lebesgue ∇-measurable. Assume that the function

$$x \to w\left(x\right) \chi_{[a,x)}\left(y\right) \left(\frac{\left|\widehat{h}_{\alpha-1}\left(x,\rho\left(y\right)\right)\right|}{K_*\left(x\right)}\right)^m$$

is Lebesgue ∇-integrable on $((a,b] \cap \mathbb{T})$ for each fixed $y \in ((a,b] \cap \mathbb{T})$.
Define v_{m*} on $((a,b] \cap \mathbb{T})$ by

$$v_{m*}\left(y\right) := \int_a^b w\left(x\right) \chi_{[a,x)}\left(y\right) \left(\frac{\left|\widehat{h}_{\alpha-1}\left(x,\rho\left(y\right)\right)\right|}{K_*\left(x\right)}\right)^m \nabla x < \infty. \tag{9.97}$$

Let $\Phi_j : \mathbb{R}_+^n \to \mathbb{R}_+$, $j = 1,\dots,m$, are convex and increasing per coordinate functions.
Then

$$\int_a^b w\left(x\right) \prod_{j=1}^m \Phi_j\left(\frac{\left|\overrightarrow{J_a^\alpha f_j\left(x\right)}\right|}{K_*\left(x\right)}\right) \nabla x \tag{9.98}$$

$$\leq \left(\prod_{j=2}^m \int_a^b \Phi_j\left(\left|\overrightarrow{f_j\left(y\right)}\right|\right) \nabla y\right) \left(\int_a^b \Phi_1\left(\left|\overrightarrow{f_1\left(y\right)}\right|\right) v_{m*}\left(y\right) \nabla y\right),$$

under the further assumptions:
  (i) $\overrightarrow{f_j}, \Phi_j\left(\left|\overrightarrow{f_j}\right|\right)$ are both $\chi_{[a,x)}\left(y\right) \left|\widehat{h}_{\alpha-1}\left(x,\rho\left(y\right)\right)\right|$ ∇y-integrable, ∇-a.e. in $x \in ((a,b] \cap \mathbb{T})$, $j = 1,\dots,m$, and
  (ii) $v_{m*}\Phi_1\left(\left|\overrightarrow{f_1}\right|\right), \Phi_2\left(\left|\overrightarrow{f_2}\right|\right), \Phi_3\left(\left|\overrightarrow{f_3}\right|\right),\dots,\Phi_m\left(\left|\overrightarrow{f_m}\right|\right)$, are all ∇-Lebesgue integrable.

Proof. As in [5], and Theorem 9.7 here. $\qquad\qquad\qquad\qquad\qquad\square$

We continue with

Theorem 9.15. *Let $\mathbb{T}$ be a time scale, and $a, b \in \mathbb{T}$, $a < b$, with $\sigma\left(a\right) = a$. Let $\alpha \geq 1$, h_α as in (9.73), and $K_a^\alpha f_{ji}$ as in (9.77), where $f_{ji} \in L_1\left([a,b) \cap \mathbb{T}\right)$, $j = 1,\dots,m \in \mathbb{N}$, $i = 1,\dots,n \in \mathbb{N}$. Let $\overrightarrow{f_j} := \left(f_{j1},\dots,f_{jn}\right)$, $\overrightarrow{K_a^\alpha f_j} := \left(K_a^\alpha f_{j1},\dots,K_a^\alpha f_{jn}\right)$, $j = 1,\dots,m$. Assume further that $h_{\alpha-1}\left(s,\sigma\left(t\right)\right)$ is Lebesgue Δ-measurable on $\left([a,b) \cap \mathbb{T}\right)^2$ Call*

$$K^*\left(x\right) := \int_a^b \chi_{[a,x)}\left(y\right) \left|h_{\alpha-1}\left(x,\sigma\left(y\right)\right)\right| \Delta y \tag{9.99}$$

$\forall\, x \in \left([a,b] \cap \mathbb{T}\right)$. *Assume that $K^*\left(x\right) > 0$, Δ-a.e. in $x \in \left([a,b] \cap \mathbb{T}\right)$.*
  Here the weight function

$$u : \left([a,b) \cap \mathbb{T}\right) \to \mathbb{R}_+,$$

is Lebesgue Δ-measurable. Assume that the function

$$x \to u\left(x\right) \chi_{[a,x)}\left(y\right) \left(\frac{\left|h_{\alpha-1}\left(x,\sigma\left(y\right)\right)\right|}{K^*\left(x\right)}\right)$$

is Lebesgue Δ-integrable on $\left([a,b) \cap \mathbb{T}\right)$ for each fixed $y \in \left([a,b) \cap \mathbb{T}\right)$.

Define v^* on $([a, b) \cap \mathbb{T})$ by

$$v^*(y) := \int_a^b \frac{u(x)}{K^*(x)} \chi_{[a,x)}(y) |h_{\alpha-1}(x, \sigma(y))| \, \Delta x < \infty. \qquad (9.100)$$

Let $p_j > 1 : \sum_{j=1}^m \frac{1}{p_j} = 1$. Let the functions $\Phi_j : \mathbb{R}_+^n \to \mathbb{R}_+$, $j = 1, \ldots, m$, be convex and increasing per coordinate.

Then

$$\int_a^b u(x) \prod_{j=1}^m \Phi_j \left(\frac{\left| \overrightarrow{K_a^\alpha f_j(x)} \right|}{K^*(x)} \right) \Delta x \qquad (9.101)$$

$$\leq \left(\prod_{j=1}^m \int_a^b v^*(y) \, \Phi_j \left(\left| \overrightarrow{f_j(y)} \right| \right)^{p_j} \Delta y \right)^{\frac{1}{p_j}},$$

under the further assumptions:

(i) $\overrightarrow{f_j}, \Phi_j \left(\left| \overrightarrow{f_j} \right| \right)^{p_j}$ *are both* $\chi_{[a,x)}(y) |h_{\alpha-1}(x, \sigma(y))| \, \Delta y$-*integrable, Δ-a.e. in $x \in ([a, b) \cap \mathbb{T})$, for all $j = 1, \ldots, m$, and*

(ii) $v^* \Phi_j \left(\left| \overrightarrow{f_j} \right| \right)^{p_j}$ *is Δ-Lebesgue integrable, $j = 1, \ldots, m$.*

Proof. As in [6], and Theorem 9.9 here. $\qquad \qquad \qquad \square$

The counterpart of last result follows.

Theorem 9.16. *Let $\mathbb{T}$ be a time scale, and $a, b \in \mathbb{T} - \{\min \mathbb{T}\}$, $a < b$, with $\rho(a) = a$. Let $\alpha \geq 1$, $\widehat{h}_\alpha$ as in (9.80), and $J_a^\alpha f_{ji}$ as in (9.82), where $f_{ji} \in L_1((a, b] \cap \mathbb{T})$, $i = 1, \ldots, n \in \mathbb{N}$, $j = 1, \ldots, m \in \mathbb{N}$. Let $\overrightarrow{f_j} := (f_{j1}, \ldots, f_{jn})$, $\overrightarrow{J_a^\alpha f_j} := (J_a^\alpha f_{j1}, \ldots, J_a^\alpha f_{jn})$, $j = 1, \ldots, m$. Assume further that $\widehat{h}_{\alpha-1}(s, \rho(t))$ is Lebesgue ∇-measurable on $((a, b] \cap \mathbb{T})^2$. Call*

$$K_*(x) := \int_a^b \chi_{[a,x)}(y) \left| \widehat{h}_{\alpha-1}(x, \rho(y)) \right| \nabla y, \qquad (9.102)$$

$\forall \, x \in ([a, b] \cap \mathbb{T})$. Assume that $K_*(x) > 0$, ∇-a.e. in $x \in ([a, b] \cap \mathbb{T})$.

Here the weight function

$$w : ((a, b] \cap \mathbb{T}) \to \mathbb{R}_+,$$

is Lebesgue ∇-measurable. Assume that the function

$$x \to w(x) \chi_{[a,x)}(y) \left(\frac{\left| \widehat{h}_{\alpha-1}(x, \rho(y)) \right|}{K_*(x)} \right)$$

is Lebesgue ∇-integrable on $((a, b] \cap \mathbb{T})$ for each fixed $y \in ((a, b] \cap \mathbb{T})$.

Define v_* on $((a, b] \cap \mathbb{T})$ by

$$v_* (y) := \int_a^b \frac{w(x)}{K_*(x)} \chi_{[a,x)}(y) \left| \widehat{h}_{\alpha-1}(x, \rho(y)) \right| \nabla x < \infty. \tag{9.103}$$

Let $p_j > 1 : \sum_{j=1}^m \frac{1}{p_j} = 1$. Let the functions $\Phi_j : \mathbb{R}_+^n \to \mathbb{R}_+, j = 1, \ldots, m$, be convex and increasing per coordinate.

Then

$$\int_a^b w(x) \prod_{j=1}^m \Phi_j \left(\frac{\left| \overrightarrow{J_a^\alpha f_j(x)} \right|}{K_*(x)} \right) \nabla x \tag{9.104}$$

$$\leq \left(\prod_{j=1}^m \int_a^b v_*(y) \Phi_j \left(\left| \overrightarrow{f_j(y)} \right| \right)^{p_j} \nabla y \right)^{\frac{1}{p_j}},$$

under the further assumptions:

(i) $\overrightarrow{f_j}, \Phi_j \left(\left| \overrightarrow{f_j} \right| \right)^{p_j}$ are both $\chi_{[a,x)}(y) \left| \widehat{h}_{\alpha-1}(x, \rho(y)) \right| \nabla y$-integrable, ∇-a.e. in $x \in ((a, b] \cap \mathbb{T})$, for all $j = 1, \ldots, m$, and

(ii) $v_* \Phi_j \left(\left| \overrightarrow{f_j} \right| \right)^{p_j}$ is ∇-Lebesgue integrable, $j = 1, \ldots, m$.

Proof. As in [6], and Theorem 9.9 here. □

We give

Corollary 9.7. *(to Theorem 9.13) It holds*

$$\int_a^b u(x) \prod_{j=1}^m \left(\sum_{i=1}^n e^{\frac{|K_a^\alpha f_{ji}(x)|}{K^*(x)}} \right) \Delta x \tag{9.105}$$

$$\leq \left(\prod_{j=2}^m \int_a^b \sum_{i=1}^n e^{|f_{ji}(y)|} \Delta y \right) \left(\int_a^b \left(\sum_{i=1}^n e^{|f_{1i}(y)|} \right) v_m^*(y) \Delta y \right),$$

under the assumptions:

(i) $\overrightarrow{f_j}, \sum_{i=1}^n e^{|f_{ji}(y)|}$ are $\chi_{[a,x)}(y) |h_{\alpha-1}(x, \sigma(y))| \Delta y$-integrable, Δ-a.e. in $x \in ([a, b) \cap \mathbb{T}), j = 1, \ldots, m$, and

(ii) $v_m^* \left(\sum_{i=1}^n e^{|f_{1i}(y)|} \right), \sum_{i=1}^n e^{|f_{ji}(y)|}, j = 2, \ldots, m$, are all Δ-integrable.

Corollary 9.8. *(to Theorem 9.16) It holds*

$$\int_a^b w(x) \prod_{j=1}^m \left(\sum_{i=1}^n e^{\frac{|J_a^\alpha f_{ji}(x)|}{K_*(x)}} \right) \nabla x \tag{9.106}$$

$$\leq \left(\prod_{j=1}^m \int_a^b v_*(y) \left(\sum_{i=1}^n e^{|f_{ji}(y)|} \right)^{p_j} \nabla y \right)^{\frac{1}{p_j}},$$

under the assumptions:

(i) $\overrightarrow{f_j}, \left(\sum_{i=1}^n e^{|f_{ji}(y)|} \right)^{p_j}$ are both $\chi_{[a,x)}(y) \left| \widehat{h}_{\alpha-1}(x, \rho(y)) \right| \nabla y$-integrable, ∇-a.e. in $x \in ((a, b] \cap \mathbb{T})$, for all $j = 1, \ldots, m$, and

(ii) $v_*(y) \left(\sum_{i=1}^n e^{|f_{ji}(y)|} \right)^{p_j}$ is ∇-Lebesgue integrable, $j = 1, \ldots, m$.

We continue with

Theorem 9.17. *Let* $\mathbb{T}$ *be a time scale, and* $a, b \in \mathbb{T}$, $a < b$, *with* $\sigma(a) = a$. *Let all as in Definition 9.3 and for* $f_{ji} \in C_{rd}^{m}([a,b] \cap \mathbb{T})$, $j = 1, \ldots, m_{*} \in \mathbb{N}$, $i = 1, \ldots, n \in \mathbb{N}$; $h_{\tilde{\nu}}$ *as in (9.73). Let* $\overrightarrow{f_{j}}^{\Delta^{m}} := \left(f_{j1}^{\Delta^{m}}, \ldots, f_{jn}^{\Delta^{m}} \right)$, $\overrightarrow{\Delta_{a*}^{\mu-1} f_{j}} :=$ $\left(\Delta_{a*}^{\mu-1} f_{j1}, \ldots, \Delta_{a*}^{\mu-1} f_{jn} \right)$, $j = 1, \ldots, m$. *Call*

$$K_1(x) := \int_a^b \chi_{[a,x)}(y) \left| h_{\tilde{\nu}}(x, \sigma(y)) \right| \Delta y \qquad (9.107)$$

$\forall \, x \in ([a,b] \cap \mathbb{T})$. *Assume that* $K_1(x) > 0$, Δ*-a.e. in* $x \in ([a,b] \cap \mathbb{T})$.
Here the weight function

$$u : ([a,b) \cap \mathbb{T}) \to \mathbb{R}_{+},$$

is Lebesgue Δ*-measurable. Assume that the function*

$$x \to u(x) \chi_{[a,x)}(y) \left(\frac{|h_{\tilde{\nu}}(x, \sigma(y))|}{K_1(x)} \right)^{m_{*}}$$

is Lebesgue Δ*-integrable on* $([a,b) \cap \mathbb{T})$ *for each fixed* $y \in ([a,b) \cap \mathbb{T})$.
Define $\varphi_{m_{*}}$ *on* $([a,b) \cap \mathbb{T})$ *by*

$$\varphi_{m_{*}}(y) := \int_a^b u(x) \chi_{[a,x)}(y) \left(\frac{|h_{\tilde{\nu}}(x, \sigma(y))|}{K_1(x)} \right)^{m_{*}} \Delta x < \infty. \qquad (9.108)$$

Here $\Phi_j : \mathbb{R}_{+}^{n} \to \mathbb{R}_{+}$, $j = 1, \ldots, m_{*}$, *are convex and increasing per coordinate functions.*
Then

$$\int_a^b u(x) \prod_{j=1}^{m_{*}} \Phi_j \left(\frac{\left| \overrightarrow{\Delta_{a*}^{\mu-1} f_{j}(x)} \right|}{K_1(x)} \right) \Delta x \qquad (9.109)$$

$$\leq \left(\prod_{j=2}^{m_{*}} \int_a^b \Phi_j \left(\left| \overrightarrow{f_{j}^{\Delta^{m}}(y)} \right| \right) \Delta y \right) \left(\int_a^b \Phi_1 \left(\left| \overrightarrow{f_{1}^{\Delta^{m}}(y)} \right| \right) \varphi_{m_{*}}(y) \Delta y \right),$$

under the assumption that $\varphi_{m_{*}}$ *is* Δ*-Lebesgue integrable on* $([a,b) \cap \mathbb{T})$.

Proof. By Theorem 9.13.   □

We also derive

Theorem 9.18. *Let* $\mathbb{T}$ *be a time scale, and* $a, b \in \mathbb{T} - \{\min \mathbb{T}\}$, $a < b$, *with* $\rho(a) = a$. *Let all as in Definition 9.6 and for* $f_{ji} \in C_{ld}^{m}([a,b] \cap \mathbb{T})$, $j = 1, \ldots, m_{*} \in \mathbb{N}$, $i = 1, \ldots, n \in \mathbb{N}$; $h_{\tilde{\nu}}$ *as in (9.80). Let* $\overrightarrow{f_{j}}^{\nabla^{m}} := \left(f_{j1}^{\nabla^{m}}, \ldots, f_{jn}^{\nabla^{m}} \right)$, $\overrightarrow{\nabla_{a*}^{\mu-1} f_{j}} :=$ $\left(\nabla_{a*}^{\mu-1} f_{j1}, \ldots, \nabla_{a*}^{\mu-1} f_{jn} \right)$, $j = 1, \ldots, m$. *Call*

$$K_2(x) := \int_a^b \chi_{[a,x)}(y) \left| \hat{h}_{\tilde{\nu}}(x, \rho(y)) \right| \nabla y, \qquad (9.110)$$

$\forall \, x \in ([a, b] \cap \mathbb{T})$. *Assume that $K_2(x) > 0$, ∇-a.e. in $x \in ([a, b] \cap \mathbb{T})$.*
 Here the weight function

$$w : ((a, b] \cap \mathbb{T}) \to \mathbb{R}_+,$$

is Lebesgue ∇-measurable. Assume that the function

$$x \to w(x) \chi_{[a,x)}(y) \left(\frac{\left| \widehat{h}_{\widetilde{\nu}}(x, \rho(y)) \right|}{K_2(x)} \right)$$

is Lebesgue ∇-integrable on $((a, b] \cap \mathbb{T})$ for each fixed $y \in ((a, b] \cap \mathbb{T})$.
 Define ψ on $((a, b] \cap \mathbb{T})$ by

$$\psi(y) := \int_a^b \frac{w(x)}{K_2(x)} \chi_{[a,x)}(y) \left| \widehat{h}_{\widetilde{\nu}}(x, \rho(y)) \right| \nabla x < \infty. \tag{9.111}$$

Let $p_j > 1 : \sum_{j=1}^{m_*} \frac{1}{p_j} = 1$. Let the functions $\Phi_j : \mathbb{R}_+^n \to \mathbb{R}_+$, $j = 1, \ldots, m_*$, be convex and increasing per coordinate.
 Then

$$\int_a^b w(x) \prod_{j=1}^m \Phi_j \left(\frac{\left| \overrightarrow{\nabla_{a*}^{\mu-1} f_j(x)} \right|}{K_2(x)} \right) \nabla x \tag{9.112}$$

$$\leq \left(\prod_{j=1}^m \int_a^b \psi(y) \, \Phi_j \left(\left| \overrightarrow{f_j^{\nabla^m}(y)} \right| \right)^{p_j} \nabla y \right)^{\frac{1}{p_j}},$$

under the assumption that ψ is ∇-Lebesgue integrable on $((a, b] \cap \mathbb{T})$.

Proof. By Theorem 9.16. $\qquad\qquad\qquad\qquad\qquad\qquad\qquad\qquad\qquad\qquad$ □

We finish with

Remark 9.3. (to Theorem 9.5)
 (i) Let $\mathbb{T}_1 = \mathbb{R}$, $\mathbb{T}_2 = \mathbb{Z}$; $0 \leq \alpha \leq 1$. Then

$$\int_a^b \cdot \Diamond_\alpha x = \int_a^b \cdot dx, \tag{9.113}$$

and

$$\int_c^d \cdot \Diamond_\alpha y = \alpha \sum_{y=c}^{d-1} \cdot + (1 - \alpha) \sum_{y=c+1}^d \cdot . \tag{9.114}$$

Assume $k : [a, b] \times [c, d]_{\mathbb{Z}} \to \mathbb{R}_+$, a continuous function. So here

$$K(x) = \alpha \sum_{y=c}^{d-1} k(x, y) + (1 - \alpha) \sum_{y=c+1}^d k(x, y) > 0, \tag{9.115}$$

$\forall \, x \in [a, b]$, and

$f_i : [c,d]_{\mathbb{Z}} \to \mathbb{R}$, with

$$g_i(x) = \alpha \sum_{y=c}^{d-1} k(x,y) f_i(y) + (1-\alpha) \sum_{y=c+1}^{d} k(x,y) f_i(y), \tag{9.116}$$

$\forall\, x \in [a,b]$, $i = 1, \ldots, n$.

Here $u : [a,b] \to \mathbb{R}_+$ continuous, and

$$v(y) = \int_a^b \frac{u(x) k(x,y)}{K(x)} dx, \tag{9.117}$$

$\forall\, y \in [c,d]_{\mathbb{Z}}$.

Let $\Phi : \mathbb{R}_+^n \to \mathbb{R}$ convex and increasing per coordinate function. Then, by (9.44), we obtain

$$\int_a^b u(x) \Phi\left(\frac{\left|\overrightarrow{g(x)}\right|}{K(x)}\right) dx \tag{9.118}$$

$$\leq \alpha \sum_{y=c}^{d-1} v(y) \Phi\left(\left|\overrightarrow{f(y)}\right|\right) + (1-\alpha) \sum_{y=c+1}^{d} v(y) \Phi\left(\left|\overrightarrow{f(y)}\right|\right).$$

(ii) Let $\mathbb{T}_1 = \mathbb{Z}$, $\mathbb{T}_2 = \mathbb{R}$. Then

$$\int_a^b \cdot \diamondsuit_\alpha x = \alpha \sum_{x=a}^{b-1} \cdot + (1-\alpha) \sum_{x=a+1}^{b} \cdot, \tag{9.119}$$

and

$$\int_c^d \cdot \diamondsuit_\alpha y = \int_c^d \cdot dy. \tag{9.120}$$

Assume $k : [a,b]_{\mathbb{Z}} \times [c,d] \to \mathbb{R}_+$, a continuous function. So here

$$K(x) = \int_c^d k(x,y)\, dy > 0, \quad \forall x \in [a,b]_{\mathbb{Z}}. \tag{9.121}$$

Consider $f_i : [c,d] \to \mathbb{R}$ continuous and

$$g_i(x) = \int_c^d k(x,y) f_i(y)\, dy, \quad \forall x \in [a,b]_{\mathbb{Z}}, \; i = 1, \ldots, n. \tag{9.122}$$

Let $u : [a,b]_{\mathbb{Z}} \to \mathbb{R}_+$. Here it is

$$v(y) = \alpha \sum_{x=a}^{b-1} \frac{u(x) k(x,y)}{K(x)} + (1-\alpha) \sum_{x=a+1}^{b} \frac{u(x) k(x,y)}{K(x)}, \tag{9.123}$$

$\forall\, y \in [c,d]$.

Let $\Phi : \mathbb{R}_+^n \to \mathbb{R}$ convex and increasing per coordinate function. Then, by (9.44), we derive

$$\alpha \sum_{x=a}^{b-1} u(x) \Phi\left(\frac{\left|\overrightarrow{g(x)}\right|}{K(x)}\right) + (1-\alpha) \sum_{x=a+1}^{b} u(x) \Phi\left(\frac{\left|\overrightarrow{g(x)}\right|}{K(x)}\right) \tag{9.124}$$

$$\leq \int_c^d v(y) \Phi\left(\left|\overrightarrow{f(y)}\right|\right) dy.$$

9.3   Appendix

We give

Proposition 9.1. *Let* $t_n, t, \tau \in ([a, b] \cap \mathbb{T})$, $\mathbb{T}$ *is time scale, and* $t_n \to t$, *then* $\chi_{[a,t_n)}(\tau) \to \chi_{[a,t)}(\tau)$, *delta Lebesgue measure* Δ*-a.e. in* τ.

Proof.   Indeed we have the cases:

A) $t_n \le t$.

We distinguish the subcases:

A1) $a \le \tau < t_n$, then $\chi_{[a,t_n)}(\tau) = \chi_{[a,t)}(\tau) = 1$,

A2) $t \le \tau \le b$, then $\chi_{[a,t_n)}(\tau) = \chi_{[a,t)}(\tau) = 0$,

A3) $t_n \le \tau < t$, then $\chi_{[a,t_n)}(\tau) = 0$, but $\chi_{[a,t)}(\tau) = 1$, and convergence fails.

As $n \to \infty$, the delta Lebesgue measure $\mu_\Delta([t_n,t)) = (t - t_n) \to 0$. Thus $\chi_{[a,t_n)}(\tau) \to \chi_{[a,t)}(\tau)$, Δ-a.e. in τ, when $t_n \to t$ from the left.

B) $t < t_n$.

We distinguish the subcases:

B1) $a \le \tau < t$, then $\chi_{[a,t_n)}(\tau) = \chi_{[a,t)}(\tau) = 1$,

B2) $t_n \le \tau \le b$, then $\chi_{[a,t_n)}(\tau) = \chi_{[a,t)}(\tau) = 0$,

B3) $t \le \tau < t_n$, then $\chi_{[a,t_n)}(\tau) = 1$, but $\chi_{[a,t)}(\tau) = 0$, and convergence fails.

As $n \to \infty$, $\mu_\Delta([t,t_n)) = (t_n - t) \to 0$. Thus $\chi_{[a,t_n)}(\tau) \to \chi_{[a,t)}(\tau)$, Δ-a.e. in τ, when $t_n \to t$ from the right, proving the claim.   □

Proposition 9.2. *Let* $y_n, y, x \in ([a, b) \cap \mathbb{T})$, $\mathbb{T}$ *is time scale, such that* $y_n \to y$, *as* $n \to \infty$. *Then* $\chi_{(y_n,b)}(x) \to \chi_{(y,b)}(x)$, Δ*-a.e. in* $x \in [a, b) \cap \mathbb{T}$.

Proof.   Indeed we have the cases:

A) Case of $y_n \le y$, we distinguish the subcases:

A1) if $a \le x \le y_n$, then $\chi_{(y_n,b)}(x) = \chi_{(y,b)}(x) = 0$,

A2) if $y < x < b$, then $\chi_{(y_n,b)}(x) = \chi_{(y,b)}(x) = 1$,

A3) if $y_n < x \le y$, then $\chi_{(y_n,b)}(x) = 1$, but $\chi_{(y,b)}(x) = 0$, and convergence fails.

We observe that (see [18]) $0 \le \mu_\Delta((y_n,y]) = \sigma(y) - \sigma(y_n) = y - \sigma(y_n) \le y - y_n \to 0$, proving $\chi_{(y_n,b)}(x) \to \chi_{(y,b)}(x)$, Δ-a.e. in $x \in [a, b) \cap \mathbb{T}$, when $y_n \to y$ from the left.

B) Case of $y < y_n$, we distinguish the subcases:

B1) if $a \le x \le y$, then $\chi_{(y_n,b)}(x) = \chi_{(y,b)}(x) = 0$,

B2) if $y_n < x < b$, then $\chi_{(y_n,b)}(x) = \chi_{(y,b)}(x) = 1$,

B3) if $y < x \le y_n$, then $\chi_{(y_n,b)}(x) = 0$, but $\chi_{(y,b)}(x) = 1$, and convergence fails.

But (by [18]) we have $\mu_\Delta((y,y_n]) = \sigma(y_n) - \sigma(y) = \sigma(y_n) - y \to 0$, since σ is rd-continuous and y is a right dense point ([13]). So we proved that $\chi_{(y_n,b)}(x) \to \chi_{(y,b)}(x)$, Δ-a.e. in $x \in [a, b) \cap \mathbb{T}$, when $y_n \to y$ from the right. That is proving the claim.   □

Bibliography

1. R. Agarwal, M. Bohner, D. O'Regan, A. Peterson, *Dynamic equations on time scales: a survey*, J. Comput. Appl. Math. 141 (1-2) (2002), 1-26.
2. R. Agarwal, M. Bohner, A. Peterson, *Inequalities on time scales: a survey*, Math. Inequal. Appl. 4 (4) (2001), 535-557.
3. G.A. Anastassiou, *Principles of delta fractional calculus on time scales and inequalities*, Math. Comput. Modelling 52 (3-4) (2010), 556-566.
4. G.A. Anastassiou, *Foundations of nabla fractional calculus on time scales and inequalities*, Comput. Math. Appl. 59 (12) (2010), 3750-3762.
5. G.A. Anastassiou, *Vectorial Hardy type fractional inequalities*, Bull. Tbilisi Internat. Centre Math. Informatics 16 (2) (2012), 21-57.
6. G.A. Anastassiou, *Integral operator inequalities on time scales*, J. Difference Equations 7 (2) (2012), 111-137.
7. G.A. Anastassiou, *Rational inequalities for integral operators under convexity*, Commun. Appl. Anal. 16 (2) (2012), 179-210.
8. G.A. Anastassiou, *Vectorial fractional integral inequalities with convexity*, Central European J. Phys. 11 (10) (2013), 1194-1211.
9. G.A. Anastassiou, *Vectorial integral operator convexity inequalities on time scales*, J. Concrete Applicable Math. 11 (1) (2013), 47-80.
10. M. Bohner, G.S. Guseinov, *Multiple Lebesgue integration on time scales*, Adv. Difference Equations 2006 (2006), Article ID 26391, 1-12.
11. M. Bohner, G.S. Guseinov, *Double integral calculus of variations on time scales*, Comput. Math. Appl. 54 (2007), 45-57.
12. M. Bohner, H. Luo, *Singular second-order multipoint dynamic boundary value problems with mixed derivatives*, Adv. Difference Equations 2006 (2006), Article ID 54989, 1-15.
13. M. Bohner, A. Peterson, *Dynamic Equations on Time Scales: An Introduction with Applications*, Birkhäuser, Boston, 2001.
14. M. Bohner, A. Peterson, *Advances in Dynamic Equations on Time Scales*, Birkhäuser, Boston, 2003.
15. C.C. Dinu, *Inequalities on time scales*, Ph.D. thesis, U. Craiova, Romania (2008).
16. T.S. Fergurson, *Mathematical Statistics*, Academic Press, New York, 1967.
17. R.A.C. Ferreira, M.R. Sidi Ammi, D.F.M. Torres, *Diamond-alpha Integral inequalities on time scales*, Int. J. Math. Stat. 5 (A09) (2009), 52-59.
18. G.S. Guseinov, *Integration on time scales*, J. Math. Anal. Appl. 285 (2003), 107-127.
19. H.G. Hardy, *Notes on some points in the integral calculus*, Messenger Math. 47 (10) (1918), 145-150.

20. S. Hilger, *Analysis on measure chains — a unified approach to continuous and discrete calculus*, Results Math. 18 (1-2) (1990), 18-56.
21. S. Hilger, *Differential and difference calculus — unified!*, Nonlinear Anal. 30 (5) (1997), 2683-2694.
22. S. Iqbal, K. Krulic, J. Pecaric, *On an inequality of H.G. Hardy*, J. Inequalities Appl. 2010 (2010), Article ID 264347, 23 pages.
23. A.A. Kilbas, H.M. Srivastava, J.J. Trujillo, *Theory and Applications of Fractional Differential Equations*, North-Holland Mathematics Studies, Vol. 204, Elsevier, New York, 2006.
24. D. Mozyrska, D.F.M. Torres, *Diamond-alpha polynomial series on time scales*, Int. J. Math. Stat. 5 (A09) (2009), 92-101.
25. M. Perlman, *Jensen's inequality for a convex vector valued function on an infinite-dimensional space*, J. Multivariate Anal. 4 (1974), 52-65.
26. J.W. Rogers, Jr., Q. Sheng, *Notes on the diamond-α dynamic derivative on time scales*, J. Math. Anal. Appl. 326 (1) (2007), 228-241.
27. S.G. Samko, A.A. Kilbas, O.I. Marichev, *Fractional Integral and Derivatives: Theory and Applications*, Gordon and Breach Science Publishers, Yverdon, Switzerland, 1993.
28. Q. Sheng, *Hybrid approximations via second order combined dynamic derivatives on time scales*, Electron. J. Qual. Theory Differ. Equ. 2007 (17) (2007), 13 pp. (electronic).
29. Q. Sheng, M. Fadag, J. Henderson, J.M. Davis, *An exploration of combined dynamic derivatives on time scales and their applications*, Nonlinear Anal. Real World Appl. 7 (3) (2006), 395-413.
30. M.R. Sidi Ammi, R.A.C. Ferreira, D.F.M. Torres, *Diamond-α Jensen's inequality on time scales*, J. Inequalities Appl. 2008 (2008), Article ID 576876, 13 pages.
31. M.R. Sidi Ammi, D.F.M. Torres, *Combined dynamic Grüss inequalities on time scales*, J. Math. Sci. 161 (6) (2012), 792-802.
32. W.-S. Cheung, *Generalizations of Hölder's inequality*, Internat. J. Math. Math. Sci. 26 (1) (2001), 7-10.

Chapter 10

General Grüss and Ostrowski Inequalities Using s-Convexity

Using the well known representation formula for functions due to Fink [8], we establish a series of general Grüss and Ostrowski type inequalities involving s-convexity and s-concavity in the second sense, acting to all possible directions. It follows [4].

10.1 Background

We are motivated by the following famous inequality due to G. Grüss of 1935.

Theorem 10.1. ([9]) Let f, g be integrable functions from $[a, b]$ into $\mathbb{R}$, that satisfy the conditions

$$m \leq f(x) \leq M, \quad n \leq g(x) \leq N, \quad x \in [a, b],$$

where $m, M, n, N \in \mathbb{R}$. Then

$$\left| \frac{1}{b-a} \int_a^b f(x) g(x) \, dx - \left(\frac{1}{b-a} \int_a^b f(x) \, dx \right) \left(\frac{1}{b-a} \int_a^b g(x) \, dx \right) \right| \qquad (10.1)$$

$$\leq \frac{1}{4} (M - m)(N - n).$$

We are also motivated by [3], Chapter 25, pp. 305-317.

Another motivation is [2].

We are strongly motivated in our method by

Proposition 10.1. Let $f : [a, b] \to \mathbb{R}$ be continuous on $[a, b]$ and differentiable on (a, b). If $f' \in L_1([a, b])$, then

$$f(x) = \frac{1}{b-a} \int_a^b f(u) \, du + (a - b) \int_0^1 p(x, t) f'(ta + (1-t)b) \, dt, \qquad (10.2)$$

for all $x \in [a, b]$, where

$$p(x, t) = \begin{cases} t, & t \in \left[0, \frac{b-x}{b-a}\right), \\ t - 1, & t \in \left[\frac{b-x}{b-a}, 1\right]. \end{cases} \qquad (10.3)$$

185

Proof. We use Montgomery identity ([6])

$$f(x) = \frac{1}{b-a} \int_a^b f(t)\, dt + \frac{1}{b-a} \int_a^b q(x,t) f'(t)\, dt, \tag{10.4}$$

for all $x \in [a,b]$, where

$$q(x,t) = \begin{cases} t-a, & t \in [a,x], \\ t-b, & t \in (x,b]. \end{cases} \tag{10.5}$$

We observe by change of variable that

$$\frac{1}{b-a} \int_a^b q(x,u) f'(u)\, du = \int_0^1 q(x, \lambda a + (1-\lambda) b) f'(\lambda a + (1-\lambda) b)\, d\lambda. \tag{10.6}$$

We notice that

$$q(x, \lambda a + (1-\lambda) b) = \begin{cases} (1-\lambda)(b-a), & \lambda a + (1-\lambda) b \in [a,x], \\ -\lambda(b-a), & \lambda a + (1-\lambda) b \in (x,b] \end{cases} \tag{10.7}$$

$$= \begin{cases} (1-\lambda)(b-a), & \lambda \in \left[\frac{b-x}{b-a}, 1\right], \\ -\lambda(b-a), & \lambda \in \left[0, \frac{b-x}{b-a}\right) \end{cases}$$

$$= (a-b) \begin{cases} \lambda, & \lambda \in \left[0, \frac{b-x}{b-a}\right), \\ \lambda-1, & \lambda \in \left[\frac{b-x}{b-a}, 1\right] \end{cases} = (a-b)\, p(x,\lambda). \tag{10.8}$$

Therefore it holds

$$\int_0^1 q(x, \lambda a + (1-\lambda) b) f'(\lambda a + (1-\lambda) b)\, d\lambda$$

$$= (a-b) \int_0^1 p(x,\lambda) f'(\lambda a + (1-\lambda) b)\, d\lambda, \tag{10.9}$$

that is

$$\frac{1}{b-a} \int_a^b q(x,u) f'(u)\, du = (a-b) \int_0^1 p(x,\lambda) f'(\lambda a + (1-\lambda) b)\, d\lambda, \tag{10.10}$$

proving the claim. $\qquad\square$

Note. From (10.3) we notice that

$$|p(x,t)| \le 1, \text{ all } x \in [a,b] \text{ and } t \in [0,1]. \tag{10.11}$$

Assume now that $|f'|$ is convex on $[a,b]$ with $f'(a), f'(b) \in \mathbb{R}$. Then $g(t) = |f'(ta + (1-t)b)| \ge 0$ is also a convex function in $t \in [0,1]$, as a composition of a convex and an affine function.

It is clear now by comparing the areas under g and the related trapezium that

$$\int_0^1 |f'(ta + (1-t)b)|\, dt \le \frac{|f'(a)| + |f'(b)|}{2}. \tag{10.12}$$

Hence when $|f'|$ is convex with $f'(a), f'(b) \in \mathbb{R}$, we get

$$\int_0^1 |p(x,t)| \, |f'(ta + (1-t)b)| \, dt \leq \frac{|f'(a)| + |f'(b)|}{2}. \qquad (10.13)$$

We need

Theorem 10.2. *([8], Fink) Let $a, b \in \mathbb{R}$, $a < b$, $f : [a,b] \to \mathbb{R}$, $n \in \mathbb{N}$, $f^{(n-1)}$ is absolutely continuous on $[a,b]$, $x \in [a,b]$. Then*

$$f(x) = \frac{n}{b-a} \int_a^b f(t) \, dt \qquad (10.14)$$

$$-\sum_{k=1}^{n-1} \left(\frac{n-k}{k!} \right) \left(\frac{f^{(k-1)}(a)(x-a)^k - f^{(k-1)}(b)(x-b)^k}{b-a} \right)$$

$$+\frac{1}{(n-1)!\,(b-a)} \int_a^b (x-t)^{n-1} q(x,t) f^{(n)}(t) \, dt,$$

where

$$q(x,t) = \begin{cases} t-a, & t \in [a,x], \\ t-b, & t \in (x,b]. \end{cases} \qquad (10.15)$$

When $n = 1$ the sum $\sum_{k=1}^{n-1}$ is zero.

We make

Remark 10.1. (on Theorem 10.2) Here we transform the remainder of (10.14) as follows

$$\frac{1}{(n-1)!\,(b-a)} \int_a^b (x-t)^{n-1} q(x,t) f^{(n)}(t) \, dt \qquad (10.16)$$

$$= \frac{1}{(n-1)!} \int_0^1 (x - \lambda a - (1-\lambda)b)^{n-1} q(x, \lambda a + (1-\lambda)b) f^{(n)}(\lambda a + (1-\lambda)b) \, d\lambda$$

$$\overset{(10.8)}{=} \frac{a-b}{(n-1)!} \int_0^1 ((x-b) + \lambda(b-a))^{n-1} p(x,\lambda) f^{(n)}(\lambda a + (1-\lambda)b) \, d\lambda. \qquad (10.17)$$

We have established the following result which is our main tool here

Theorem 10.3. *Let $a, b \in \mathbb{R}$, $a < b$, $f : [a,b] \to \mathbb{R}$, $n \in \mathbb{N}$, $f^{(n-1)}$ is absolutely continuous on $[a,b]$, $x \in [a,b]$. Then*

$$f(x) = \frac{n}{b-a} \int_a^b f(t) \, dt - \sum_{k=1}^{n-1} \left(\frac{n-k}{k!} \right) \left(\frac{f^{(k)}(a)(x-a)^k - f^{(k-1)}(b)(x-b)^k}{b-a} \right)$$

$$+\frac{a-b}{(n-1)!} \int_0^1 ((x-b) + \lambda(b-a))^{n-1} p(x,\lambda) f^{(n)}(\lambda a + (1-\lambda)b) \, d\lambda, \qquad (10.18)$$

where

$$p(x,\lambda) = \begin{cases} \lambda, & \lambda \in \left[0, \frac{b-x}{b-a}\right), \\ \lambda - 1, & \lambda \in \left[\frac{b-x}{b-a}, 1\right]. \end{cases} \qquad (10.19)$$

When $n = 1$ the sum $\sum_{k=1}^{n-1}$ is zero.

Note. When $n = 1$ formula (10.18) collapses to (10.2).

We call the remainder of (10.18) as

$$\text{Rem}(10.18) = \frac{a-b}{(n-1)!} \int_0^1 ((x-b) + \lambda(b-a))^{n-1} p(x,\lambda) f^{(n)}(\lambda a + (1-\lambda) b) \, d\lambda.$$
$$(10.20)$$

We have that

$$|\text{Rem}(10.18)| \leq \frac{(b-a)^n}{(n-1)!} \int_0^1 \left| f^{(n)}(\lambda a + (1-\lambda) b) \right| d\lambda. \qquad (10.21)$$

Assume now that $\left| f^{(n)} \right|$ is convex on $[a,b]$ with $f^{(n)}(a), f^{(n)}(b) \in \mathbb{R}$, then

$$\int_0^1 \left| f^{(n)}(\lambda a + (1-\lambda) b) \right| d\lambda \leq \frac{\left| f^{(n)}(a) \right| + \left| f^{(n)}(b) \right|}{2}. \qquad (10.22)$$

So given that $\left| f^{(n)} \right|$ is convex on $[a,b]$ with $f^{(n)}(a), f^{(n)}(b) \in \mathbb{R}$, we derive

$$|\text{Rem}(10.18)| \leq \frac{(b-a)^n}{(n-1)!} \frac{\left(\left| f^{(n)}(a) \right| + \left| f^{(n)}(b) \right| \right)}{2}. \qquad (10.23)$$

We need

Definition 10.1. ([10]) A function $f : \mathbb{R}_+ = [0, \infty) \to \mathbb{R}$ is said to be s-convex in the second sense if

$$f(\lambda x + (1-\lambda) y) \leq \lambda^s f(x) + (1-\lambda)^s f(y), \qquad (10.24)$$

for all $x, y \in [0, \infty)$, $\lambda \in [0,1]$ and for some fixed $s \in (0,1]$.

When $s = 1$, s-convexity in the second sense reduces to ordinary convexity. If "$\geq$" holds in (10.24), we talk about s-concavity in the second sense.

We also need

Definition 10.2. (see also [1])Let I be a subinterval of $\mathbb{R}_+$ and $f : I \to (0, \infty)$. We call f s-logarithmically convex (s-log-convex) in the second sense, iff $\log f(x)$ is s-convex in the second sense, iff

$$f(\lambda x + (1-\lambda) y) \leq (f(x))^{\lambda^s} (f(y))^{(1-\lambda)^s}, \qquad (10.25)$$

for all $x, y \in I$, $\lambda \in [0,1]$ and for some fixed $s \in (0,1]$.

When $s = 1$, s-log-convexity in the second sense reduces to usual log-convexity. If "$\geq$" holds in (10.25), we talk about s-log-concavity in the second sense.

We also need the s-convex Hadamard's inequality

Theorem 10.4. ([7]) *Suppose that $f : [0, \infty) \to [0, \infty)$ is an s-convex function in the second sense, where $s \in (0,1]$ and let $a, b \in [0, \infty)$, $a < b$. If $f \in L_1([a,b])$, then*

$$2^{s-1} f\left(\frac{a+b}{2} \right) \leq \frac{1}{b-a} \int_a^b f(x) \, dx \leq \frac{f(a) + f(b)}{s+1}. \qquad (10.26)$$

The constant $K = \frac{1}{s+1}$ is the best possible in the second inequality (10.26). The above inequalities are sharp.

We mention also the s-convex Ostrowski type inequality

Theorem 10.5. *([2]) Let $f : I \subset [0, \infty) \to \mathbb{R}$ be a differentiable mapping on I^0 such that $f' \in L_1([a, b])$, where $a, b \in I$ with $a < b$. If $|f'|$ is s-convex in the second sense on $[a, b]$ for some fixed $s \in (0, 1]$ and $|f'(x)| \leq M$, $x \in [a, b]$, then it holds*

$$\left| f(x) - \frac{1}{b-a} \int_a^b f(u)\, du \right| \leq \frac{M}{b-a} \left(\frac{(x-a)^2 + (b-x)^2}{s+1} \right), \qquad (10.27)$$

for each $x \in [a, b]$.

We are also motivated by

Theorem 10.6. *([5],Čebyšev 1882) Let $f, g : [a, b] \to \mathbb{R}$ be absolutely continuous functions. If $f', g' \in L_\infty([a, b])$, then*

$$\left| \frac{1}{b-a} \int_a^b f(x) g(x)\, dx - \left(\frac{1}{b-a} \int_a^b f(x)\, dx \right) \left(\frac{1}{b-a} \int_a^b g(x)\, dx \right) \right|$$

$$\leq \frac{1}{12} (b-a)^2 \, \|f'\|_\infty \, \|g'\|_\infty. \qquad (10.28)$$

Next we derive general inequalities similar to (10.28) and (10.27).

10.2 Main Results

We present our first general main result that is about Grüss type inequalities and involves s-convexity and s-concavity in the second sense.

Theorem 10.7. *Let $a, b \in \mathbb{R}$, $a < b$, $f, g : [a, b] \to \mathbb{R}$, $n \in \mathbb{N}$; $f^{(n-1)}$, $g^{(n-1)}$ are absolutely continuous on $[a, b]$, $x \in [a, b]$. Whenever we consider s-convexity or s-concavity of functions we take $a \geq 0$.*
Denote by

$$F_{n-1}^f(x) := \sum_{k=1}^{n-1} \left(\frac{n-k}{k!} \right) \left(\frac{f^{(k-1)}(b)(x-b)^k - f^{(k-1)}(a)(x-a)^k}{b-a} \right), \qquad (10.29)$$

with $F_0^f(x) := 0$, and

$$\Delta_{(f,g)} := \int_a^b f(x) g(x)\, dx - \frac{n}{b-a} \left(\int_a^b f(x)\, dx \right) \left(\int_a^b g(x)\, dx \right)$$

$$- \frac{1}{2} \left[\int_a^b \left(g(x) F_{n-1}^f(x) + f(x) F_{n-1}^g(x) \right) dx \right]. \qquad (10.30)$$

We distinguish the cases:

1) Here $\left|f^{(n)}\right|$, $\left|g^{(n)}\right|$ are s-convex in the second sense and

$$\left|f^{(n)}\left(x\right)\right| \leq M, \quad \left|g^{(n)}\left(x\right)\right| \leq K, \quad x \in [a,b]. \tag{10.31}$$

Then

$$\left|\Delta_{(f,g)}\right| \leq \frac{2\left(b-a\right)^{n+1}}{\left(n-1\right)!\left(s+2\right)\left(s+3\right)}\left(K\left\|f\right\|_{\infty} + M\left\|g\right\|_{\infty}\right). \tag{10.32}$$

2) Let $p_i > 1 : \sum_{i=1}^{4}\frac{1}{p_i} = 1$, and $f^{(n)}$, $g^{(n)} \in L_{p_4}\left([a,b]\right)$. Then

$$\left|\Delta_{(f,g)}\right|$$

$$\leq \frac{\left(b-a\right)^{n+\frac{1}{p_2}+\frac{1}{p_3}}\left[\left\|f\right\|_{p_1}\left\|g^{(n)}\right\|_{p_4} + \left\|g\right\|_{p_1}\left\|f^{(n)}\right\|_{p_4}\right]}{2^{\frac{1}{p_1}+\frac{1}{p_4}}\left(n-1\right)!\left(\left(p_2\left(n-1\right)+1\right)\left(p_2\left(n-1\right)+2\right)\right)^{\frac{1}{p_2}}\left(\left(p_3+1\right)\left(p_3+2\right)\right)^{\frac{1}{p_3}}}. \tag{10.33}$$

3) Let $p_i > 1 : \sum_{i=1}^{4}\frac{1}{p_i} = 1$, with $\left|f^{(n)}\right|$, $\left|g^{(n)}\right|$ are s-convex in the second sense and $\left|f^{(n)}\left(x\right)\right| \leq M$, $\left|g^{(n)}\left(x\right)\right| \leq K$, $x \in [a,b]$. Then

$$\left|\Delta_{(f,g)}\right|$$

$$\leq \frac{2^{\frac{1}{p_2}+\frac{1}{p_3}}\left(b-a\right)^{n+1-\frac{1}{p_1}}\left[\left\|f\right\|_{p_1}K + \left\|g\right\|_{p_1}M\right]}{\left(n-1\right)!\left(\left(p_2\left(n-1\right)+1\right)\left(p_2\left(n-1\right)+2\right)\right)^{\frac{1}{p_2}}\left(\left(p_3+1\right)\left(p_3+2\right)\right)^{\frac{1}{p_3}}\left(p_4 s+1\right)^{\frac{1}{p_4}}}. \tag{10.34}$$

4) Let $p_i > 1 : \sum_{i=1}^{4}\frac{1}{p_i} = 1$, with $\left|f^{(n)}\right|^{p_4}$, $\left|g^{(n)}\right|^{p_4}$ are s-convex in the second sense and $\left|f^{(n)}\left(x\right)\right| \leq M$, $\left|g^{(n)}\left(x\right)\right| \leq K$, $x \in [a,b]$. Then

$$\left|\Delta_{(f,g)}\right|$$

$$\leq \frac{\left(b-a\right)^{n+1-\frac{1}{p_1}}\left[\left\|f\right\|_{p_1}K + \left\|g\right\|_{p_1}M\right]}{2^{\frac{1}{p_1}}\left(n-1\right)!\left(\left(p_2\left(n-1\right)+1\right)\left(p_2\left(n-1\right)+2\right)\right)^{\frac{1}{p_2}}\left(\left(p_3+1\right)\left(p_3+2\right)\right)^{\frac{1}{p_3}}\left(s+1\right)^{\frac{1}{p_4}}}. \tag{10.35}$$

5) Here $p_i > 1 : \sum_{i=1}^{4}\frac{1}{p_i} = 1$. Let $f^{(n)}, g^{(n)} \in L_{p_4}\left([a,b]\right)$, with $\left|f^{(n)}\right|^{p_4}$, $\left|g^{(n)}\right|^{p_4}$ being s-concave in the second sense. Then

$$\left|\Delta_{(f,g)}\right|$$

$$\leq \frac{\left(b-a\right)^{n+1-\frac{1}{p_1}}\left[\left\|f\right\|_{p_1}\left|g^{(n)}\left(\frac{a+b}{2}\right)\right| + \left\|g\right\|_{p_1}\left|f^{(n)}\left(\frac{a+b}{2}\right)\right|\right]}{2^{\left(\frac{1}{p_1}+\frac{2-s}{p_4}\right)}\left(n-1\right)!\left(\left(p_2\left(n-1\right)+1\right)\left(p_2\left(n-1\right)+2\right)\right)^{\frac{1}{p_2}}\left(\left(p_3+1\right)\left(p_3+2\right)\right)^{\frac{1}{p_3}}}. \tag{10.36}$$

Proof. Since $f, g : [a, b] \to \mathbb{R}$, $n \in \mathbb{N}$, with $f^{(n-1)}, g^{(n-1)}$ are absolutely continuous on $[a, b]$, $x \in [a, b]$, by Theorem 10.3 we obtain

$$f(x) = \frac{n}{b-a} \int_a^b f(t)\, dt + F_{n-1}^f(x) + R_n^f(x), \tag{10.37}$$

where

$$F_{n-1}^f(x) := \sum_{k=1}^{n-1} \left(\frac{n-k}{k!} \right) \left(\frac{f^{(k-1)}(b)(x-b)^k - f^{(k-1)}(a)(x-a)^k}{b-a} \right), \tag{10.38}$$

and

$$R_n^f(x) := \frac{a-b}{(n-1)!} \int_0^1 ((x-b) + \lambda(b-a))^{n-1} p(x, \lambda) f^{(n)}(\lambda a + (1-\lambda) b)\, d\lambda$$

$$= \frac{1}{(n-1)!(b-a)} \int_a^b (x-t)^{n-1} q(x,t) f^{(n)}(t)\, dt. \tag{10.39}$$

Similarly we have

$$g(x) = \frac{n}{b-a} \int_a^b g(t)\, dt + F_{n-1}^g(x) + R_n^g(x). \tag{10.40}$$

Then

$$f(x) g(x) = \frac{n}{b-a} g(x) \int_a^b f(t)\, dt + g(x) F_{n-1}^f(x) + g(x) R_n^f(x), \tag{10.41}$$

$$f(x) g(x) = \frac{n}{b-a} f(x) \int_a^b g(t)\, dt + f(x) F_{n-1}^g(x) + f(x) R_n^g(x).$$

Then by integrating we obtain

$$\int_a^b f(x) g(x)\, dx = \frac{n}{b-a} \left(\int_a^b g(x)\, dx \right) \left(\int_a^b f(x)\, dx \right) \tag{10.42}$$

$$+ \int_a^b g(x) F_{n-1}^f(x)\, dx + \int_a^b g(x) R_n^f(x)\, dx,$$

$$\int_a^b f(x) g(x)\, dx = \frac{n}{b-a} \left(\int_a^b f(x)\, dx \right) \left(\int_a^b g(x)\, dx \right)$$

$$+ \int_a^b f(x) F_{n-1}^g(x)\, dx + \int_a^b f(x) R_n^g(x)\, dx.$$

That is

$$\int_a^b f(x) g(x)\, dx - \frac{n}{b-a} \left(\int_a^b f(x)\, dx \right) \left(\int_a^b g(x)\, dx \right)$$

$$= \int_a^b g(x) F_{n-1}^f(x) dx + \int_a^b g(x) R_n^f(x) dx \qquad (10.43)$$

$$= \int_a^b f(x) F_{n-1}^g(x) dx + \int_a^b f(x) R_n^g(x) dx.$$

Adding the last we derive

$$\Delta_{(f,g)} := \int_a^b f(x) g(x) dx - \frac{n}{b-a} \left(\int_a^b f(x) dx \right) \left(\int_a^b g(x) dx \right)$$

$$- \frac{1}{2} \left[\int_a^b \left(g(x) F_{n-1}^f(x) + f(x) F_{n-1}^g(x) \right) dx \right]$$

$$= \frac{1}{2} \left[\int_a^b \left(f(x) R_n^g(x) + g(x) R_n^f(x) \right) dx \right]. \qquad (10.44)$$

Next we estimate $\Delta_{(f,g)}$.

1) Estimate with respect to $\|\cdot\|_\infty$ and s-convexity in the second sense. We have that

$$|\Delta_{(f,g)}| \le \frac{1}{2} \left[\|f\|_\infty \int_a^b |R_n^g(x)| dx + \|g\|_\infty \int_a^b |R_n^f(x)| dx \right]. \qquad (10.45)$$

We notice under the assumption $|f^{(n)}|$ is s-convex in the second sense and $|f^{(n)}(x)| \le M$, $x \in [a, b]$, that

$$|R_n^f(x)| \le \frac{(b-a)^n}{(n-1)!} \int_0^1 |p(x, \lambda)| \left| f^{(n)}(\lambda a + (1-\lambda) b) \right| d\lambda \qquad (10.46)$$

$$\le \frac{(b-a)^n}{(n-1)!} \int_0^1 |p(x, \lambda)| \left(\lambda^s \left| f^{(n)}(a) \right| + (1-\lambda)^s \left| f^{(n)}(b) \right| \right) d\lambda$$

$$\le \frac{M(b-a)^n}{(n-1)!} \int_0^1 |p(x, \lambda)| (\lambda^s + (1-\lambda)^s) d\lambda \qquad (10.47)$$

$$= \frac{M(b-a)^n}{(n-1)!} \left[\int_0^{\frac{b-x}{b-a}} \lambda (\lambda^s + (1-\lambda)^s) d\lambda + \int_{\frac{b-x}{b-a}}^1 (1-\lambda)(\lambda^s + (1-\lambda)^s) d\lambda \right]$$

$$= \frac{M(b-a)^n}{(n-1)!} \left[\int_0^{\frac{b-x}{b-a}} \lambda^{s+1} d\lambda + \int_0^{\frac{b-x}{b-a}} \lambda(1-\lambda)^s d\lambda \right. \qquad (10.48)$$

$$\left. + \int_{\frac{b-x}{b-a}}^1 (\lambda^s - \lambda^{s+1}) d\lambda + \int_{\frac{b-x}{b-a}}^1 (1-\lambda)^{s+1} d\lambda \right]$$

$$= \frac{M\,(b-a)^n}{(n-1)!}\left[\left(\frac{1}{s+2}\left(\frac{b-x}{b-a}\right)^{s+2}\right)\right.$$

$$+\left(\frac{1}{s+2}\left(\frac{x-a}{b-a}\right)^{s+2} - \frac{1}{s+1}\left(\frac{x-a}{b-a}\right)^{s+1} + \frac{1}{(s+1)(s+2)}\right)$$

$$+\left(\frac{1}{s+2}\left(\frac{b-x}{b-a}\right)^{s+2} - \frac{1}{s+1}\left(\frac{b-x}{b-a}\right)^{s+1} + \frac{1}{(s+1)(s+2)}\right)$$

$$\left.+\left(\frac{1}{s+2}\left(\frac{x-a}{b-a}\right)^{s+2}\right)\right]$$

$$= \frac{M\,(b-a)^n}{(n-1)!}\left[\frac{2}{s+2}\left(\frac{b-x}{b-a}\right)^{s+2} + \frac{2}{s+2}\left(\frac{x-a}{b-a}\right)^{s+2}\right.$$

$$\left.-\frac{1}{s+1}\left(\frac{x-a}{b-a}\right)^{s+1} - \frac{1}{s+1}\left(\frac{b-x}{b-a}\right)^{s+1} + \frac{2}{(s+1)(s+2)}\right] \qquad (10.49)$$

$$= \frac{M\,(b-a)^n}{(n-1)!}\left[\frac{2}{s+2}\left(\frac{(b-x)^{s+2}+(x-a)^{s+2}}{(b-a)^{s+2}}\right)\right.$$

$$\left.-\frac{1}{s+1}\left(\frac{(b-x)^{s+1}+(x-a)^{s+1}}{(b-a)^{s+1}}\right) + \frac{2}{(s+1)(s+2)}\right] =: \frac{M\,(b-a)^n}{(n-1)!}\varphi\,(x,s)\,.$$

$$\qquad (10.50)$$

We have proved that

$$\left|R_n^f\,(x)\right| \le \frac{M\,(b-a)^n}{(n-1)!}\varphi\,(x,s)\,, \quad x\in[a,b]\,. \qquad (10.51)$$

Given that $\left|g^{(n)}\right|$ is s-convex in the second sense and $\left|g^{(n)}\,(x)\right| \le K$, $x\in[a,b]$, we similarly get

$$\left|R_n^g\,(x)\right| \le \frac{K\,(b-a)^n}{(n-1)!}\varphi\,(x,s)\,, \quad x\in[a,b]\,. \qquad (10.52)$$

We notice that

$$\int_a^b \varphi\,(x,s)\,dx = \frac{4\,(b-a)}{(s+2)(s+3)}\,. \qquad (10.53)$$

Hence it holds

$$\int_a^b \left|R_n^f\,(x)\right|\,dx \le \frac{4M\,(b-a)^{n+1}}{(n-1)!\,(s+2)(s+3)}\,, \qquad (10.54)$$

and similarly

$$\int_a^b |R_n^g(x)|\, dx \le \frac{4K(b-a)^{n+1}}{(n-1)!(s+2)(s+3)}. \tag{10.55}$$

Therefore we proved that

$$|\Delta_{(f,g)}| \le \frac{1}{2}\left[\|f\|_\infty \frac{4K(b-a)^{n+1}}{(n-1)!(s+2)(s+3)} + \|g\|_\infty \frac{4M(b-a)^{n+1}}{(n-1)!(s+2)(s+3)}\right]. \tag{10.56}$$

Hence we derive

$$|\Delta_{(f,g)}| \le \frac{2(b-a)^{n+1}}{(n-1)!(s+2)(s+3)}\left(K\|f\|_\infty + M\|g\|_\infty\right). \tag{10.57}$$

2) Estimate with respect to $\|\cdot\|_{p_i}$ norms, $i = 1, 2, 3, 4$; $p_i > 1$, $\sum_{i=1}^4 \frac{1}{p_i} = 1$.

We use Hölder's inequality for four functions, and we use the old remainder form, see (10.14).

So here

$$R_n^f(x) = \frac{1}{(n-1)!(b-a)}\int_a^b (x-t)^{n-1} q(x,t) f^{(n)}(t)\, dt, \tag{10.58}$$

where $q(x,t)$ as in (10.15).

We have that

$$\left|\int_a^b g(x) R_n^f(x)\, dx\right| \le \int_a^b |g(x)|\, |R_n^f(x)|\, dx \tag{10.59}$$

$$\le \frac{1}{(n-1)!(b-a)}\int_a^b\int_a^b |g(x)|\,|x-t|^{n-1}\,|q(x,t)|\,\left|f^{(n)}(t)\right|\, dt\, dx$$

$$\le \frac{1}{(n-1)!(b-a)}\left(\int_a^b\int_a^b |g(x)|^{p_1}\, dt\, dx\right)^{\frac{1}{p_1}}\left(\int_a^b\int_a^b |x-t|^{(n-1)p_2}\, dt\, dx\right)^{\frac{1}{p_2}}$$

$$\cdot\left(\int_a^b\int_a^b |q(x,t)|^{p_3}\, dt\, dx\right)^{\frac{1}{p_3}}\left(\int_a^b\int_a^b \left|f^{(n)}(t)\right|^{p_4}\, dt\, dx\right)^{\frac{1}{p_4}} \tag{10.60}$$

$$= \frac{1}{(n-1)!(b-a)^{\frac{1}{p_2}+\frac{1}{p_3}}}\|g\|_{p_1}\left\|f^{(n)}\right\|_{p_4}\left(\int_a^b\left(\int_a^b |x-t|^{(n-1)p_2}\, dt\right) dx\right)^{\frac{1}{p_2}}$$

$$\cdot\left(\int_a^b\left(\int_a^b |q(x,t)|^{p_3}\, dt\right) dx\right)^{\frac{1}{p_3}}$$

$$= \frac{\|g\|_{p_1}\left\|f^{(n)}\right\|_{p_4}}{(n-1)!(b-a)^{\frac{1}{p_2}+\frac{1}{p_3}}}\cdot\frac{2^{\frac{1}{p_2}}(b-a)^{n-1+\frac{2}{p_2}}}{((p_2(n-1)+1)(p_2(n-1)+2))^{\frac{1}{p_2}}} \tag{10.61}$$

$$\cdot \frac{2^{\frac{1}{p_3}} (b-a)^{1+\frac{2}{p_3}}}{((p_3+1)(p_3+2))^{\frac{1}{p_3}}}$$

$$= \frac{2^{\frac{1}{p_2}+\frac{1}{p_3}} (b-a)^{n+\frac{1}{p_2}+\frac{1}{p_3}} \|g\|_{p_1} \|f^{(n)}\|_{p_4}}{(n-1)! ((p_2(n-1)+1)(p_2(n-1)+2))^{\frac{1}{p_2}} ((p_3+1)(p_3+2))^{\frac{1}{p_3}}}. \tag{10.62}$$

We have established that

$$\left| \int_a^b g(x) R_n^f(x) dx \right|$$

$$\leq \frac{2^{\frac{1}{p_2}+\frac{1}{p_3}} (b-a)^{n+\frac{1}{p_2}+\frac{1}{p_3}} \|g\|_{p_1} \|f^{(n)}\|_{p_4}}{(n-1)! ((p_2(n-1)+1)(p_2(n-1)+2))^{\frac{1}{p_2}} ((p_3+1)(p_3+2))^{\frac{1}{p_3}}}. \tag{10.63}$$

Similarly we obtain

$$\left| \int_a^b f(x) R_n^g(x) dx \right|$$

$$\leq \frac{2^{\frac{1}{p_2}+\frac{1}{p_3}} (b-a)^{n+\frac{1}{p_2}+\frac{1}{p_3}} \|f\|_{p_1} \|g^{(n)}\|_{p_4}}{(n-1)! ((p_2(n-1)+1)(p_2(n-1)+2))^{\frac{1}{p_2}} ((p_3+1)(p_3+2))^{\frac{1}{p_3}}}. \tag{10.64}$$

Consequently by (10.44), (10.63), (10.64), we get

$$|\Delta_{(f,g)}|$$

$$\leq \frac{(b-a)^{n+\frac{1}{p_2}+\frac{1}{p_3}} \left[\|f\|_{p_1} \|g^{(n)}\|_{p_4} + \|g\|_{p_1} \|f^{(n)}\|_{p_4} \right]}{2^{\frac{1}{p_1}+\frac{1}{p_4}} (n-1)! ((p_2(n-1)+1)(p_2(n-1)+2))^{\frac{1}{p_2}} ((p_3+1)(p_3+2))^{\frac{1}{p_3}}}. \tag{10.65}$$

3) We notice also that if $|f^{(n)}|$, $|g^{(n)}|$ are s-convex in the second sense, then

$$\|g^{(n)}\|_{p_4}^{p_4} = \int_a^b |g^{(n)}(t)|^{p_4} dt$$

$$= (b-a) \int_0^1 |g^{(n)}(\lambda a + (1-\lambda)b)|^{p_4} d\lambda \tag{10.66}$$

$$\leq (b-a) \int_0^1 \left(\lambda^s |g^{(n)}(a)| + (1-\lambda)^s |g^{(n)}(b)| \right)^{p_4} d\lambda$$

$$\leq K^{p_4} (b-a) \int_0^1 (\lambda^s + (1-\lambda)^s)^{p_4} d\lambda$$

$$\leq 2^{p_4-1} K^{p_4} (b-a) \left[\int_0^1 (\lambda^{p_4 s} + (1-\lambda)^{p_4 s}) d\lambda \right] = \frac{(2K)^{p_4} (b-a)}{(p_4 s + 1)}. \tag{10.67}$$

Hence it holds

$$\left\|g^{(n)}\right\|_{p_4} \le 2K \left(\frac{b-a}{p_4 s + 1}\right)^{\frac{1}{p_4}}. \tag{10.68}$$

Similarly we find

$$\left\|f^{(n)}\right\|_{p_4} \le 2M \left(\frac{b-a}{p_4 s + 1}\right)^{\frac{1}{p_4}}. \tag{10.69}$$

So if $\left|f^{(n)}\right|$, $\left|g^{(n)}\right|$ are s-convex in the second sense and $\left|f^{(n)}(x)\right| \le M$, $\left|g^{(n)}(x)\right| \le K$, $x \in [a, b]$, then by (10.65), (10.68), (10.69), we get

$$\left|\Delta_{(f,g)}\right|$$

$$\le \frac{2^{\frac{1}{p_2}+\frac{1}{p_3}} (b-a)^{n+1-\frac{1}{p_1}} \left[\|f\|_{p_1} K + \|g\|_{p_1} M\right]}{(n-1)! \left((p_2(n-1)+1)(p_2(n-1)+2)\right)^{\frac{1}{p_2}} \left((p_3+1)(p_3+2)\right)^{\frac{1}{p_3}} (p_4 s + 1)^{\frac{1}{p_4}}}. \tag{10.70}$$

4) Next assuming that $\left|f^{(n)}\right|^{p_4}$, $\left|g^{(n)}\right|^{p_4}$ are s-convex in the second sense, then

$$\left\|g^{(n)}\right\|_{p_4}^{p_4} = (b-a) \int_0^1 \left|g^{(n)}(\lambda a + (1-\lambda)b)\right|^{p_4} d\lambda$$

$$\le (b-a) \int_0^1 \left(\lambda^s \left|g^{(n)}(a)\right|^{p_4} + (1-\lambda)^s \left|g^{(n)}(b)\right|^{p_4}\right) d\lambda \tag{10.71}$$

$$\le K^{p_4}(b-a) \int_0^1 \left(\lambda^s + (1-\lambda)^s\right) d\lambda = \frac{2K^{p_4}(b-a)}{s+1}. \tag{10.72}$$

That is

$$\left\|g^{(n)}\right\|_{p_4} \le \frac{2^{\frac{1}{p_4}} K (b-a)^{\frac{1}{p_4}}}{(s+1)^{\frac{1}{p_4}}}. \tag{10.73}$$

Similarly we get

$$\left\|f^{(n)}\right\|_{p_4} \le \frac{2^{\frac{1}{p_4}} M (b-a)^{\frac{1}{p_4}}}{(s+1)^{\frac{1}{p_4}}}. \tag{10.74}$$

So if $\left|f^{(n)}\right|^{p_4}$, $\left|g^{(n)}\right|^{p_4}$ are s-convex in the second sense and $\left|f^{(n)}(x)\right| \le M$, $\left|g^{(n)}(x)\right| \le K$, $x \in [a, b]$, then by (10.65), (10.73), (10.74), we derive

$$\left|\Delta_{(f,g)}\right|$$

$$\le \frac{(b-a)^{n+1-\frac{1}{p_1}} \left[\|f\|_{p_1} K + \|g\|_{p_1} M\right]}{2^{\frac{1}{p_1}} (n-1)! \left((p_2(n-1)+1)(p_2(n-1)+2)\right)^{\frac{1}{p_2}} \left((p_3+1)(p_3+2)\right)^{\frac{1}{p_3}} (s+1)^{\frac{1}{p_4}}}. \tag{10.75}$$

5) Assume finally that $\left|g^{(n)}\right|^{p_4}$, $\left|f^{(n)}\right|^{p_4}$ are s-concave in the second sense, then by (10.26) we get

$$\int_a^b \left|g^{(n)}(t)\right|^{p_4} dt \leq 2^{s-1} \left|g^{(n)}\left(\frac{a+b}{2}\right)\right|^{p_4} (b-a) \tag{10.76}$$

and

$$\int_a^b \left|f^{(n)}(t)\right|^{p_4} dt \leq 2^{s-1} \left|f^{(n)}\left(\frac{a+b}{2}\right)\right|^{p_4} (b-a). \tag{10.77}$$

Therefore

$$\left\|g^{(n)}\right\|_{p_4} \leq 2^{\frac{s-1}{p_4}} \left|g^{(n)}\left(\frac{a+b}{2}\right)\right| (b-a)^{\frac{1}{p_4}}, \tag{10.78}$$

and

$$\left\|f^{(n)}\right\|_{p_4} \leq 2^{\frac{s-1}{p_4}} \left|f^{(n)}\left(\frac{a+b}{2}\right)\right| (b-a)^{\frac{1}{p_4}}. \tag{10.79}$$

Then by using (10.65), (10.78), (10.79), we get

$$\left|\Delta_{(f,g)}\right|$$

$$\leq \frac{(b-a)^{n+1-\frac{1}{p_1}} \left[\|f\|_{p_1} \left|g^{(n)}\left(\frac{a+b}{2}\right)\right| + \|g\|_{p_1} \left|f^{(n)}\left(\frac{a+b}{2}\right)\right|\right]}{2^{\frac{1}{p_1}+\frac{2-s}{p_4}} (n-1)! \left((p_2(n-1)+1)(p_2(n-1)+2)\right)^{\frac{1}{p_2}} \left((p_3+1)(p_3+2)\right)^{\frac{1}{p_3}}}. \tag{10.80}$$

The proof of the theorem is complete. $\qquad\square$

We apply Theorem 10.7 for $n = 1$.

Proposition 10.2. *Let* $a, b \in \mathbb{R}$, $a < b$, $f, g : [a, b] \to \mathbb{R}$; f, g *are absolutely continuous on* $[a, b]$, $x \in [a, b]$. *Whenever we consider* s-*convexity or* s-*concavity of functions we take* $a \geq 0$.

Denote by

$$\Delta^*_{(f,g)} := \int_a^b f(x)g(x)\,dx - \frac{1}{b-a}\left(\int_a^b f(x)\,dx\right)\left(\int_a^b g(x)\,dx\right). \tag{10.81}$$

We distinguish the cases:

1) Here $|f'|$, $|g'|$ *are* s-*convex in the second sense and*

$$|f'(x)| \leq M, \quad |g'(x)| \leq K, \quad x \in [a, b]. \tag{10.82}$$

Then

$$\left|\Delta^*_{(f,g)}\right| \leq \frac{2(b-a)^2}{(s+2)(s+3)}\left(K\|f\|_\infty + M\|g\|_\infty\right). \tag{10.83}$$

2) Let $p_i > 1 : \sum_{i=1}^4 \frac{1}{p_i} = 1$, *and* f', $g' \in L_{p_4}([a,b])$. *Then*

$$\left|\Delta^*_{(f,g)}\right| \leq \frac{(b-a)^{1+\frac{1}{p_2}+\frac{1}{p_3}} \left[\|f\|_{p_1}\|g'\|_{p_4} + \|g\|_{p_1}\|f'\|_{p_4}\right]}{2^{\frac{1}{p_1}+\frac{1}{p_2}+\frac{1}{p_4}} \left((p_3+1)(p_3+2)\right)^{\frac{1}{p_3}}}. \tag{10.84}$$

3) Let $p_i > 1 : \sum_{i=1}^{4} \frac{1}{p_i} = 1$, with $|f'|$, $|g'|$ are s-convex in the second sense and $|f'(x)| \leq M$, $|g'(x)| \leq K$, $x \in [a,b]$. Then

$$\left| \Delta^*_{(f,g)} \right| \leq \frac{2^{\frac{1}{p_3}} (b-a)^{2-\frac{1}{p_1}} \left[\|f\|_{p_1} K + \|g\|_{p_1} M \right]}{((p_3+1)(p_3+2))^{\frac{1}{p_3}} (p_4 s + 1)^{\frac{1}{p_4}}}. \tag{10.85}$$

4) Let $p_i > 1 : \sum_{i=1}^{4} \frac{1}{p_i} = 1$, with $|f'|^{p_4}$, $|g'|^{p_4}$ are s-convex in the second sense and $|f'(x)| \leq M$, $|g'(x)| \leq K$, $x \in [a,b]$. Then

$$\left| \Delta^*_{(f,g)} \right| \leq \frac{(b-a)^{2-\frac{1}{p_1}} \left[\|f\|_{p_1} K + \|g\|_{p_1} M \right]}{2^{\frac{1}{p_1}+\frac{1}{p_2}} ((p_3+1)(p_3+2))^{\frac{1}{p_3}} (s+1)^{\frac{1}{p_4}}}. \tag{10.86}$$

5) Here $p_i > 1 : \sum_{i=1}^{4} \frac{1}{p_i} = 1$. Let $f', g' \in L_{p_4}([a,b])$, with $|f'|^{p_4}$, $|g'|^{p_4}$ being s-concave in the second sense. Then

$$\left| \Delta^*_{(f,g)} \right| \leq \frac{(b-a)^{2-\frac{1}{p_1}} \left[\|f\|_{p_1} \left| g' \left(\frac{a+b}{2} \right) \right| + \|g\|_{p_1} \left| f' \left(\frac{a+b}{2} \right) \right| \right]}{2^{\left(\frac{1}{p_1}+\frac{1}{p_2}+\frac{2-s}{p_4} \right)} ((p_3+1)(p_3+2))^{\frac{1}{p_3}}}. \tag{10.87}$$

We continue with

Theorem 10.8. *Let $0 \leq a < b$, $f, g : [a,b] \to \mathbb{R}$, $n \in \mathbb{N}$; $f^{(n-1)}, g^{(n-1)}$ are absolutely continuous on $[a,b]$, $x \in [a,b]$. We assume that $|f^{(n)}|, |g^{(n)}|$ are s-logarithmically convex in the second sense, and $|f^{(n)}(a)|, |f^{(n)}(b)|, |g^{(n)}(a)|, |g^{(n)}(b)| \in (0,1]$. Call $A := \frac{|f^{(n)}(a)|}{|f^{(n)}(b)|}$, $B := \frac{|g^{(n)}(a)|}{|g^{(n)}(b)|}$, $s \in (0,1]$, and*

$$\psi_s(z) := \begin{cases} \frac{z^s - 1}{s \ln z}, & \text{if } z \in (0, \infty) - \{1\}, \\ 1, & \text{if } z = 1. \end{cases} \tag{10.88}$$

Denote by

$$F^f_{n-1}(x) := \sum_{k=1}^{n-1} \left(\frac{n-k}{k!} \right) \left(\frac{f^{(k-1)}(b)(x-b)^k - f^{(k-1)}(a)(x-a)^k}{b-a} \right), \tag{10.89}$$

with $F^f_0(x) := 0$, and

$$\Delta_{(f,g)} := \int_a^b f(x) g(x) \, dx - \frac{n}{b-a} \left(\int_a^b f(x) \, dx \right) \left(\int_a^b g(x) \, dx \right)$$

$$- \frac{1}{2} \left[\int_a^b \left(g(x) F^f_{n-1}(x) + f(x) F^g_{n-1}(x) \right) dx \right]. \tag{10.90}$$

We distinguish the cases:
1) Estimate with respect to $\|\cdot\|_\infty$. It holds

$$\left| \Delta_{(f,g)} \right| \leq \frac{(b-a)^{n+1}}{2(n-1)!} \left[\|f\|_\infty \left| g^{(n)}(b) \right|^s \psi_s(B) + \|g\|_\infty \left| f^{(n)}(b) \right|^s \psi_s(A) \right]. \tag{10.91}$$

2) Let $p_i > 1 : \sum_{i=1}^{4} \frac{1}{p_i} = 1$. Then

$$\left| \Delta_{(f,g)} \right|$$

$$\leq \frac{(b-a)^{n+1-\frac{1}{p_1}} \left[\|f\|_{p_1} \left| g^{(n)}(b) \right|^s (\psi_s (B^{p_4}))^{\frac{1}{p_4}} + \|g\|_{p_1} \left| f^{(n)}(b) \right|^s (\psi_s (A^{p_4}))^{\frac{1}{p_4}} \right]}{2^{\frac{1}{p_1}+\frac{1}{p_4}} (n-1)! ((p_2 (n-1)+1)(p_2 (n-1)+2))^{\frac{1}{p_2}} ((p_3 + 1)(p_3 + 2))^{\frac{1}{p_3}}}.$$

$$(10.92)$$

Proof. 1) See also [1]. Here we assume that $0 < \left| f^{(n)}(a) \right|, \left| f^{(n)}(b) \right| \leq 1$, set $A := \frac{\left| f^{(n)}(a) \right|}{\left| f^{(n)}(b) \right|}$, $\left| f^{(n)} \right|$ is s-logarithmically convex in the second sense, $s \in (0,1]$, $\lambda \in [0,1]$.

We observe that

$$\left| R_n^f (x) \right| \overset{\text{(by (10.46), (10.11))}}{\leq} \frac{(b-a)^n}{(n-1)!} \int_0^1 \left| f^{(n)} (\lambda a + (1-\lambda) b) \right| d\lambda \qquad (10.93)$$

$$\leq \frac{(b-a)^n}{(n-1)!} \int_0^1 \left| f^{(n)}(a) \right|^{\lambda^s} \left| f^{(n)}(b) \right|^{(1-\lambda)^s} d\lambda$$

$$\leq \frac{(b-a)^n}{(n-1)!} \int_0^1 \left| f^{(n)}(a) \right|^{\lambda s} \left| f^{(n)}(b) \right|^{(1-\lambda)s} d\lambda$$

$$= \frac{(b-a)^n \left| f^{(n)}(b) \right|^s}{(n-1)!} \int_0^1 \left(\frac{\left| f^{(n)}(a) \right|}{\left| f^{(n)}(b) \right|} \right)^{\lambda s} d\lambda \qquad (10.94)$$

$$= \frac{(b-a)^n \left| f^{(n)}(b) \right|^s}{(n-1)!} \int_0^1 A^{\lambda s} d\lambda =: (*).$$

If $A = 1$, then

$$(*) = \frac{(b-a)^n \left| f^{(n)}(b) \right|^s}{(n-1)!}. \qquad (10.95)$$

If $A \neq 1$ we get

$$(*) = \frac{(b-a)^n \left| f^{(n)}(b) \right|^s}{(n-1)!} \int_0^1 e^{s(\ln A)\lambda} d\lambda = \frac{(b-a)^n \left| f^{(n)}(b) \right|^s}{(n-1)!} \frac{1}{s \ln A} \left(A^{s\lambda} \big|_0^1 \right)$$

$$= \frac{(b-a)^n \left| f^{(n)}(b) \right|^s}{(n-1)!} \left(\frac{A^s - 1}{s \ln A} \right). \qquad (10.96)$$

Therefore we derive that

$$\left| R_n^f (x) \right| \leq \frac{(b-a)^n \left| f^{(n)}(b) \right|^s}{(n-1)!} \psi_s (A). \qquad (10.97)$$

We further assume $0 < \left|g^{(n)}(a)\right|, \left|g^{(n)}(b)\right| \leq 1$, set $B := \frac{\left|g^{(n)}(a)\right|}{\left|g^{(n)}(b)\right|}$, $\left|g^{(n)}\right|$ is s-log-convex in the second sense. Then, similarly, we obtain that

$$\left|R_n^g(x)\right| \leq \frac{(b-a)^n \left|g^{(n)}(b)\right|^s}{(n-1)!} \psi_s(B). \tag{10.98}$$

Notice that

$$\int_a^b \left|R_n^f(x)\right| dx \leq \frac{(b-a)^{n+1} \left|f^{(n)}(b)\right|^s}{(n-1)!} \psi_s(A) \tag{10.99}$$

and

$$\int_a^b \left|R_n^g(x)\right| dx \leq \frac{(b-a)^{n+1} \left|g^{(n)}(b)\right|^s}{(n-1)!} \psi_s(B). \tag{10.100}$$

It is clear now

$$\left|\Delta_{(f,g)}\right| \leq \frac{1}{2}\left[\|f\|_\infty \left|g^{(n)}(b)\right|^s \psi_s(B) + \|g\|_\infty \left|f^{(n)}(b)\right|^s \psi_s(A)\right] \frac{(b-a)^{n+1}}{(n-1)!}. \tag{10.101}$$

2) Here we assume that $0 < \left|f^{(n)}(a)\right|, \left|f^{(n)}(b)\right|, \left|g^{(n)}(a)\right|, \left|g^{(n)}(b)\right| \leq 1$, set $C := \left(\frac{\left|f^{(n)}(a)\right|}{\left|f^{(n)}(b)\right|}\right)^{p_4}$ and $D := \left(\frac{\left|g^{(n)}(a)\right|}{\left|g^{(n)}(b)\right|}\right)^{p_4}$; $\left|f^{(n)}\right|, \left|g^{(n)}\right|$ are s-log-convex in the second sense, $s \in (0,1], \lambda \in [0,1]$.

We have as in (10.66) that

$$\left\|g^{(n)}\right\|_{p_4}^{p_4} = (b-a)\int_0^1 \left|g^{(n)}(\lambda a + (1-\lambda)b)\right|^{p_4} d\lambda \tag{10.102}$$

$$\leq (b-a)\int_0^1 \left(\left|g^{(n)}(a)\right|^{\lambda^s}\right)^{p_4} \left(\left|g^{(n)}(b)\right|^{(1-\lambda)^s}\right)^{p_4} d\lambda$$

$$= (b-a)\int_0^1 \left(\left|g^{(n)}(a)\right|^{p_4}\right)^{\lambda^s} \left(\left|g^{(n)}(b)\right|^{p_4}\right)^{(1-\lambda)^s} d\lambda$$

$$\leq (b-a)\int_0^1 \left(\left|g^{(n)}(a)\right|^{p_4}\right)^{\lambda s} \left(\left|g^{(n)}(b)\right|^{p_4}\right)^{(1-\lambda)s} d\lambda$$

$$= (b-a)\left|g^{(n)}(b)\right|^{p_4 s} \int_0^1 \left(\left(\frac{\left|g^{(n)}(a)\right|}{\left|g^{(n)}(b)\right|}\right)^{p_4}\right)^{s\lambda} d\lambda$$

$$= (b-a)\left|g^{(n)}(b)\right|^{p_4 s} \int_0^1 D^{s\lambda} d\lambda. \tag{10.103}$$

Hence

$$\left\|g^{(n)}\right\|_{p_4}^{p_4} \leq (b-a)\left|g^{(n)}(b)\right|^{p_4 s} \psi_s(D), \tag{10.104}$$

and

$$\left\| g^{(n)} \right\|_{p_4} \le (b-a)^{\frac{1}{p_4}} \left| g^{(n)}(b) \right|^s (\psi_s(D))^{\frac{1}{p_4}}. \tag{10.105}$$

Similarly we obtain

$$\left\| f^{(n)} \right\|_{p_4} \le (b-a)^{\frac{1}{p_4}} \left| f^{(n)}(b) \right|^s (\psi_s(C))^{\frac{1}{p_4}}. \tag{10.106}$$

Finally using (10.33) we find

$$\left| \Delta_{(f,g)} \right|$$

$$\le \frac{(b-a)^{n+1-\frac{1}{p_1}} \left[\|f\|_{p_1} \left| g^{(n)}(b) \right|^s (\psi_s(D))^{\frac{1}{p_4}} + \|g\|_{p_1} \left| f^{(n)}(b) \right|^s (\psi_s(C))^{\frac{1}{p_4}} \right]}{2^{\frac{1}{p_1}+\frac{1}{p_4}} (n-1)! \left((p_2(n-1)+1)(p_2(n-1)+2) \right)^{\frac{1}{p_2}} \left((p_3+1)(p_3+2) \right)^{\frac{1}{p_3}}}. \tag{10.107}$$

The proof of the theorem now is complete. $\qquad\square$

We apply Theorem 10.8 for $n = 1$.

Proposition 10.3. *Let* $0 \le a < b$, $f, g : [a, b] \to \mathbb{R}$ *are absolutely continuous on* $[a, b]$, $x \in [a, b]$. *Assume that* $|f'|, |g'|$ *are s-log-convex in the second sense, and* $|f'(a)|, |f'(b)|, |g'(a)|, |g'(b)| \in (0, 1]$. *Call* $A^* := \frac{|f'(a)|}{|f'(b)|}$, $B^* := \frac{|g'(a)|}{|g'(b)|}$, $s \in (0, 1]$, *and*

$$\psi_s(z) := \begin{cases} \frac{z^s - 1}{s \ln z}, & \text{if } z \in (0, \infty) - \{1\}, \\ 1, & \text{if } z = 1. \end{cases} \tag{10.108}$$

Denote by

$$\Delta^*_{(f,g)} := \int_a^b f(x)g(x)\,dx - \frac{1}{b-a} \left(\int_a^b f(x)\,dx \right) \left(\int_a^b g(x)\,dx \right). \tag{10.109}$$

We distinguish the cases:
1) It holds

$$\left| \Delta^*_{(f,g)} \right| \le \frac{(b-a)^2}{2} \left[\|f\|_\infty \left| g'(b) \right|^s \psi_s(B^*) + \|g\|_\infty \left| f'(b) \right|^s \psi_s(A^*) \right]. \tag{10.110}$$

2) Let $p_i > 1 : \sum_{i=1}^4 \frac{1}{p_i} = 1$. *Then*

$$\left| \Delta^*_{(f,g)} \right|$$

$$\le \frac{(b-a)^{2-\frac{1}{p_1}} \left[\|f\|_{p_1} \left| g'(b) \right|^s (\psi_s(B^{*p_4}))^{\frac{1}{p_4}} + \|g\|_{p_1} \left| f'(b) \right|^s (\psi_s(A^{*p_4}))^{\frac{1}{p_4}} \right]}{2^{1-\frac{1}{p_3}} \left((p_3+1)(p_3+2) \right)^{\frac{1}{p_3}}}. \tag{10.111}$$

Next we present s-convexity general Ostrowski type inequalities.

Theorem 10.9. *Let $a, b \in \mathbb{R}$, $a < b$, $f : [a, b] \to \mathbb{R}$, $n \in \mathbb{N}$; $f^{(n-1)}$ is absolutely continuous on $[a, b]$, $x \in [a, b]$. When we consider s-convexity or s-concavity of functions we take $a \geq 0$.*
Denote by

$$F_{n-1}^f(x) := \sum_{k=1}^{n-1} \left(\frac{n-k}{k!} \right) \left(\frac{f^{(k-1)}(b)(x-b)^k - f^{(k-1)}(a)(x-a)^k}{b-a} \right), \quad (10.112)$$

with $F_0^f(x) := 0$, and

$$R_n^f(x) := f(x) - \frac{n}{b-a} \int_a^b f(t) \, dt - F_{n-1}^f(x). \quad (10.113)$$

We distinguish the cases:
1) Here $\left| f^{(n)} \right|$ is s-convex in the second sense and $\left| f^{(n)}(x) \right| \leq M$, $x \in [a, b]$. Then

$$\left| R_n^f(x) \right| \leq \frac{M(b-a)^n}{(n-1)!} \left[\frac{2}{(s+2)} \left(\frac{(b-x)^{s+2} + (x-a)^{s+2}}{(b-a)^{s+2}} \right) \right. \quad (10.114)$$

$$\left. - \frac{1}{(s+1)} \left(\frac{(b-x)^{s+1} + (x-a)^{s+1}}{(b-a)^{s+1}} \right) + \frac{2}{(s+1)(s+2)} \right].$$

2) Let $p_i > 1 : \sum_{i=1}^3 \frac{1}{p_i} = 1$, and $f^{(n)} \in L_{p_3}([a, b])$. Then

$$\left| R_n^f(x) \right| \leq \frac{\left\| f^{(n)} \right\|_{p_3}}{(n-1)!(b-a)} \left(\frac{(b-x)^{p_1(n-1)+1} + (x-a)^{p_1(n-1)+1}}{p_1(n-1)+1} \right)^{\frac{1}{p_1}}$$

$$\cdot \left(\frac{(b-x)^{p_2+1} + (x-a)^{p_2+1}}{p_2+1} \right)^{\frac{1}{p_2}}. \quad (10.115)$$

3) Let $p_i > 1 : \sum_{i=1}^3 \frac{1}{p_i} = 1$. If $\left| f^{(n)} \right|$ is s-convex in the second sense and $\left| f^{(n)}(x) \right| \leq M$, $x \in [a, b]$, then

$$\left| R_n^f(x) \right| \leq \frac{2M(b-a)^{\frac{1}{p_3}-1}}{(p_3 s+1)^{\frac{1}{p_3}}(n-1)!} \quad (10.116)$$

$$\cdot \left(\frac{(b-x)^{p_1(n-1)+1} + (x-a)^{p_1(n-1)+1}}{p_1(n-1)+1} \right)^{\frac{1}{p_1}} \left(\frac{(b-x)^{p_2+1} + (x-a)^{p_2+1}}{p_2+1} \right)^{\frac{1}{p_2}}.$$

4) Let $p_i > 1 : \sum_{i=1}^3 \frac{1}{p_i} = 1$. If $\left| f^{(n)} \right|^{p_3}$ is s-convex in the second sense and $\left| f^{(n)}(x) \right| \leq M$, $x \in [a, b]$, then

$$\left| R_n^f(x) \right| \leq \frac{2^{\frac{1}{p_3}} M(b-a)^{\frac{1}{p_3}-1}}{(s+1)^{\frac{1}{p_3}}(n-1)!} \quad (10.117)$$

$$\cdot \left(\frac{(b-x)^{p_1(n-1)+1} + (x-a)^{p_1(n-1)+1}}{p_1(n-1)+1} \right)^{\frac{1}{p_1}} \left(\frac{(b-x)^{p_2+1} + (x-a)^{p_2+1}}{p_2+1} \right)^{\frac{1}{p_2}}.$$

5) Let $p_i > 1 : \sum_{i=1}^{3} \frac{1}{p_i} = 1$, $f^{(n)} \in L_{p_3}([a,b])$, with $|f^{(n)}|^{p_3}$ being s-concave in the second sense. Then

$$\left| R_n^f(x) \right| \le \frac{2^{\frac{s-1}{p_3}} \left| f^{(n)} \left(\frac{a+b}{2} \right) \right| (b-a)^{\frac{1}{p_3}-1}}{(n-1)!} \tag{10.118}$$

$$\cdot \left(\frac{(b-x)^{p_1(n-1)+1} + (x-a)^{p_1(n-1)+1}}{p_1(n-1)+1} \right)^{\frac{1}{p_1}} \left(\frac{(b-x)^{p_2+1} + (x-a)^{p_2+1}}{p_2+1} \right)^{\frac{1}{p_2}}.$$

Proof. 1) As in (10.50), (10.51), we get

$$\left| R_n^f(x) \right| \le \frac{M(b-a)^n}{(n-1)!} \varphi(x,s), \quad x \in [a,b]. \tag{10.119}$$

2) Let $p_i > 1, \sum_{i=1}^{3} \frac{1}{p_i} = 1$. Then

$$\left| R_n^f(x) \right| \le \frac{1}{(n-1)!(b-a)} \int_a^b |x-t|^{n-1} |q(x,t)| \left| f^{(n)}(t) \right| dt$$

$$\le \frac{1}{(n-1)!(b-a)} \left(\int_a^b |x-t|^{p_1(n-1)} dt \right)^{\frac{1}{p_1}} \left(\int_a^b |q(x,t)|^{p_2} dt \right)^{\frac{1}{p_2}} \left\| f^{(n)} \right\|_{p_3} \tag{10.120}$$

$$= \frac{\left\| f^{(n)} \right\|_{p_3}}{(n-1)!(b-a)} \left(\frac{(b-x)^{p_1(n-1)+1} + (x-a)^{p_1(n-1)+1}}{p_1(n-1)+1} \right)^{\frac{1}{p_1}}$$

$$\cdot \left(\frac{(b-x)^{p_2+1} + (x-a)^{p_2+1}}{p_2+1} \right)^{\frac{1}{p_2}}. \tag{10.121}$$

3) If $\left| f^{(n)} \right|$ is s-convex in the second sense and $\left| f^{(n)}(x) \right| \le M$, $x \in [a,b]$, we get

$$\left\| f^{(n)} \right\|_{p_3} \le 2M \left(\frac{b-a}{p_3 s + 1} \right)^{\frac{1}{p_3}}, \tag{10.122}$$

see also (10.69).

4) If $\left| f^{(n)} \right|^{p_3}$ is s-convex in the second sense and $\left| f^{(n)}(x) \right| \le M$, $x \in [a,b]$, we get

$$\left\| f^{(n)} \right\|_{p_3} \le \frac{2^{\frac{1}{p_3}} M (b-a)^{\frac{1}{p_3}}}{(s+1)^{\frac{1}{p_3}}}, \tag{10.123}$$

see also (10.74).

5) If $\left| f^{(n)} \right|^{p_3}$ is s-concave in the second sense, then by (10.26) we find

$$\left\| f^{(n)} \right\|_{p_3} \le 2^{\frac{s-1}{p_3}} \left| f^{(n)} \left(\frac{a+b}{2} \right) \right| (b-a)^{\frac{1}{p_3}}, \tag{10.124}$$

see also (10.79). $\qquad \square$

We apply Theorem 10.9 for $n = 1$.

Proposition 10.4. *Let* $a, b \in \mathbb{R}$, $a < b$, $f : [a,b] \to \mathbb{R}$ *is absolutely continuous on* $[a,b]$, $x \in [a,b]$. *In case of s-convexity or s-concavity we take* $a \geq 0$.
Denote by

$$R_1^f(x) = f(x) - \frac{1}{b-a} \int_a^b f(t)\, dt. \tag{10.125}$$

We distinguish the cases:
1) Here $|f'|$ *is s-convex in the second sense and* $|f'(x)| \leq M$, $x \in [a,b]$. *Then*

$$\left| R_1^f(x) \right| \leq M(b-a) \left[\frac{2}{(s+2)} \left(\frac{(b-x)^{s+2} + (x-a)^{s+2}}{(b-a)^{s+2}} \right) \right. \tag{10.126}$$

$$\left. - \frac{1}{(s+1)} \left(\frac{(b-x)^{s+1} + (x-a)^{s+1}}{(b-a)^{s+1}} \right) + \frac{2}{(s+1)(s+2)} \right].$$

2) Let $p_i > 1 : \sum_{i=1}^{3} \frac{1}{p_i} = 1$, *and* $f' \in L_{p_3}([a,b])$. *Then*

$$\left| R_1^f(x) \right| \leq \|f'\|_{p_3} (b-a)^{\frac{1}{p_1}-1} \left(\frac{(b-x)^{p_2+1} + (x-a)^{p_2+1}}{p_2+1} \right)^{\frac{1}{p_2}}. \tag{10.127}$$

3) Let $p_i > 1 : \sum_{i=1}^{3} \frac{1}{p_i} = 1$. *If* $|f'|$ *is s-convex in the second sense and* $|f'(x)| \leq M$, $x \in [a,b]$, *then*

$$\left| R_1^f(x) \right| \leq \frac{2M}{(p_3 s + 1)^{\frac{1}{p_3}} (b-a)^{\frac{1}{p_2}}} \left(\frac{(b-x)^{p_2+1} + (x-a)^{p_2+1}}{p_2+1} \right)^{\frac{1}{p_2}}. \tag{10.128}$$

4) Let $p_i > 1 : \sum_{i=1}^{3} \frac{1}{p_i} = 1$. *If* $|f'|^{p_3}$ *is s-convex in the second sense and* $|f'(x)| \leq M$, $x \in [a,b]$, *then*

$$\left| R_1^f(x) \right| \leq \frac{2^{\frac{1}{p_3}} M}{(s+1)^{\frac{1}{p_3}} (b-a)^{\frac{1}{p_2}}} \left(\frac{(b-x)^{p_2+1} + (x-a)^{p_2+1}}{p_2+1} \right)^{\frac{1}{p_2}}. \tag{10.129}$$

5) Let $p_i > 1 : \sum_{i=1}^{3} \frac{1}{p_i} = 1$, $f' \in L_{p_3}([a,b])$, *with* $|f'|^{p_3}$ *being s-concave in the second sense. Then*

$$\left| R_1^f(x) \right| \leq \frac{2^{\frac{s-1}{p_3}} \left| f'\left(\frac{a+b}{2}\right) \right|}{(b-a)^{\frac{1}{p_2}}} \left(\frac{(b-x)^{p_2+1} + (x-a)^{p_2+1}}{p_2+1} \right)^{\frac{1}{p_2}}. \tag{10.130}$$

We continue with

Theorem 10.10. *Let* $0 \leq a < b$, $f : [a,b] \to \mathbb{R}$, $n \in \mathbb{N}$; $f^{(n-1)}$ *is absolutely continuous on* $[a,b]$, $x \in [a,b]$. *We assume that* $\left| f^{(n)} \right|$ *is s-log-convex in the second sense, and* $\left| f^{(n)}(a) \right|, \left| f^{(n)}(b) \right| \in (0,1]$. *Call* $A := \frac{\left| f^{(n)}(a) \right|}{\left| f^{(n)}(b) \right|}$, $s \in (0,1]$, *and*

$$\psi_s(z) := \begin{cases} \frac{z^s - 1}{s \ln z}, & \text{if } z \in (0, \infty) - \{1\}, \\ 1, & \text{if } z = 1. \end{cases} \tag{10.131}$$

Denote by

$$F_{n-1}^f(x) := \sum_{k=1}^{n-1} \left(\frac{n-k}{k!} \right) \left(\frac{f^{(k-1)}(b)(x-b)^k - f^{(k-1)}(a)(x-a)^k}{b-a} \right), \quad (10.132)$$

with $F_0^f(x) := 0$, *and*

$$R_n^f(x) := f(x) - \frac{n}{b-a} \int_a^b f(t)\,dt - F_{n-1}^f(x). \quad (10.133)$$

We distinguish the cases:
1) It holds generally

$$\left| R_n^f(x) \right| \leq \frac{(b-a)^n \left| f^{(n)}(b) \right|^s}{(n-1)!} \psi_s(A). \quad (10.134)$$

2) Let $p_i > 1 : \sum_{i=1}^3 \frac{1}{p_i} = 1$. *Then*

$$\left| R_n^f(x) \right| \leq \frac{(b-a)^{\frac{1}{p_3}-1}}{(n-1)!} \left| f^{(n)}(b) \right|^s (\psi_s(A^{p_3}))^{\frac{1}{p_3}} \quad (10.135)$$

$$\cdot \left(\frac{(b-x)^{p_1(n-1)+1} + (x-a)^{p_1(n-1)+1}}{p_1(n-1)+1} \right)^{\frac{1}{p_1}} \left(\frac{(b-x)^{p_2+1} + (x-a)^{p_2+1}}{p_2+1} \right)^{\frac{1}{p_2}}.$$

Proof. 1) As in (10.97) we get

$$\left| R_n^f(x) \right| \leq \frac{(b-a)^n \left| f^{(n)}(b) \right|^s}{(n-1)!} \psi_s(A). \quad (10.136)$$

2) As in (10.106) we obtain

$$\left\| f^{(n)} \right\|_{p_3} \leq (b-a)^{\frac{1}{p_3}} \left| f^{(n)}(b) \right|^s (\psi_s(A^{p_3}))^{\frac{1}{p_3}}. \quad (10.137)$$

Then we use (10.115). $\qquad\qquad\qquad\qquad\qquad\qquad\qquad\qquad\qquad\qquad \square$

At last we apply Theorem 10.10 for $n = 1$.

Proposition 10.5. *Let* $0 \leq a < b$, $f : [a,b] \to \mathbb{R}$ *absolutely continuous on* $[a,b]$, $x \in [a,b]$. *We assume that* $|f'|$ *is s-log-convex in the second sense, and* $|f'(a)|, |f'(b)| \in (0,1]$. *Call* $A^* := \frac{|f'(a)|}{|f'(b)|}$, $s \in (0,1]$, *and* $\psi_s(z)$ *as in (10.131).*
Denote by

$$R_1^f(x) := f(x) - \frac{1}{b-a} \int_a^b f(t)\,dt. \quad (10.138)$$

We distinguish the cases:
1) It holds

$$\left| R_1^f(x) \right| \leq (b-a) \left| f'(b) \right|^s \psi_s(A^*). \quad (10.139)$$

2) Let $p_i > 1 : \sum_{i=1}^3 \frac{1}{p_i} = 1$. *Then*

$$\left| R_1^f(x) \right| \leq (b-a)^{-\frac{1}{p_2}} \left| f'(b) \right|^s (\psi_s(A^{*p_3}))^{\frac{1}{p_3}} \left(\frac{(b-x)^{p_2+1} + (x-a)^{p_2+1}}{p_2+1} \right)^{\frac{1}{p_2}}.$$

$$(10.140)$$

Bibliography

1. A.O. Akdemir, M. Tunc, *On some integral inequalities for s-logarithmically convex functions and their applications*, arXiv: 1212.1584v1[math.FA] 7 Dec 2012.
2. M. Alomari, M. Darus, S.S. Dragomir, P. Cerone, *Ostrowski type inequalities for functions whose derivatives are s-convex in the second sense*, Appl. Math. Lett. 23 (2010), 1071-1076.
3. G. Anastassiou, *Advanced Inequalities*, World Scientific, Singapore, 2011.
4. G. Anastassiou, *General Grüss and Ostrowski type inequalities involving s-convexity*, Bull. Allahabad Math. Soc. 28 (Part 1) (2013), 101-129.
5. P.L. Chebyshev, *Sur les expressions approximatives des integrales definies par les autres prises entre les mêmes limites*, Proc. Math. Soc. Charkov 2 (1882), 93-98.
6. S.S. Dragomir, N.S. Barnett, *An Ostrowski type inequality for mappings whose second derivatives are bounded and applications*, RGMIA Research Report Collection 1 (2) (1998), article no. 9, pp. 69-77, http://ajmaa.org/RGMIA/v1n2.php.
7. S.S. Dragomir, S. Fitzpatrick, *The Hadamard's inequality for s-convex functions in the second sense*, Demonstratio Math. 32 (4) (1999), 687-696.
8. A.M. Fink, *Bounds on the deviation of a function from its averages*, Czechoslovak Math. J. 42 (117) (2) (1992), 289-310.
9. G. Grüss, *Über das maximum des absoluten Betrages von* $\left[\left(\frac{1}{b-a}\right)\int_a^b f(x)g(x)\,dx - \left(\frac{1}{(b-a)^2}\int_a^b f(x)\,dx\int_a^b g(x)\,dx\right)\right]$, Math. Z. 39 (1935), 215-226.
10. H. Hudzik, L. Maligranda, *Some remarks on s-convex functions*, Aequationes Math. 48 (1994), 100-111.

Chapter 11

Essential and *s*-Convexity Ostrowski and Grüss Inequalities Using Several Functions

Using the harmonic polynomials representation formula due to Dedic, Pečaric and Ujevic [6], we establish Ostrowski and Grüss type inequalities involving several functions. The estimates are with respect all norms $\|\cdot\|_p$, $1 \leq p \leq \infty$, and also take into account the *s*-convexity and *s*-concavity in the second sense of the involved functions. It follows [3].

11.1 Introduction

The problem of estimating the difference of a value of a function from its average is a paramount one. The answer to it are the Ostrowski type inequalities. Ostrowski type inequalities are very useful among others in Numerical Analysis for approximating integrals.

The problem of estimating the difference between the average of a product of functions from the product of their averages is also a very important one. The answer to it are the Grüss type inequalities. Grüss type inequalities are very useful among others in Probability for estimating expected values, etc. There exists a vast literature about Ostrowski and Grüss type inequalities to all possible directions. Mathematical community is very much interested to these inequalities due to their applications.

So here we produce Ostrowski and Grüss type inequalities for several functions, acting to all possible directions, including *s*-convexity and *s*-concavity in the second sense complete study. Our results are univariate.

The following results motivate our work.

Theorem 11.1. *(1938, Ostrowski [12]) Let $f : [a, b] \to \mathbb{R}$ be continuous on $[a, b]$ and differentiable on (a, b) whose derivative $f' : (a, b) \to \mathbb{R}$ is bounded on (a, b),* *i.e.,* $\|f'\|_\infty^{\sup} := \sup\limits_{t \in (a,b)} |f'(t)| < +\infty.$ *Then*

$$\left| \frac{1}{b-a} \int_a^b f(t)\,dt - f(x) \right| \leq \left[\frac{1}{4} + \frac{\left(x - \frac{a+b}{2}\right)^2}{(b-a)^2} \right] (b-a)\,\|f'\|_\infty^{\sup}, \qquad (11.1)$$

for any $x \in [a,b]$. *The constant* $\frac{1}{4}$ *is the best possible.*

Theorem 11.2. *(1935, Grüss [10]) Let* f, g *be integrable functions from* $[a,b]$ *into* $\mathbb{R}$, *that satisfy the conditions*

$$m \leq f(x) \leq M, \quad n \leq g(x) \leq N, \quad x \in [a,b],$$

where $m, M, n, N \in \mathbb{R}$. *Then*

$$\left| \frac{1}{b-a} \int_a^b f(x)\,g(x)\,dx - \left(\frac{1}{b-a} \int_a^b f(x)\,dx \right) \left(\frac{1}{b-a} \int_a^b g(x)\,dx \right) \right|$$

$$\leq \frac{1}{4} (M - m)(N - n). \qquad (11.2)$$

Theorem 11.3. *(1998, Dragomir and Wang [8]) Let* $f : [a,b] \to \mathbb{R}$ *is absolutely continuous function with* $f' \in L_p([a,b])$, $p > 1$, $\frac{1}{p} + \frac{1}{q} = 1$, $x \in [a,b]$. *Then*

$$\left| f(x) - \frac{1}{b-a} \int_a^b f(t)\,dt \right|$$

$$\leq \frac{1}{(q+1)^{\frac{1}{q}}} \left[\left(\frac{x-a}{b-a} \right)^{q+1} + \left(\frac{b-x}{b-a} \right)^{q+1} \right]^{\frac{1}{q}} (b-a)^{\frac{1}{q}} \|f'\|_p. \qquad (11.3)$$

Theorem 11.4. *(1882, Čebyšev [4]) Let* $f, g : [a,b] \to \mathbb{R}$ *absolutely continuous functions with* $f', g' \in L_\infty([a,b])$. *Then*

$$\left| \frac{1}{b-a} \int_a^b f(x)\,g(x)\,dx - \left(\frac{1}{b-a} \int_a^b f(x)\,dx \right) \left(\frac{1}{b-a} \int_a^b g(x)\,dx \right) \right|$$

$$\leq \frac{1}{12} (b-a)^2 \|f'\|_\infty \|g'\|_\infty. \qquad (11.4)$$

Above is also assumed that the involved integrals exist.

We are also greatly inspired by the great work of B.G. Pachpatte, see [13], [14], [15].

11.2   Background

Let $(P_n)_{n \in \mathbb{N}}$ be a harmonic sequence of polynomials, that is $P_n' = P_{n-1}$, $n \geq 1$, $P_0 = 1$. Furthermore, let $[a, b] \subset \mathbb{R}$, $a \neq b$, and $h : [a, b] \to \mathbb{R}$ be such that $h^{(n-1)}$ is absolutely continuous function for some fixed $n \geq 1$. We use the notation

$$L_n \left[h\left(x\right) \right] = \frac{1}{n} \left[h\left(x\right) + \sum_{k=1}^{n-1} (-1)^k P_k\left(x\right) h^{(k)}\left(x\right) \right. \tag{11.5}$$

$$\left. + \sum_{k=1}^{n-1} \frac{(-1)^k (n-k)}{b-a} \left[P_k\left(a\right) h^{(k-1)}\left(a\right) - P_k\left(b\right) h^{(k-1)}\left(b\right) \right] \right],$$

$x \in [a, b]$, for convenience.

For $n = 1$ the above sums are defined to be zero, that is $L_1 \left[h\left(x\right) \right] = h\left(x\right)$.

Dedic, Pečaric and Ujevic, see [6], [5], established the following identity,

$$L_n \left[h\left(x\right) \right] - \frac{1}{b-a} \int_a^b h\left(t\right) dt = \frac{(-1)^{n+1}}{n\left(b-a\right)} \int_a^b P_{n-1}\left(t\right) q\left(x, t\right) h^{(n)}\left(t\right) dt, \tag{11.6}$$

where

$$q\left(x, t\right) = \begin{cases} t - a, & \text{if } t \in [a, x], \\ t - b, & \text{if } t \in (x, b], \end{cases} \quad x \in [a, b]. \tag{11.7}$$

For the harmonic sequence of polynomials $P_k\left(t\right) = \frac{(t-x)^k}{k!}$, $k \geq 0$, the identity (11.6) reduces to the Fink identity in [9], (see also [6], p. 177).

11.3   Main Results

We present our first main result, a set of very general Ostrowski type inequalities involving several functions.

Theorem 11.5. *Let* $n_j \in \mathbb{N}$, $j = 1, \dots, r \in \mathbb{N} - \{1\}$, $n_1 \leq n_2 \leq \dots \leq n_r$ *and* $f_j : [a, b] \to \mathbb{R}$ *be such that* $f_j^{(n_j - 1)}$ *is absolutely continuous function. Denote*

$$S_1\left(f_1, \dots, f_r\right) := \sum_{j=1}^r \left[\left(\prod_{\substack{i=1 \\ i \neq j}}^r f_i\left(x\right) \right) \left[L_{n_j}\left[f_j\left(x\right) \right] - \frac{1}{b-a} \int_a^b f_j\left(t\right) dt \right] \right], \tag{11.8}$$

$$S_2\left(f_1, \dots, f_r\right) := \sum_{j=1}^r \left[\left(\prod_{\substack{i=1 \\ i \neq j}}^r L_{n_i}\left[f_i\left(x\right) \right] \right) \left[L_{n_j}\left[f_j\left(x\right) \right] - \frac{1}{b-a} \int_a^b f_j\left(t\right) dt \right] \right], \tag{11.9}$$

$x \in [a, b]$. *Then*

1)

$$|S_1(f_1,\ldots,f_r)| \le \frac{1}{n_1}\left[\frac{1}{2} + \frac{|a+b-2x|}{2(b-a)}\right]$$

$$\cdot \left[\sum_{j=1}^{r}\left[\left(\prod_{\substack{i=1\\i\neq j}}^{r}|f_i(x)|\right)\left\|P_{n_j-1}\right\|_{\infty,[a,b]}\left\|f_j^{(n_j)}\right\|_{1,[a,b]}\right]\right], \qquad (11.10)$$

and

$$|S_2(f_1,\ldots,f_r)| \le \frac{1}{n_1}\left[\frac{1}{2} + \frac{|a+b-2x|}{2(b-a)}\right]$$

$$\cdot \left[\sum_{j=1}^{r}\left[\left(\prod_{\substack{i=1\\i\neq j}}^{r}|L_{n_i}[f_i(x)]|\right)\left\|P_{n_j-1}\right\|_{\infty,[a,b]}\left\|f_j^{(n_j)}\right\|_{1,[a,b]}\right]\right], \qquad (11.11)$$

2) *let* $p_{lj} > 1 : \sum_{lj=1}^{3}\frac{1}{p_{lj}} = 1$, *with* $f_j^{(n_j)} \in L_{p_{3j}}([a,b])$, $j=1,\ldots,r$, *it holds*

$$|S_1(f_1,\ldots,f_r)| \le \frac{1}{n_1(b-a)}\left[\sum_{j=1}^{r}\left[\left(\prod_{\substack{i=1\\i\neq j}}^{r}|f_i(x)|\right)\left\|P_{n_j-1}\right\|_{p_{1j},[a,b]}\right.\right.$$

$$\left.\left.\cdot\left\|f_j^{(n_j)}\right\|_{p_{3j},[a,b]}\left(\frac{(b-x)^{p_{2j}+1}+(x-a)^{p_{2j}+1}}{p_{2j}+1}\right)^{\frac{1}{p_{2j}}}\right]\right], \qquad (11.12)$$

and

$$|S_2(f_1,\ldots,f_r)| \le \frac{1}{n_1(b-a)}\left[\sum_{j=1}^{r}\left[\left(\prod_{\substack{i=1\\i\neq j}}^{r}|L_{n_i}[f_i(x)]|\right)\left\|P_{n_j-1}\right\|_{p_{1j},[a,b]}\right.\right.$$

$$\left.\left.\cdot\left\|f_j^{(n_j)}\right\|_{p_{3j},[a,b]}\left(\frac{(b-x)^{p_{2j}+1}+(x-a)^{p_{2j}+1}}{p_{2j}+1}\right)^{\frac{1}{p_{2j}}}\right]\right], \qquad (11.13)$$

3) *assuming* $f_j^{(n_j)} \in L_{\infty}([a,b])$, $j=1,\ldots,r$, *we get*

$$|S_1(f_1,\ldots,f_r)| \le \left(\frac{(b-x)^2+(x-a)^2}{2n_1(b-a)}\right)$$

$$\cdot\left[\sum_{j=1}^{r}\left[\left(\prod_{\substack{i=1\\i\neq j}}^{r}|f_i(x)|\right)\left\|P_{n_j-1}\right\|_{\infty,[a,b]}\left\|f_j^{(n_j)}\right\|_{\infty,[a,b]}\right]\right], \qquad (11.14)$$

and

$$|S_2(f_1,\ldots,f_r)| \le \left(\frac{(b-x)^2+(x-a)^2}{2n_1(b-a)}\right)$$

$$\cdot\left[\sum_{j=1}^{r}\left[\left(\prod_{\substack{i=1\\i\neq j}}^{r}|L_{n_i}[f_i(x)]|\right)\left\|P_{n_j-1}\right\|_{\infty,[a,b]}\left\|f_j^{(n_j)}\right\|_{\infty,[a,b]}\right]\right]. \qquad (11.15)$$

Proof. For $j = 1, \ldots, r$, $r \in \mathbb{N} - \{1\}$, we have

$$L_{n_j}[f_j(x)] = \frac{1}{n_j}\left[f_j(x) + \sum_{k=1}^{n_j-1}(-1)^k P_k(x) f_j^{(k)}(x)\right.$$

$$\left. + \sum_{k=1}^{n_j-1}\frac{(-1)^k(n_j-k)}{b-a}\left[P_k(a)f_j^{(k-1)}(a) - P_k(b)f_j^{(k-1)}(b)\right]\right], \qquad (11.16)$$

and

$$L_{n_j}[f_j(x)] - \frac{1}{b-a}\int_a^b f_j(t)\,dt \overset{(11.6)}{=} \frac{(-1)^{n_j+1}}{n_j(b-a)}\int_a^b P_{n_j-1}(t)\,q(x,t)\,f_j^{(n_j)}(t)\,dt. \tag{11.17}$$

Hence it holds

$$\left(\prod_{\substack{i=1\\i\neq j}}^{r}f_i(x)\right)\left[L_{n_j}[f_j(x)] - \frac{1}{b-a}\int_a^b f_j(t)\,dt\right] \tag{11.18}$$

$$= \left(\prod_{\substack{i=1\\i\neq j}}^{r}f_i(x)\right)\left[\frac{(-1)^{n_j+1}}{n_j(b-a)}\int_a^b P_{n_j-1}(t)\,q(x,t)\,f_j^{(n_j)}(t)\,dt\right], \quad \text{for all } j = 1, \ldots, r.$$

Therefore by addition of (11.18), we derive the identity

$$S_1(f_1, \ldots, f_r) := \sum_{j=1}^{r}\left[\left(\prod_{\substack{i=1\\i\neq j}}^{r}f_i(x)\right)\left[L_{n_j}[f_j(x)] - \frac{1}{b-a}\int_a^b f_j(t)\,dt\right]\right] \tag{11.19}$$

$$= \sum_{j=1}^{r}\left[\left(\prod_{\substack{i=1\\i\neq j}}^{r}f_i(x)\right)\left[\frac{(-1)^{n_j+1}}{n_j(b-a)}\int_a^b P_{n_j-1}(t)\,q(x,t)\,f_j^{(n_j)}(t)\,dt\right]\right].$$

Similarly we produce the identity

$$S_2(f_1, \ldots, f_r) := \sum_{j=1}^{r}\left[\left(\prod_{\substack{i=1\\i\neq j}}^{r}L_{n_i}[f_i(x)]\right)\left[L_{n_j}[f_j(x)] - \frac{1}{b-a}\int_a^b f_j(t)\,dt\right]\right] \tag{11.20}$$

$$= \sum_{j=1}^{r}\left[\left(\prod_{\substack{i=1\\i\neq j}}^{r}L_{n_i}[f_i(x)]\right)\left[\frac{(-1)^{n_j+1}}{n_j(b-a)}\int_a^b P_{n_j-1}(t)\,q(x,t)\,f_j^{(n_j)}(t)\,dt\right]\right].$$

Consequently we have

$$|S_1(f_1, \ldots, f_r)|$$

$$\leq \sum_{j=1}^{r} \left[\left(\prod_{\substack{i=1 \\ i \neq j}}^{r} |f_i(x)| \right) \left[\frac{1}{n_j(b-a)} \int_a^b |P_{n_j-1}(t)| \, |q(x,t)| \left| f_j^{(n_j)}(t) \right| dt \right] \right], \quad (11.21)$$

and

$$|S_2(f_1,\ldots,f_r)|$$

$$\leq \sum_{j=1}^{r} \left[\left(\prod_{\substack{i=1 \\ i \neq j}}^{r} |L_{n_i}[f_i(x)]| \right) \left[\frac{1}{n_j(b-a)} \int_a^b |P_{n_j-1}(t)| \, |q(x,t)| \left| f_j^{(n_j)}(t) \right| dt \right] \right].$$

$$(11.22)$$

Furthermore it holds

$$|S_1(f_1,\ldots,f_r)|$$

$$\leq \sum_{j=1}^{r} \left[\left(\prod_{\substack{i=1 \\ i \neq j}}^{r} |f_i(x)| \right) \left[\frac{\|P_{n_j-1}\|_{\infty,[a,b]}}{n_j(b-a)} \int_a^b |q(x,t)| \left| f_j^{(n_j)}(t) \right| dt \right] \right], \quad (11.23)$$

and

$$|S_2(f_1,\ldots,f_r)|$$

$$\leq \sum_{j=1}^{r} \left[\left(\prod_{\substack{i=1 \\ i \neq j}}^{r} |L_{n_i}[f_i(x)]| \right) \left[\frac{\|P_{n_j-1}\|_{\infty,[a,b]}}{n_j(b-a)} \int_a^b |q(x,t)| \left| f_j^{(n_j)}(t) \right| dt \right] \right]. \quad (11.24)$$

Since

$$|q(x,t)| \leq \max(x-a, b-x) = \frac{(b-a) + |a+b-2x|}{2}, \quad (11.25)$$

we get

$$|S_1(f_1,\ldots,f_r)|$$

$$\leq \sum_{j=1}^{r} \left[\left(\prod_{\substack{i=1 \\ i \neq j}}^{r} |f_i(x)| \right) \left[\frac{\|P_{n_j-1}\|_{\infty,[a,b]}}{n_j(b-a)} \left[\frac{(b-a) + |a+b-2x|}{2} \right] \left\| f_j^{(n_j)} \right\|_{1,[a,b]} \right] \right],$$

$$(11.26)$$

and

$$|S_2(f_1,\ldots,f_r)|$$

$$\leq \sum_{j=1}^{r} \left[\left(\prod_{\substack{i=1 \\ i \neq j}}^{r} |L_{n_i}[f_i(x)]| \right) \left[\frac{\|P_{n_j-1}\|_{\infty,[a,b]}}{n_j(b-a)} \left[\frac{(b-a) + |a+b-2x|}{2} \right] \left\| f_j^{(n_j)} \right\|_{1,[a,b]} \right] \right].$$

$$(11.27)$$

Let now $p_{lj} > 1 : \sum_{lj=1}^{3} \frac{1}{p_{lj}} = 1$, with $f_j^{(n_j)} \in L_{p_{3j}}([a, b])$, $j = 1, \ldots, r$.
Hence, by Hölder's inequality for three functions, it holds

$$\int_a^b \left|P_{n_j-1}(t)\right| \left|q(x, t)\right| \left|f_j^{(n_j)}(t)\right| dt$$

$$\leq \left\|P_{n_j-1}\right\|_{p_{1j},[a,b]} \left(\int_a^b |q(x, t)|^{p_{2j}} dt\right)^{\frac{1}{p_{2j}}} \left\|f_j^{(n_j)}\right\|_{p_{3j},[a,b]}$$

$$= \left\|P_{n_j-1}\right\|_{p_{1j},[a,b]} \left(\frac{(b-x)^{p_{2j}+1} + (x-a)^{p_{2j}+1}}{p_{2j}+1}\right)^{\frac{1}{p_{2j}}} \left\|f_j^{(n_j)}\right\|_{p_{3j},[a,b]}. \quad (11.28)$$

Consequently we derive

$$|S_1(f_1, \ldots, f_r)| \leq \sum_{j=1}^{r} \left[\left(\prod_{\substack{i=1 \\ i \neq j}}^{r} |f_i(x)|\right) \left[\frac{1}{n_j(b-a)} \left\|P_{n_j-1}\right\|_{p_{1j},[a,b]}\right.\right. \quad (11.29)$$

$$\left.\left.\cdot \left(\frac{(b-x)^{p_{2j}+1} + (x-a)^{p_{2j}+1}}{p_{2j}+1}\right)^{\frac{1}{p_{2j}}} \left\|f_j^{(n_j)}\right\|_{p_{3j},[a,b]}\right]\right],$$

and

$$|S_2(f_1, \ldots, f_r)| \leq \sum_{j=1}^{r} \left[\left(\prod_{\substack{i=1 \\ i \neq j}}^{r} |L_{n_i}[f_i(x)]|\right) \left[\frac{1}{n_j(b-a)} \left\|P_{n_j-1}\right\|_{p_{1j},[a,b]}\right.\right. \quad (11.30)$$

$$\left.\left.\cdot \left(\frac{(b-x)^{p_{2j}+1} + (x-a)^{p_{2j}+1}}{p_{2j}+1}\right)^{\frac{1}{p_{2j}}} \left\|f_j^{(n_j)}\right\|_{p_{3j},[a,b]}\right]\right].$$

Assuming that $f_j^{(n_j)} \in L_\infty([a, b])$, $j = 1, \ldots, r$, we find

$$|S_1(f_1, \ldots, f_r)| \leq \sum_{j=1}^{r} \left[\left(\prod_{\substack{i=1 \\ i \neq j}}^{r} |f_i(x)|\right) \left[\frac{1}{n_j(b-a)} \left\|P_{n_j-1}\right\|_{\infty,[a,b]}\right.\right. \quad (11.31)$$

$$\left.\left.\cdot \left\|f_j^{(n_j)}\right\|_{\infty,[a,b]} \left(\frac{(b-x)^2 + (x-a)^2}{2}\right)\right]\right],$$

and

$$|S_2(f_1, \ldots, f_r)| \leq \sum_{j=1}^{r} \left[\left(\prod_{\substack{i=1 \\ i \neq j}}^{r} |L_{n_i}[f_i(x)]|\right) \left[\frac{1}{n_j(b-a)} \left\|P_{n_j-1}\right\|_{\infty,[a,b]}\right.\right. \quad (11.32)$$

$$\left.\left.\cdot \left\|f_j^{(n_j)}\right\|_{\infty,[a,b]} \left(\frac{(b-x)^2 + (x-a)^2}{2}\right)\right]\right].$$

The proof of the theorem is now complete. $\qquad\square$

We need

Definition 11.1. ([11]) A function $f : \mathbb{R}_+ = [0, \infty) \to \mathbb{R}$ is said to be s-convex in the second sense if

$$f(\lambda x + (1 - \lambda) y) \le \lambda^s f(x) + (1 - \lambda)^s f(y), \tag{11.33}$$

for all $x, y \in [0, \infty)$, $\lambda \in [0, 1]$ and for some fixed $s \in (0, 1]$.

When $s = 1$, s-convexity in the second sense reduces to ordinary convexity. If "$\ge$" holds in (11.33), we talk about s-concavity in the second sense.
We also need

Definition 11.2. (see also [1]) Let I be a subinterval of $\mathbb{R}_+$ and $f : I \to (0, \infty)$. We call f s-logarithmically convex (s-log-convex) in the second sense, iff $\log f(x)$ is s-convex in the second sense, iff

$$f(\lambda x + (1 - \lambda) y) \le (f(x))^{\lambda^s} (f(y))^{(1-\lambda)^s}, \tag{11.34}$$

for all $x, y \in I$, $\lambda \in [0, 1]$ and for some fixed $s \in (0, 1]$.

When $s = 1$, s-log-convexity in the second sense reduces to usual log-convexity. If "$\ge$" holds in (11.34), we talk about s-log-concavity in the second sense.
We also need the s-convex Hadamard's inequality.

Theorem 11.6. ([7]) Suppose that $f : [0, \infty) \to [0, \infty)$ is an s-convex function in the second sense, where $s \in (0, 1]$ and let $a, b \in [0, \infty)$, $a < b$. If $f \in L_1([a, b])$, then

$$2^{s-1} f\left(\frac{a+b}{2}\right) \le \frac{1}{b-a} \int_a^b f(x)\, dx \le \frac{f(a) + f(b)}{s+1}. \tag{11.35}$$

The constant $K = \frac{1}{s+1}$ is the best possible in the second inequality (11.35). The above inequalities are sharp.

Next we present general Ostrowski type inequalities for several functions under s-convexity and s-concavity in the second sense.

Theorem 11.7. Same terms and assumptions as in Theorem 11.5. Assume that $a \ge 0$.

1) Suppose $\left| f_j^{(n_j)} \right|$ is s-convex in the second sense and $\left| f_j^{(n_j)}(x) \right| \le M_j$, $x \in [a, b]$, $j = 1, \ldots, r$. Then

$$|S_1(f_1, \ldots, f_r)| \le \frac{(b-a)}{n_1} \left[\frac{2}{s+2} \left(\frac{(b-x)^{s+2} + (x-a)^{s+2}}{(b-a)^{s+2}} \right) \right.$$

$$\left. - \frac{1}{s+1} \left(\frac{(b-x)^{s+1} + (x-a)^{s+1}}{(b-a)^{s+1}} \right) + \frac{2}{(s+1)(s+2)} \right]$$

$$\left[\sum_{j=1}^{r} \left[\left(\prod_{\substack{i=1 \\ i\neq j}}^{r} |f_i(x)| \right) \|P_{n_j-1}\|_{\infty,[a,b]} M_j \right] \right], \tag{11.36}$$

and

$$|S_2(f_1,\ldots,f_r)| \leq \frac{(b-a)}{n_1} \left[\frac{2}{s+2} \left(\frac{(b-x)^{s+2} + (x-a)^{s+2}}{(b-a)^{s+2}} \right) \right.$$

$$\left. -\frac{1}{s+1} \left(\frac{(b-x)^{s+1} + (x-a)^{s+1}}{(b-a)^{s+1}} \right) + \frac{2}{(s+1)(s+2)} \right]$$

$$\cdot \left[\sum_{j=1}^{r} \left[\left(\prod_{\substack{i=1 \\ i\neq j}}^{r} |L_{n_i}[f_i(x)]| \right) \|P_{n_j-1}\|_{\infty,[a,b]} M_j \right] \right], \tag{11.37}$$

$x \in [a,b]$.

2) Let $p_{lj} > 1 : \sum_{lj=1}^{3} \frac{1}{p_{lj}} = 1$, with $f_j^{(n_j)} \in L_{p_{3j}}([a,b])$, $j = 1,\ldots,r$.

2i) Assume again $\left| f_j^{(n_j)} \right|$ is s-convex in the second sense, and $\left| f_j^{(n_j)}(x) \right| \leq M_j$, $j = 1,\ldots,r$, $x \in [a,b]$. Then

$$|S_1(f_1,\ldots,f_r)| \leq \frac{2}{n_1(b-a)} \left[\sum_{j=1}^{r} \left[\left(\prod_{\substack{i=1 \\ i\neq j}}^{r} |f_i(x)| \right) \|P_{n_j-1}\|_{p_{1j},[a,b]} \right. \right.$$

$$\left. \left. \cdot M_j \left(\frac{b-a}{p_{3j}s+1} \right)^{\frac{1}{p_{3j}}} \left(\frac{(b-x)^{p_{2j}+1} + (x-a)^{p_{2j}+1}}{p_{2j}+1} \right)^{\frac{1}{p_{2j}}} \right] \right], \tag{11.38}$$

and

$$|S_2(f_1,\ldots,f_r)| \leq \frac{2}{n_1(b-a)} \left[\sum_{j=1}^{r} \left[\left(\prod_{\substack{i=1 \\ i\neq j}}^{r} |L_{n_i}[f_i(x)]| \right) \|P_{n_j-1}\|_{p_{1j},[a,b]} \right. \right.$$

$$\left. \left. \cdot M_j \left(\frac{b-a}{p_{3j}s+1} \right)^{\frac{1}{p_{3j}}} \left(\frac{(b-x)^{p_{2j}+1} + (x-a)^{p_{2j}+1}}{p_{2j}+1} \right)^{\frac{1}{p_{2j}}} \right] \right]. \tag{11.39}$$

2ii) Assume that $\left| f_j^{(n_j)} \right|^{p_{3j}}$ is s-convex in the second sense, and $\left| f_j^{(n_j)}(x) \right| \leq M_j$, $j = 1,\ldots,r$, $x \in [a,b]$. Then

$$|S_1(f_1,\ldots,f_r)| \leq \frac{1}{n_1} \left[\sum_{j=1}^{r} \left[\left(\prod_{\substack{i=1 \\ i\neq j}}^{r} |f_i(x)| \right) \|P_{n_j-1}\|_{p_{1j},[a,b]} \right. \right.$$

$$\cdot \frac{2^{\frac{1}{p_{3j}}} M_j \, (b-a)^{\frac{1}{p_{3j}}-1}}{(s+1)^{\frac{1}{p_{3j}}}} \left(\frac{(b-x)^{p_{2j}+1} + (x-a)^{p_{2j}+1}}{p_{2j}+1} \right)^{\frac{1}{p_{2j}}} \bigg] \bigg], \qquad (11.40)$$

and

$$|S_2 \, (f_1, \dots, f_r)| \leq \frac{1}{n_1} \left[\sum_{j=1}^{r} \left[\left(\prod_{\substack{i=1 \\ i \neq j}}^{r} |L_{n_i} \, [f_i \, (x)]| \right) \|P_{n_j-1}\|_{p_{1j},[a,b]} \right. \right.$$

$$\cdot \frac{2^{\frac{1}{p_{3j}}} M_j \, (b-a)^{\frac{1}{p_{3j}}-1}}{(s+1)^{\frac{1}{p_{3j}}}} \left(\frac{(b-x)^{p_{2j}+1} + (x-a)^{p_{2j}+1}}{p_{2j}+1} \right)^{\frac{1}{p_{2j}}} \bigg] \bigg]. \qquad (11.41)$$

2iii) Assume that $\left| f_j^{(n_j)} \right|^{p_{3j}}$ *is s-concave in the second sense. Then*

$$|S_1 \, (f_1, \dots, f_r)| \leq \frac{1}{n_1} \left[\sum_{j=1}^{r} \left[\left(\prod_{\substack{i=1 \\ i \neq j}}^{r} |f_i \, (x)| \right) \|P_{n_j-1}\|_{p_{1j},[a,b]} \right. \right.$$

$$\cdot 2^{\frac{s-1}{p_{3j}}} \left| f_j^{(n_j)} \left(\frac{a+b}{2} \right) \right| (b-a)^{\frac{1}{p_{3j}}-1} \left(\frac{(b-x)^{p_{2j}+1} + (x-a)^{p_{2j}+1}}{p_{2j}+1} \right)^{\frac{1}{p_{2j}}} \bigg] \bigg], \qquad (11.42)$$

and

$$|S_2 \, (f_1, \dots, f_r)| \leq \frac{1}{n_1} \left[\sum_{j=1}^{r} \left[\left(\prod_{\substack{i=1 \\ i \neq j}}^{r} |L_{n_i} \, [f_i \, (x)]| \right) \|P_{n_j-1}\|_{p_{1j},[a,b]} \right. \right.$$

$$\cdot 2^{\frac{s-1}{p_{3j}}} \left| f_j^{(n_j)} \left(\frac{a+b}{2} \right) \right| (b-a)^{\frac{1}{p_{3j}}-1} \left(\frac{(b-x)^{p_{2j}+1} + (x-a)^{p_{2j}+1}}{p_{2j}+1} \right)^{\frac{1}{p_{2j}}} \bigg] \bigg]. \qquad (11.43)$$

Proof.   As in (11.23) and (11.24) we have that

$$|S_1 \, (f_1, \dots, f_r)| \leq \frac{1}{n_1 \, (b-a)}$$

$$\cdot \left[\sum_{j=1}^{r} \left[\left(\prod_{\substack{i=1 \\ i \neq j}}^{r} |f_i \, (x)| \right) \|P_{n_j-1}\|_{\infty,[a,b]} \int_a^b |q \, (x,t)| \left| f_j^{(n_j)} \, (t) \right| dt \right] \right], \qquad (11.44)$$

and

$$|S_2 \, (f_1, \dots, f_r)| \leq \frac{1}{n_1 \, (b-a)}$$

$$\cdot \left[\sum_{j=1}^{r} \left[\left(\prod_{\substack{i=1 \\ i \neq j}}^{r} |L_{n_i}[f_i(x)]| \right) \|P_{n_j-1}\|_{\infty,[a,b]} \int_a^b |q(x,t)| \left| f_j^{(n_j)}(t) \right| dt \right] \right]. \quad (11.45)$$

Set

$$p(x,t) := \begin{cases} t, & t \in \left[0, \frac{b-x}{b-a}\right), \\ t-1, & t \in \left[\frac{b-x}{b-a}, 1\right]. \end{cases} \quad (11.46)$$

In [2], for $\lambda \in [0,1]$, we proved that

$$q(x, \lambda a + (1-\lambda)b) = (a-b)p(x,\lambda), \quad (11.47)$$

that is

$$|q(x, \lambda a + (1-\lambda)b)| = (b-a)|p(x,\lambda)|. \quad (11.48)$$

One can write

$$\int_a^b |q(x,t)| \left| f_j^{(n_j)}(t) \right| dt \quad (11.49)$$

$$= (b-a) \int_0^1 |q(x, \lambda a + (1-\lambda)b)| \left| f_j^{(n_j)}(\lambda a + (1-\lambda)b) \right| d\lambda$$

$$\overset{(11.48)}{=} (b-a)^2 \int_0^1 |p(x,\lambda)| \left| f_j^{(n_j)}(\lambda a + (1-\lambda)b) \right| d\lambda =: (*). \quad (11.50)$$

We notice under the assumption that $\left| f_j^{(n_j)} \right|$ is s-convex in the second sense and $\left| f_j^{(n_j)}(x) \right| \leq M_j$, $x \in [a,b]$, that

$$(*) \leq (b-a)^2 \int_0^1 |p(x,t)| \left(\lambda^s \left| f_j^{(n_j)}(a) \right| + (1-\lambda)^s \left| f_j^{(n_j)}(b) \right| \right) d\lambda \quad (11.51)$$

$$\leq M_j(b-a)^2 \int_0^1 |p(x,\lambda)| (\lambda^s + (1-\lambda)^s) d\lambda$$

(as in [2])

$$= M_j(b-a)^2 \left[\frac{2}{s+2} \left(\frac{(b-x)^{s+2} + (x-a)^{s+2}}{(b-a)^{s+2}} \right) \right.$$

$$\left. - \frac{1}{s+1} \left(\frac{(b-x)^{s+1} + (x-a)^{s+1}}{(b-a)^{s+1}} \right) + \frac{2}{(s+1)(s+2)} \right]. \quad (11.52)$$

So we got that

$$\int_a^b |q(x,t)| \left| f_j^{(n_j)}(t) \right| dt \leq M_j(b-a)^2 \left[\frac{2}{s+2} \left(\frac{(b-x)^{s+2} + (x-a)^{s+2}}{(b-a)^{s+2}} \right) \right.$$

$$-\frac{1}{s+1}\left(\frac{(b-x)^{s+1}+(x-a)^{s+1}}{(b-a)^{s+1}}\right)+\frac{2}{(s+1)(s+2)}\right], \quad j=1,\ldots,r. \quad (11.53)$$

Using (11.53) into (11.44) and (11.45) we derive (11.36) and (11.37).

Next we elaborate on (11.12) and (11.13).

Assume that $\left|f_j^{(n_j)}\right|$ is s-convex in the second sense, acting as in [2], we obtain

$$\left\|f_j^{(n_j)}\right\|_{p_{3j},[a,b]}\leq 2M_j\left(\frac{b-a}{p_{3j}s+1}\right)^{\frac{1}{p_{3j}}}, \quad (11.54)$$

$j=1,\ldots,r$, with $\left|f_j^{(n_j)}(x)\right|\leq M_j$, $x\in[a,b]$.

Next suppose that $\left|f_j^{(n_j)}\right|^{p_{3j}}$ is s-convex in the second sense. As in [2] we get

$$\left\|f_j^{(n_j)}\right\|_{p_{3j},[a,b]}\leq\frac{2^{\frac{1}{p_{3j}}}M_j(b-a)^{\frac{1}{p_{3j}}}}{(s+1)^{\frac{1}{p_{3j}}}}, \quad (11.55)$$

$j=1,\ldots,r$, with $\left|f_j^{(n_j)}(x)\right|\leq M_j$, $x\in[a,b]$.

Finally assume that $\left|f_j^{(n_j)}\right|^{p_{3j}}$ is s-concave in the second sense. Based on Theorem 11.6 and acting as in [2], we derive

$$\left\|f_j^{(n_j)}\right\|_{p_{3j},[a,b]}\leq 2^{\frac{s-1}{p_{3j}}}\left|f_j^{(n_j)}\left(\frac{a+b}{2}\right)\right|(b-a)^{\frac{1}{p_{3j}}}. \quad (11.56)$$

The proof is completed. $\qquad\qquad\qquad\qquad\qquad\qquad\qquad\qquad\qquad\quad \square$

Ostrowski type inequalities for several functions under s-log-convexity in the second sense follow.

Theorem 11.8. *Same terms and assumptions as in Theorem 11.5. Assume that $a\geq 0$. We further suppose that $\left|f_j^{(n_j)}\right|\neq 0$ is s-log-convex in the second sense, and $\left|f_j^{(n_j)}(a)\right|,\left|f_j^{(n_j)}(b)\right|\in(0,1]$, $j=1,\ldots,r$. Call $A_j:=\left|\dfrac{f_j^{(n_j)}(a)}{f_j^{(n_j)}(b)}\right|$, $s\in(0,1]$, $j=1,\ldots,r$, and*

$$\psi_s(z):=\begin{cases}\frac{z^s-1}{s\ln z}, & \text{if } z\in(0,\infty)-\{1\},\\ 1, & \text{if } z=1.\end{cases} \quad (11.57)$$

1) It holds

$$|S_1(f_1,\ldots,f_r)|\leq\left[\frac{(b-a)+|a+b-2x|}{2n_1}\right]$$

$$\cdot\left[\sum_{j=1}^{r}\left[\left(\prod_{\substack{i=1\\i\neq j}}^{r}|f_i(x)|\right)\|P_{n_j-1}\|_{\infty,[a,b]}\left|f_j^{(n_j)}(b)\right|^s\psi_s(A_j)\right]\right], \quad (11.58)$$

and

$$|S_2(f_1,\ldots,f_r)| \le \left[\frac{(b-a)+|a+b-2x|}{2n_1}\right]$$

$$\cdot \left[\sum_{j=1}^{r}\left[\left(\prod_{\substack{i=1\\i\neq j}}^{r}|L_{n_i}[f_i(x)]|\right)\|P_{n_j-1}\|_{\infty,[a,b]}\left|f_j^{(n_j)}(b)\right|^{s}\psi_s(A_j)\right]\right]. \qquad (11.59)$$

2) Let $p_{lj} > 1 : \sum_{lj=1}^{3}\frac{1}{p_{lj}} = 1$, with $f_j^{(n_j)} \in L_{p3j}([a,b])$, and set $B_j := A_j^{p3j}$, $j = 1,\ldots,r$. Then

$$|S_1(f_1,\ldots,f_r)| \le \frac{1}{n_1}\left[\sum_{j=1}^{r}\left[\left(\prod_{\substack{i=1\\i\neq j}}^{r}|f_i(x)|\right)\|P_{n_j-1}\|_{p_{1j},[a,b]}\right.\right.$$

$$\cdot (b-a)^{\frac{1}{p_{3j}}-1}\left|f_j^{(n_j)}(b)\right|^{s}(\psi_s(B_j))^{\frac{1}{p_{3j}}}\left(\frac{(b-x)^{p_{2j}+1}+(x-a)^{p_{2j}+1}}{p_{2j}+1}\right)^{\frac{1}{p_{2j}}}\left.\left.\right]\right], \qquad (11.60)$$

and

$$|S_2(f_1,\ldots,f_r)| \le \frac{1}{n_1}\left[\sum_{j=1}^{r}\left[\left(\prod_{\substack{i=1\\i\neq j}}^{r}|L_{n_i}[f_i(x)]|\right)\|P_{n_j-1}\|_{p_{1j},[a,b]}\right.\right.$$

$$\cdot (b-a)^{\frac{1}{p_{3j}}-1}\left|f_j^{(n_j)}(b)\right|^{s}(\psi_s(B_j))^{\frac{1}{p_{3j}}}\left(\frac{(b-x)^{p_{2j}+1}+(x-a)^{p_{2j}+1}}{p_{2j}+1}\right)^{\frac{1}{p_{2j}}}\left.\left.\right]\right]. \qquad (11.61)$$

Proof. 1) See also [1]. Here we assume that $0 < \left|f_j^{(n_j)}(a)\right|, \left|f_j^{(n_j)}(b)\right| \le 1$, set $A_j := \left|\frac{f_j^{(n_j)}(a)}{f_j^{(n_j)}(b)}\right|$, $\left|f_j^{(n_j)}\right|$ is s-logarithmically convex in the second sense, $s \in (0,1]$, $\lambda \in [0,1]$, $j = 1,\ldots,r$. From (11.26) we get

$$|S_1(f_1,\ldots,f_r)| \le \sum_{j=1}^{r}\left[\left(\prod_{\substack{i=1\\i\neq j}}^{r}|f_i(x)|\right)\right.$$

$$\cdot \left[\frac{\|P_{n_j-1}\|_{\infty,[a,b]}}{n_j}\left[\frac{(b-a)+|a+b-2x|}{2}\right]\int_{0}^{1}\left|f_j^{(n_j)}(\lambda a+(1-\lambda)b)\right|d\lambda\right]\right], \qquad (11.62)$$

and from (11.27) we obtain

$$|S_2(f_1,\ldots,f_r)| \leq \sum_{j=1}^{r} \left[\left(\prod_{\substack{i=1 \\ i\neq j}}^{r} |L_{n_i}[f_i(x)]| \right) \right.$$

$$\left. \cdot \left[\frac{\|P_{n_j-1}\|_{\infty,[a,b]}}{n_j} \left[\frac{(b-a)+|a+b-2x|}{2} \right] \int_0^1 \left| f_j^{(n_j)}(\lambda a + (1-\lambda) b) \right| d\lambda \right] \right].$$
(11.63)

We study separately the integral

$$\int_0^1 \left| f_j^{(n_j)}(\lambda a + (1-\lambda) b) \right| d\lambda \overset{(11.34)}{\leq} \int_0^1 \left| f_j^{(n_j)}(a) \right|^{\lambda s} \left| f_j^{(n_j)}(b) \right|^{(1-\lambda)s} d\lambda \quad (11.64)$$

$$\leq \int_0^1 \left| f_j^{(n_j)}(a) \right|^{\lambda s} \left| f_j^{(n_j)}(b) \right|^{(1-\lambda)s} d\lambda = \left| f_j^{(n_j)}(b) \right|^s \int_0^1 \left(\frac{\left| f_j^{(n_j)}(a) \right|}{\left| f_j^{(n_j)}(b) \right|} \right)^{\lambda s} d\lambda$$

$$= \left| f_j^{(n_j)}(b) \right|^s \int_0^1 A_j^{\lambda s} d\lambda =: (**).$$
(11.65)

If $A_j = 1$, then

$$(**) = \left| f_j^{(n_j)}(b) \right|^s.$$
(11.66)

If $A_j \neq 1$, we get

$$(**) = \left| f_j^{(n_j)}(b) \right|^s \int_0^1 e^{s(\ln A_j)\lambda} d\lambda = \left| f_j^{(n_j)}(b) \right|^s \left(\frac{A_j^s - 1}{s \ln A_j} \right).$$
(11.67)

Therefore we derive

$$\int_0^1 \left| f_j^{(n_j)}(\lambda a + (1-\lambda) b) \right| d\lambda \leq \left| f_j^{(n_j)}(b) \right|^s \psi_s(A_j).$$
(11.68)

Hence by (11.68) used in (11.62) and (11.63) we derive (11.58) and (11.59).

    2) As before and as in [2], we obtain

$$\left\| f_j^{(n_j)} \right\|_{p_{3j},[a,b]} \leq (b-a)^{\frac{1}{p_{3j}}} \left| f_j^{(n_j)}(b) \right|^s (\psi_s(B_j))^{\frac{1}{p_{3j}}}.$$
(11.69)

Using (11.69) into (11.12), (11.13) we derive (11.60), (11.61).   □

Next we give applications when $n_1 = n_2 = \ldots = n_r = 1$.

Corollary 11.1. *(to Theorem 11.5) Let* $f_j : [a,b] \to \mathbb{R}$ *is absolutely continuous function,* $j = 1,\ldots,r \in \mathbb{N} - \{1\}$. *Denote*

$$S^*(f_1,\ldots,f_r) := \sum_{j=1}^{r} \left[\left(\prod_{\substack{i=1 \\ i\neq j}}^{r} f_i(x) \right) \left[f_j(x) - \frac{1}{b-a} \int_a^b f_j(t)\,dt \right] \right],$$
(11.70)

$x \in [a, b]$.

Then

1)

$$|S^* (f_1, \ldots, f_r)| \leq \left[\frac{1}{2} + \frac{|a + b - 2x|}{2 (b - a)} \right]$$

$$\cdot \left[\sum_{j=1}^{r} \left[\left(\prod_{\substack{i=1 \\ i \neq j}}^{r} |f_i (x)| \right) \|f_j'\|_{1,[a,b]} \right] \right],$$ (11.71)

2) let $p_{lj} > 1 : \sum_{lj=1}^{3} \frac{1}{p_{lj}} = 1$, with $f_j' \in L_{p_{3j}} ([a,b])$, $j = 1, \ldots, r$, it holds

$$|S^* (f_1, \ldots, f_r)|$$

$$\leq \sum_{j=1}^{r} \left[\left(\prod_{\substack{i=1 \\ i \neq j}}^{r} |f_i (x)| \right) (b - a)^{\frac{1}{p_{1j}} - 1} \|f_j'\|_{p_{3j},[a,b]} \left(\frac{(b - x)^{p_{2j}+1} + (x - a)^{p_{2j}+1}}{p_{2j} + 1} \right)^{\frac{1}{p_{2j}}} \right],$$ (11.72)

3) assuming $f_j' \in L_\infty ([a,b])$, $j = 1, \ldots, r$, we get

$$|S^* (f_1, \ldots, f_r)| \leq \left(\frac{(b - x)^2 + (x - a)^2}{2 (b - a)} \right)$$

$$\cdot \left[\sum_{j=1}^{r} \left[\left(\prod_{\substack{i=1 \\ i \neq j}}^{r} |f_i (x)| \right) \|f_j'\|_{\infty,[a,b]} \right] \right].$$ (11.73)

We continue with

Corollary 11.2. *(to Theorem 11.7) Same terms and assumptions as in Corollary 11.1. Assume that $a \geq 0$.*

1) Suppose $|f_j'|$ is s-convex in the second sense and $|f_j' (x)| \leq M_{1j}$, $x \in [a, b]$, $j = 1, \ldots, r$. Then

$$|S^* (f_1, \ldots, f_r)| \leq (b - a) \left[\frac{2}{s + 2} \left(\frac{(b - x)^{s+2} + (x - a)^{s+2}}{(b - a)^{s+2}} \right) \right.$$

$$\left. - \frac{1}{s + 1} \left(\frac{(b - x)^{s+1} + (x - a)^{s+1}}{(b - a)^{s+1}} \right) + \frac{2}{(s + 1)(s + 2)} \right]$$

$$\cdot \left[\sum_{j=1}^{r} \left[\left(\prod_{\substack{i=1 \\ i \neq j}}^{r} |f_i (x)| \right) M_{1j} \right] \right],$$ (11.74)

2) Let $p_{lj} > 1 : \sum_{l j=1}^{3} \frac{1}{p_{lj}} = 1$, with $f_j' \in L_{p_{3j}}([a,b])$, $j = 1, \ldots, r$.

2i) Assume again $\left|f_j'\right|$ is s-convex in the second sense, and $\left|f_j'(x)\right| \leq M_{1j}$, $j = 1, \ldots, r$, $x \in [a,b]$. Then

$$|S^*(f_1, \ldots, f_r)| \leq 2 \left[\sum_{j=1}^{r} \left[\left(\prod_{\substack{i=1 \\ i \neq j}}^{r} |f_i(x)| \right) (b-a)^{-\frac{1}{p_{2j}}} \right. \right.$$

$$\left. \left. \cdot M_{1j} (p_{3j}s + 1)^{-\frac{1}{p_{3j}}} \left(\frac{(b-x)^{p_{2j}+1} + (x-a)^{p_{2j}+1}}{p_{2j}+1} \right)^{\frac{1}{p_{2j}}} \right] \right], \qquad (11.75)$$

2ii) Assume that $\left|f_j'\right|^{p_{3j}}$ is s-convex in the second sense, and $\left|f_j'(x)\right| \leq M_{1j}$, $j = 1, \ldots, r$, $x \in [a,b]$. Then

$$|S^*(f_1, \ldots, f_r)| \leq \sum_{j=1}^{r} \left[\left(\prod_{\substack{i=1 \\ i \neq j}}^{r} |f_i(x)| \right) \right.$$

$$\left. \cdot \frac{2^{\frac{1}{p_{3j}}} M_{1j} (b-a)^{-\frac{1}{p_{2j}}}}{(s+1)^{\frac{1}{p_{3j}}}} \left(\frac{(b-x)^{p_{2j}+1} + (x-a)^{p_{2j}+1}}{p_{2j}+1} \right)^{\frac{1}{p_{2j}}} \right], \qquad (11.76)$$

2iii) Assume that $\left|f_j'\right|^{p_{3j}}$ is s-concave in the second sense. Then

$$|S^*(f_1, \ldots, f_r)| \leq \sum_{j=1}^{r} \left[\left(\prod_{\substack{i=1 \\ i \neq j}}^{r} |f_i(x)| \right) \right.$$

$$\left. \cdot 2^{\frac{s-1}{p_{3j}}} \left| f_j' \left(\frac{a+b}{2} \right) \right| (b-a)^{-\frac{1}{p_{2j}}} \left(\frac{(b-x)^{p_{2j}+1} + (x-a)^{p_{2j}+1}}{p_{2j}+1} \right)^{\frac{1}{p_{2j}}} \right]. \qquad (11.77)$$

We also give

Corollary 11.3. *(to Theorem 11.8) Same terms and assumptions as in Corollary 11.1. Assume that $a \geq 0$. We further suppose that $\left|f_j'\right| \neq 0$ is s-log-convex in the second sense, and $\left|f_j'(a)\right|, \left|f_j'(b)\right| \in (0,1]$, $j = 1, \ldots, r$. Call $A_{1j} := \left|\frac{f_j'(a)}{f_j'(b)}\right|$, $s \in (0,1]$, $j = 1, \ldots, r$, and ψ_s as in (11.57).*

1) It holds

$$|S^*(f_1, \ldots, f_r)| \leq \left[\frac{(b-a) + |a+b-2x|}{2} \right]$$

$$\cdot \left[\sum_{j=1}^{r} \left[\left(\prod_{\substack{i=1 \\ i \neq j}}^{r} |f_i(x)| \right) |f_j'(b)|^s \psi_s(A_{1j}) \right] \right]. \qquad (11.78)$$

2) Let $p_{lj} > 1 : \sum_{lj=1}^{3} \frac{1}{p_{lj}} = 1$, with $f'_j \in L_{p_{3j}}([a,b])$, $B_{1j} := A_{1j}^{p_{3j}}$, $j = 1, \ldots, r$. Then

$$|S^* (f_1, \ldots, f_r)| \leq \sum_{j=1}^{r} \left[\left(\prod_{\substack{i=1 \\ i \neq j}}^{r} |f_i(x)| \right) \right.$$

$$\left. \cdot (b-a)^{-\frac{1}{p_{2j}}} |f'_j(b)|^s (\psi_s(B_{1j}))^{\frac{1}{p_{3j}}} \left(\frac{(b-x)^{p_{2j}+1} + (x-a)^{p_{2j}+1}}{p_{2j}+1} \right)^{\frac{1}{p_{2j}}} \right]. \quad (11.79)$$

Next we present a set of very general Grüss type inequalities involving several functions.

Theorem 11.9. *Let $n_j \in \mathbb{N}$, $j = 1, \ldots, r \in \mathbb{N} - \{1\}$, $n_1 \leq n_2 \leq \ldots \leq n_r$ and $f_j : [a,b] \to \mathbb{R}$ be such that $f_j^{(n_j-1)}$ is absolutely continuous function. Denote*

$$\Delta(f_1, \ldots, f_r) := \sum_{j=1}^{r} \left[\left(\int_a^b \left(\prod_{\substack{i=1 \\ i \neq j}}^{r} f_i(x) \right) L_{n_j} [f_j(x)] \, dx \right) \right.$$

$$\left. - \frac{1}{b-a} \left(\int_a^b \left(\prod_{\substack{i=1 \\ i \neq j}}^{r} f_i(x) \right) dx \right) \left(\int_a^b f_j(x) \, dx \right) \right]. \quad (11.80)$$

Then
1)

$$|\Delta(f_1, \ldots, f_r)| \leq \left(\frac{(b-a) + |a+b-2x|}{2n_1} \right)$$

$$\cdot \left[\sum_{j=1}^{r} \left[\left(\prod_{\substack{i=1 \\ i \neq j}}^{r} \|f_i\|_{\infty,[a,b]} \right) \|P_{n_j-1}\|_{\infty,[a,b]} \left\| f_j^{(n_j)} \right\|_{1,[a,b]} \right] \right], \quad (11.81)$$

2)

$$|\Delta(f_1, \ldots, f_r)| \leq (b-a) \left(\frac{(b-a) + |a+b-2x|}{2n_1} \right)$$

$$\cdot \left[\sum_{j=1}^{r} \left[\left(\prod_{\substack{i=1 \\ i \neq j}}^{r} \|f_i\|_{\infty,[a,b]} \right) \|P_{n_j-1}\|_{\infty,[a,b]} \left\| f_j^{(n_j)} \right\|_{\infty,[a,b]} \right] \right], \quad (11.82)$$

3) let $p_{i,j} > 1 : \sum_{i=1}^{r+2} \frac{1}{p_{i,j}} = 1$, and $f_j^{(n_j)} \in L_{(r+2),j}([a,b])$, $j = 1, \ldots, r$, it holds

$$|\Delta(f_1, \ldots, f_r)| \leq \frac{1}{n_1} \left[\sum_{j=1}^{r} \left[\left(\frac{2^{\frac{1}{p_{r+1,j}}} (b-a)^{1+\frac{1}{p_{r+1,j}}}}{((p_{r+1,j}+1)(p_{r+1,j}+2))^{\frac{1}{p_{r+1,j}}}} \right) \right. \right.$$

$$\cdot\left(\prod_{\substack{i=1\\i\neq j}}^{r}\|f_i\|_{p_{i,j},[a,b]}\right)\|P_{n_j-1}\|_{p_{j,j},[a,b]}\left\|f_j^{(n_j)}\right\|_{p_{r+2,j},[a,b]}\Bigg]\Bigg].\qquad(11.83)$$

Proof. From (11.19) we obtain

$$\sum_{j=1}^{r}\left[\left(\prod_{\substack{i=1\\i\neq j}}^{r}f_i\left(x\right)\right)L_{n_j}\left[f_j\left(x\right)\right]-\frac{1}{b-a}\left(\prod_{\substack{i=1\\i\neq j}}^{r}f_i\left(x\right)\right)\int_a^b f_j\left(t\right)dt\right]$$

$$=\sum_{j=1}^{r}\left[\frac{(-1)^{n_j+1}}{n_j\left(b-a\right)}\left(\prod_{\substack{i=1\\i\neq j}}^{r}f_i\left(x\right)\right)\int_a^b P_{n_j-1}\left(t\right)q\left(x,t\right)f_j^{(n_j)}\left(t\right)dt\right].\qquad(11.84)$$

Hence we get

$$\Delta\left(f_1,\dots,f_r\right):=\sum_{j=1}^{r}\left[\left(\int_a^b\left(\prod_{\substack{i=1\\i\neq j}}^{r}f_i\left(x\right)\right)L_{n_j}\left[f_j\left(x\right)\right]dx\right)\right.$$

$$\left.-\frac{1}{b-a}\left(\int_a^b\left(\prod_{\substack{i=1\\i\neq j}}^{r}f_i\left(x\right)\right)dx\right)\left(\int_a^b f_j\left(t\right)dt\right)\right]$$

$$=\sum_{j=1}^{r}\left[\frac{(-1)^{n_j+1}}{n_j\left(b-a\right)}\int_a^b\int_a^b\left(\prod_{\substack{i=1\\i\neq j}}^{r}f_i\left(x\right)\right)P_{n_j-1}\left(t\right)q\left(x,t\right)f_j^{(n_j)}\left(t\right)dtdx\right].\qquad(11.85)$$

Therefore it holds

$$\left|\Delta\left(f_1,\dots,f_r\right)\right|$$

$$\leq\frac{1}{n_1\left(b-a\right)}\sum_{j=1}^{r}\left[\int_a^b\int_a^b\left(\prod_{\substack{i=1\\i\neq j}}^{r}\left|f_i\left(x\right)\right|\right)\left|P_{n_j-1}\left(t\right)\right|\left|q\left(x,t\right)\right|\left|f_j^{(n_j)}\left(t\right)\right|dtdx\right].$$

$$(11.86)$$

We first find

$$\left|\Delta\left(f_1,\dots,f_r\right)\right|\leq\left(\frac{(b-a)+|a+b-2x|}{2n_1\left(b-a\right)}\right)$$

$$\cdot\left[\sum_{j=1}^{r}\left[\prod_{\substack{i=1\\i\neq j}}^{r}\|f_i\|_{\infty,[a,b]}\|P_{n_j-1}\|_{\infty,[a,b]}\left(b-a\right)\left\|f_j^{(n_j)}\right\|_{1,[a,b]}\right]\right]$$

$$= \left(\frac{(b-a) + |a+b-2x|}{2n_1} \right) \left[\sum_{j=1}^{r} \left[\prod_{\substack{i=1 \\ i \neq j}}^{r} \|f_i\|_{\infty,[a,b]} \|P_{n_j-1}\|_{\infty,[a,b]} \left\| f_j^{(n_j)} \right\|_{1,[a,b]} \right] \right].$$

$$\tag{11.87}$$

Also it holds

$$|\Delta (f_1, \ldots, f_r)| \leq \left(\frac{(b-a) + |a+b-2x|}{2n_1 (b-a)} \right)$$

$$\cdot \left[\sum_{j=1}^{r} \left[\left(\prod_{\substack{i=1 \\ i \neq j}}^{r} \|f_i\|_{\infty,[a,b]} \right) \|P_{n_j-1}\|_{\infty,[a,b]} \left\| f_j^{(n_j)} \right\|_{\infty,[a,b]} (b-a)^2 \right] \right]$$

$$= (b-a) \left(\frac{(b-a) + |a+b-2x|}{2n_1} \right) \cdot$$

$$\left[\sum_{j=1}^{r} \left[\left(\prod_{\substack{i=1 \\ i \neq j}}^{r} \|f_i\|_{\infty,[a,b]} \right) \|P_{n_j-1}\|_{\infty,[a,b]} \left\| f_j^{(n_j)} \right\|_{\infty,[a,b]} \right] \right]. \tag{11.88}$$

Let $p_{i,j} > 1 : \sum_{i=1}^{r+2} \frac{1}{p_{i,j}} = 1$, and $f_j^{(n_j)} \in L_{(r+2),j}([a,b])$, $j = 1, \ldots, r$. Hence, by Hölder's inequality we find

$$\int_a^b \int_a^b \left(\prod_{\substack{i=1 \\ i \neq j}}^{r} |f_i(x)| \right) |P_{n_j-1}(t)| |q(x,t)| \left| f_j^{(n_j)}(t) \right| dt dx$$

$$\leq \left(\prod_{\substack{i=1 \\ i \neq j}}^{r} \left(\int_a^b \int_a^b |f_i(x)|^{p_{i,j}} dt dx \right)^{\frac{1}{p_{i,j}}} \right) \left(\int_a^b \int_a^b |P_{n_j-1}(t)|^{p_{j,j}} dt dx \right)^{\frac{1}{p_{j,j}}} \tag{11.89}$$

$$\cdot \left(\int_a^b \int_a^b |q(x,t)|^{p_{r+1,j}} dt dx \right)^{\frac{1}{p_{r+1,j}}} \left(\int_a^b \int_a^b \left| f_j^{(n_j)}(t) \right|^{p_{r+2,j}} dt dx \right)^{\frac{1}{p_{r+2,j}}}$$

$$= \left(\prod_{\substack{i=1 \\ i \neq j}}^{r} \|f_i\|_{p_{i,j},[a,b]} (b-a)^{\frac{1}{p_{i,j}}} \right) \left(\|P_{n_j-1}\|_{p_{j,j},[a,b]} (b-a)^{\frac{1}{p_{j,j}}} \right)$$

$$\cdot \left(\left\| f_j^{(n_j)} \right\|_{p_{r+2,j},[a,b]} (b-a)^{\frac{1}{p_{r+2,j}}} \right) \frac{2^{\frac{1}{p_{r+1,j}}} (b-a)^{1+\frac{2}{p_{r+1,j}}}}{((p_{r+1,j}+1)(p_{r+1,j}+2))^{\frac{1}{p_{r+1,j}}}}$$

$$= \frac{2^{\frac{1}{p_{r+1,j}}} (b-a)^{2+\frac{1}{p_{r+1,j}}}}{((p_{r+1,j}+1)(p_{r+1,j}+2))^{\frac{1}{p_{r+1,j}}}}$$

$$\cdot \left(\prod_{\substack{i=1 \\ i \neq j}}^{r} \|f_i\|_{p_{i,j},[a,b]} \right) \|P_{n_j-1}\|_{p_{j,j},[a,b]} \left\| f_j^{(n_j)} \right\|_{p_{r+2,j},[a,b]}. \tag{11.90}$$

That is we found

$$\int_a^b \int_a^b \left(\prod_{\substack{i=1 \\ i \neq j}}^{r} |f_i(x)| \right) |P_{n_j-1}(t)| |q(x,t)| \left| f_j^{(n_j)}(t) \right| dt dx$$

$$\leq \frac{2^{\frac{1}{p_{r+1,j}}} (b-a)^{2+\frac{1}{p_{r+1,j}}}}{((p_{r+1,j}+1)(p_{r+1,j}+2))^{\frac{1}{p_{r+1,j}}}}$$

$$\cdot \left(\prod_{\substack{i=1 \\ i \neq j}}^{r} \|f_i\|_{p_{i,j},[a,b]} \right) \|P_{n_j-1}\|_{p_{j,j},[a,b]} \left\| f_j^{(n_j)} \right\|_{p_{r+2,j},[a,b]}. \tag{11.91}$$

Using (11.91) into (11.86) we obtain (11.83).

The proof of the theorem now is complete. □

Next we produce Grüss type inequalities for several functions under s-convexity and s-concavity in the second sense.

Theorem 11.10. *Here all as in Theorem 11.9, with $a \geq 0$.*

1) Suppose $\left| f_j^{(n_j)} \right|$ is s-convex in the second sense and $\left| f_j^{(n_j)}(x) \right| \leq M_j$, $x \in [a,b]$, $j = 1, \ldots, r$. Then

$$|\Delta(f_1, \ldots, f_r)| \leq \frac{4(b-a)^2}{(s+2)(s+3)n_1}$$

$$\cdot \left[\sum_{j=1}^{r} \left[\left(\prod_{\substack{i=1 \\ i \neq j}}^{r} \|f_i\|_{\infty,[a,b]} \right) \|P_{n_j-1}\|_{\infty,[a,b]} M_j \right] \right]. \tag{11.92}$$

2) Let $p_{i,j} > 1 : \sum_{i=1}^{r+2} \frac{1}{p_{i,j}} = 1$, with $f_j^{(n_j)} \in L_{p_{r+2,j}}([a,b])$, $j = 1, \ldots, r$.

2i) Assume again $\left| f_j^{(n_j)} \right|$ is s-convex in the second sense, and $\left| f_j^{(n_j)}(x) \right| \leq M_j$, $j = 1, \ldots, r$, $x \in [a,b]$. Then

$$|\Delta(f_1, \ldots, f_r)| \leq \frac{1}{n_1} \left[\sum_{j=1}^{r} \left[\left(\frac{2^{1+\frac{1}{p_{r+1,j}}} (b-a)^{1+\frac{1}{p_{r+1,j}}+\frac{1}{p_{r+2,j}}}}{((p_{r+1,j}+1)(p_{r+1,j}+2))^{\frac{1}{p_{r+1,j}}}} \right) \right. \right.$$

$$\left. \left. \cdot \left(\prod_{\substack{i=1 \\ i \neq j}}^{r} \|f_i\|_{p_{i,j},[a,b]} \right) \|P_{n_j-1}\|_{p_{j,j},[a,b]} M_j (p_{r+2,j}s+1)^{-\frac{1}{p_{r+2,j}}} \right] \right]. \tag{11.93}$$

2ii) Assume that $\left|f_j^{(n_j)}\right|^{p_{r+2,j}}$ *is s-convex in the second sense, and* $\left|f_j^{(n_j)}(x)\right| \le$ M_j, $j = 1, \ldots, r$, $x \in [a, b]$. *Then*

$$|\Delta(f_1, \ldots, f_r)| \le \frac{1}{n_1} \left[\sum_{j=1}^{r} \left[\left(\frac{2^{\frac{1}{p_{r+1,j}} + \frac{1}{p_{r+2,j}}} (b-a)^{1 + \frac{1}{p_{r+1,j}} + \frac{1}{p_{r+2,j}}}}{((p_{r+1,j} + 1)(p_{r+1,j} + 2))^{\frac{1}{p_{r+1,j}}}} \right) \right. \right.$$

$$\left. \left. \cdot \frac{M_j}{(s+1)^{\frac{1}{p_{r+2,j}}}} \left(\prod_{\substack{i=1 \\ i \neq j}}^{r} \|f_i\|_{p_{i,j}, [a,b]} \right) \|P_{n_j - 1}\|_{p_{j,j}, [a,b]} \right] \right]. \tag{11.94}$$

2iii) Assume that $\left|f_j^{(n_j)}\right|^{p_{r+2,j}}$ *is s-concave in the second sense. Then*

$$|\Delta(f_1, \ldots, f_r)| \le \frac{1}{n_1} \left[\sum_{j=1}^{r} \left[\left(\frac{2^{\frac{1}{p_{r+1,j}} + \frac{s-1}{p_{r+2,j}}} (b-a)^{1 + \frac{1}{p_{r+1,j}} + \frac{1}{p_{r+2,j}}}}{((p_{r+1,j} + 1)(p_{r+1,j} + 2))^{\frac{1}{p_{r+1,j}}}} \right) \right. \right.$$

$$\left. \left. \cdot \left|f_j^{(n_j)}\left(\frac{a+b}{2}\right)\right| \left(\prod_{\substack{i=1 \\ i \neq j}}^{r} \|f_i\|_{p_{i,j}, [a,b]} \right) \|P_{n_j - 1}\|_{p_{j,j}, [a,b]} \right] \right]. \tag{11.95}$$

Proof. From (11.86) we get

$$|\Delta(f_1, \ldots, f_r)| \le \frac{1}{n_1(b-a)} \left[\sum_{j=1}^{r} \left[\left(\prod_{\substack{i=1 \\ i \neq j}}^{r} \|f_i\|_{\infty, [a,b]} \right) \|P_{n_j - 1}\|_{\infty, [a,b]} \right. \right. \tag{11.96}$$

$$\left. \left. \cdot \int_a^b \left(\int_a^b |q(x,t)| \left|f_j^{(n_j)}(t)\right| dt \right) dx \right] \right].$$

Here $\left|f_j^{(n_j)}\right|$ is s-convex in the second sense and $\left|f_j^{(n_j)}(x)\right| \le M_j$, $x \in [a, b]$, $j = 1, \ldots, r$.

Using (11.53) we obtain

$$\int_a^b \left(\int_a^b |q(x,t)| \left|f_j^{(n_j)}(t)\right| dt \right) dx \le \frac{4M_j(b-a)^3}{(s+2)(s+3)}, \tag{11.97}$$

$j = 1, \ldots, r$.

Consequently by (11.97) and (11.96) we derive (11.92).

Next we elaborate on (11.83).

Assume that $\left|f_j^{(n_j)}\right|$ is s-convex in the second sense, acting as in [2], we obtain

$$\left\|f_j^{(n_j)}\right\|_{p_{r+2,j}, [a,b]} \le 2M_j \left(\frac{b-a}{p_{r+2,j} s + 1} \right)^{\frac{1}{p_{r+2,j}}}, \tag{11.98}$$

$j = 1, \ldots, r$, with $\left| f_j^{(n_j)} (x) \right| \leq M_j$, $x \in [a, b]$.

Next suppose that $\left| f_j^{(n_j)} \right|^{p_{r+2,j}}$ is s-convex in the second sense. As in [2] we get

$$\left\| f_j^{(n_j)} \right\|_{p_{r+2,j},[a,b]} \leq \frac{2^{\frac{1}{p_{r+2,j}}} M_j (b-a)^{\frac{1}{p_{r+2,j}}}}{(s+1)^{\frac{1}{p_{r+2,j}}}}, \tag{11.99}$$

$j = 1, \ldots, r$, with $\left| f_j^{(n_j)} (x) \right| \leq M_j$, $x \in [a, b]$.

Finally assume that $\left| f_j^{(n_j)} \right|^{p_{r+2,j}}$ is s-concave in the second sense. Based on Theorem 11.6 and acting as in [2], we derive

$$\left\| f_j^{(n_j)} \right\|_{p_{r+2,j},[a,b]} \leq 2^{\frac{s-1}{p_{r+2,j}}} \left| f_j^{(n_j)} \left(\frac{a+b}{2} \right) \right| (b-a)^{\frac{1}{p_{r+2,j}}} . \tag{11.100}$$

The proof is done.   □

Grüss type inequalities for several functions under s-log-convexity in the second sense follow.

Theorem 11.11. *Same terms and assumptions as in Theorem 11.9, $a \geq 0$. We further suppose that $\left| f_j^{(n_j)} \right| \neq 0$ is s-log-convex in the second sense, and $\left| f_j^{(n_j)} (a) \right|$,*

$\left| f_j^{(n_j)} (b) \right| \in (0, 1]$, $j = 1, \ldots, r$. *Call* $A_j := \left| \frac{f_j^{(n_j)} (a)}{f_j^{(n_j)} (b)} \right|$, $s \in (0, 1]$, $j = 1, \ldots, r$, *and*

$\psi_s (z)$ *as in (11.57).*

1) It holds

$$|\Delta (f_1, \ldots, f_r)| \leq (b-a) \left(\frac{(b-a) + |a+b-2x|}{2n_1} \right)$$

$$\cdot \left[\sum_{j=1}^{r} \left[\left(\prod_{\substack{i=1 \\ i \neq j}}^{r} \| f_i \|_{\infty,[a,b]} \right) \| P_{n_j - 1} \|_{\infty,[a,b]} \left| f_j^{(n_j)} (b) \right|^s \psi_s (A_j) \right] \right]. \tag{11.101}$$

2) Let $p_{i,j} > 1 : \sum_{i=1}^{r+2} \frac{1}{p_{i,j}} = 1$, *and* $f_j^{(n_j)} \in L_{(r+2),j} ([a,b])$, $B_j^* := A_j^{p_{r+2,j}}$, $j = 1, \ldots, r$. *Then*

$$|\Delta (f_1, \ldots, f_r)| \leq \frac{1}{n_1} \left[\sum_{j=1}^{r} \left[\left(\frac{2^{\frac{1}{p_{r+1,j}}} (b-a)^{1 + \frac{1}{p_{r+1,j}} + \frac{1}{p_{r+2,j}}}}{((p_{r+1,j} + 1)(p_{r+1,j} + 2))^{\frac{1}{p_{r+1,j}}}} \right) \right. \right.$$

$$\left. \left. \cdot \prod_{\substack{i=1 \\ i \neq j}}^{r} \| f_i \|_{p_{i,j},[a,b]} \| P_{n_j - 1} \|_{p_{j,j},[a,b]} \left| f_j^{(n_j)} (b) \right|^s (\psi_s (B_j^*))^{\frac{1}{p_{r+2,j}}} \right] \right]. \tag{11.102}$$

Proof.   1) From (11.81) we get

$$\left| \Delta \left(f_1, \ldots, f_r \right) \right| \leq (b-a) \left(\frac{(b-a) + |a+b-2x|}{2n_1} \right)$$

$$\cdot \left[\sum_{j=1}^{r} \left[\left(\prod_{\substack{i=1 \\ i \neq j}}^{r} \| f_i \|_{\infty,[a,b]} \right) \| P_{n_j-1} \|_{\infty,[a,b]} \int_0^1 \left| f_j^{(n_j)} \left(\lambda a + (1-\lambda) b \right) \right| d\lambda \right] \right]$$

$$\tag{11.103}$$

$$\overset{\text{(by (11.68))}}{\leq} (b-a) \left(\frac{(b-a) + |a+b-2x|}{2n_1} \right)$$

$$\cdot \left[\sum_{j=1}^{r} \left[\left(\prod_{\substack{i=1 \\ i \neq j}}^{r} \| f_i \|_{\infty,[a,b]} \right) \| P_{n_j-1} \|_{\infty,[a,b]} \left| f_j^{(n_j)}(b) \right|^s \psi_s (A_j) \right] \right]. \tag{11.104}$$

That is proving (11.101).

2) As in (11.69) we get

$$\left\| f_j^{(n_j)} \right\|_{p_{r+2,j},[a,b]} \leq (b-a)^{\frac{1}{p_{r+2,j}}} \left| f_j^{(n_j)}(b) \right|^s \left(\psi_s \left(B_j^* \right) \right)^{\frac{1}{p_{r+2,j}}}. \tag{11.105}$$

Using (11.105) into (11.83), we derive (11.102).   □

Finally we give applications to Grüss type inequalities for several functions when $n_1 = n_2 = \ldots = n_r = 1$.

Corollary 11.4. *(to Theorem 11.9) Let* $f_j : [a,b] \to \mathbb{R}$ *be absolutely continuous,* $j = 1, \ldots, r \in \mathbb{N} - \{1\}$. *Denote*

$$\Delta^* \left(f_1, \ldots, f_r \right) := r \int_a^b \left(\prod_{i=1}^{r} f_i(x) \right) dx$$

$$- \frac{1}{b-a} \left[\sum_{j=1}^{r} \left[\left(\int_a^b \left(\prod_{\substack{i=1 \\ i \neq j}}^{r} f_i(x) \right) dx \right) \left(\int_a^b f_j(x) \, dx \right) \right] \right]. \tag{11.106}$$

Then
1)

$$\left| \Delta^* \left(f_1, \ldots, f_r \right) \right| \leq \left(\frac{(b-a) + |a+b-2x|}{2} \right)$$

$$\cdot \left[\sum_{j=1}^{r} \left[\left(\prod_{\substack{i=1 \\ i \neq j}}^{r} \| f_i \|_{\infty,[a,b]} \right) \| f_j' \|_{1,[a,b]} \right] \right], \tag{11.107}$$

2)

$$|\Delta^* (f_1, \ldots, f_r)| \le (b-a) \left(\frac{(b-a) + |a+b-2x|}{2} \right)$$

$$\cdot \left[\sum_{j=1}^{r} \left[\left(\prod_{\substack{i=1 \\ i \neq j}}^{r} \|f_i\|_{\infty,[a,b]} \right) \|f_j'\|_{\infty,[a,b]} \right] \right], \tag{11.108}$$

3) let $p_{i,j} > 1 : \sum_{i=1}^{r+2} \frac{1}{p_{i,j}} = 1$, and $f_j' \in L_{(r+2),j}([a,b])$, $j = 1, \ldots, r$, it holds

$$|\Delta^* (f_1, \ldots, f_r)| \le \sum_{j=1}^{r} \left[\left(\frac{2^{\frac{1}{p_{r+1,j}}} (b-a)^{1 + \frac{1}{p_{r+1,j}} + \frac{1}{p_{j,j}}}}{((p_{r+1,j}+1)(p_{r+1,j}+2))^{\frac{1}{p_{r+1,j}}}} \right) \right.$$

$$\left. \cdot \left(\prod_{\substack{i=1 \\ i \neq j}}^{r} \|f_i\|_{p_{i,j},[a,b]} \right) \|f_j'\|_{p_{r+2,j},[a,b]} \right]. \tag{11.109}$$

Corollary 11.5. *(to Theorem 11.10) Here all as in Corollary 11.4, with $a \ge 0$.*
1) Suppose $|f_j'|$ is s-convex in the second sense and $|f_j'(x)| \le M_{1j}$, $x \in [a,b]$, $j = 1, \ldots, r$. Then

$$|\Delta^* (f_1, \ldots, f_r)| \le \frac{4(b-a)^2}{(s+2)(s+3)} \left[\sum_{j=1}^{r} \left[\left(\prod_{\substack{i=1 \\ i \neq j}}^{r} \|f_i\|_{\infty,[a,b]} \right) M_{1j} \right] \right]. \tag{11.110}$$

2) Let $p_{i,j} > 1 : \sum_{i=1}^{r+2} \frac{1}{p_{i,j}} = 1$, with $f_j' \in L_{p_{r+2,j}}([a,b])$, $j = 1, \ldots, r$.
2i) Assume again $|f_j'|$ is s-convex in the second sense, and $|f_j'(x)| \le M_{1j}$, $j = 1, \ldots, r$, $x \in [a,b]$. Then

$$|\Delta^* (f_1, \ldots, f_r)| \le \sum_{j=1}^{r} \left[\left(\frac{2^{1 + \frac{1}{p_{r+1,j}}} (b-a)^{1 + \frac{1}{p_{r+1,j}} + \frac{1}{p_{r+2,j}} + \frac{1}{p_{j,j}}}}{((p_{r+1,j}+1)(p_{r+1,j}+2))^{\frac{1}{p_{r+1,j}}}} \right) \right.$$

$$\left. \cdot \left(\prod_{\substack{i=1 \\ i \neq j}}^{r} \|f_i\|_{p_{i,j},[a,b]} \right) M_{1j} (p_{r+2,j}s + 1)^{-\frac{1}{p_{r+2,j}}} \right]. \tag{11.111}$$

2ii) Assume that $|f_j'|^{p_{r+2,j}}$ is s-convex in the second sense, and $|f_j'(x)| \le M_{1j}$, $j = 1, \ldots, r$, $x \in [a,b]$. Then

$$|\Delta^* (f_1, \ldots, f_r)| \le \sum_{j=1}^{r} \left[\left(\frac{2^{\frac{1}{p_{r+1,j}} + \frac{1}{p_{r+2,j}}} (b-a)^{1 + \frac{1}{p_{r+1,j}} + \frac{1}{p_{r+2,j}} + \frac{1}{p_{j,j}}}}{((p_{r+1,j}+1)(p_{r+1,j}+2))^{\frac{1}{p_{r+1,j}}}} \right) \right.$$

$$\left. \cdot \frac{M_{1j}}{(s+1)^{\frac{1}{p_{r+2,j}}}} \left(\prod_{\substack{i=1 \\ i \neq j}}^{r} \|f_i\|_{p_{i,j},[a,b]} \right) \right]. \tag{11.112}$$

2iii) Assume that $\left|f_j'\right|^{p_{r+2,j}}$ *is s-concave in the second sense. Then*

$$|\Delta^*(f_1,\ldots,f_r)| \leq \sum_{j=1}^r \left[\left(\frac{2^{\frac{1}{p_{r+1,j}}+\frac{s-1}{p_{r+2,j}}}(b-a)^{1+\frac{1}{p_{r+1,j}}+\frac{1}{p_{r+2,j}}+\frac{1}{p_{j,j}}}}{((p_{r+1,j}+1)(p_{r+1,j}+2))^{\frac{1}{p_{r+1,j}}}} \right) \right.$$

$$\left. \cdot \left|f_j'\left(\frac{a+b}{2}\right)\right| \left(\prod_{\substack{i=1 \\ i\neq j}}^r \|f_i\|_{p_{i,j},[a,b]} \right) \right]. \qquad (11.113)$$

Corollary 11.6. *(to Theorem 11.11) Here all as in Corollary 11.4, with $a \geq 0$. We further suppose that $\left|f_j'\right| \neq 0$ is s-log-convex in the second sense, and $\left|f_j'(a)\right|, \left|f_j'(b)\right| \in (0,1], j = 1,\ldots,r$. Call $A_{1j} := \left|\frac{f_j'(a)}{f_j'(b)}\right|, s \in (0,1], j = 1,\ldots,r,$ and $\psi_s(z)$ as in (11.57).*

1) It holds

$$|\Delta^*(f_1,\ldots,f_r)| \leq (b-a)\left(\frac{(b-a)+|a+b-2x|}{2}\right)$$

$$\cdot \left[\sum_{j=1}^r \left[\left(\prod_{\substack{i=1 \\ i\neq j}}^r \|f_i\|_{\infty,[a,b]} \right) \left|f_j'(b)\right|^s \psi_s(A_{1j}) \right] \right]. \qquad (11.114)$$

2) Let $p_{i,j} > 1 : \sum_{i=1}^{r+2} \frac{1}{p_{i,j}} = 1$, and $f_j' \in L_{(r+2),j}([a,b])$, $B_{1j}^* := A_{1j}^{p_{r+2,j}}$, $j = 1,\ldots,r$. Then

$$|\Delta^*(f_1,\ldots,f_r)| \leq \sum_{j=1}^r \left[\left(\frac{2^{\frac{1}{p_{r+1,j}}}(b-a)^{1+\frac{1}{p_{r+1,j}}+\frac{1}{p_{r+2,j}}+\frac{1}{p_{j,j}}}}{((p_{r+1,j}+1)(p_{r+1,j}+2))^{\frac{1}{p_{r+1,j}}}} \right) \right.$$

$$\left. \cdot \prod_{\substack{i=1 \\ i\neq j}}^r \|f_i\|_{p_{i,j},[a,b]} \left|f_j'(b)\right|^s (\psi_s(B_{1j}^*))^{\frac{1}{p_{r+2,j}}} \right]. \qquad (11.115)$$

Remark 11.1. From (11.20) one can work out the analogous Grüss type inequalities general theory involving the functions $L_{n_j}[f_j(x)]$, for $j = 1,\ldots,r \in \mathbb{N} - \{1\}$. The results will be very similar to the results of Theorems 11.9-11.11, and when $n_1 = n_2 = \ldots = n_r = 1$ their applications will be identical to Corollaries 11.4-11.6. We choose to omit this study.

Bibliography

1. A.O. Akdemir, M. Tunc, *On some integral inequalities for s-logarithmically convex functions and their applications*, arXiv:1212.1584v1[math.FA] 7 Dec 2012.
2. G.A. Anastassiou, *General Grüss and Ostrowski type inequalities involving s-convexity*, Bull. Allahabad Math. Soc. 28 (Part 1) (2013), 101-129.
3. G.A. Anastassiou, *Basic and s-convexity Ostrowski and Grüss type inequalities involving several functions*, Commun. Appl. Anal. 17(2) (2013), 189-212.
4. P.L. Čebyšev, *Sur les expressions approximatives des intégrales définies par les aures prises entre les mêmes limites*, Proc. Math. Soc. Charkov 2 (1882), 93-98.
5. Lj. Dedic, M. Matic, J. Pečaric, *On some generalizations of Ostrowski' inequality for Lipschitz functions and functions of bounded variation*, Math. Inequal. Appl. 3(1) (2000), 1-14.
6. Lj. Dedic, J. Pečaric, N. Ujevic, *On generalizations of Ostrowski inequality and some related results*, Czechoslovak Math. J. 53(128) (2003), 173-189.
7. S.S. Dragomir, S. Fitzpatrick, *The Hadamard's inequality for s-convex functions in the second sense*, Demonstratio Math. 32(4) (1999), 687-696.
8. S.S. Dragomir, S. Wang, *A new inequality of Ostrowski type in L_p norm*, Indian J. Math. 40 (1998), 299-304.
9. A.M. Fink, *Bounds of the deviation of a function from its averages*, Czechoslovak. Math. J. 42(117) (1992), 289-310.
10. G. Grüss, *Über das maximum des absoluten Betrages von $\frac{1}{b-a} \int_a^b f(x) g(x) dx - \frac{1}{(b-a)^2} \int_a^b f(x) dx \int_a^b g(x) dx$*, Math. Z. 39 (1935), 215-226.
11. H. Hudzik, L. Maligranda, *Some remarks on s-convex functions*, Aequationes Math. 48 (1994), 100-111.
12. A. Ostrowski, *Über die Absolutabweichung einer differentiabaren Funktion von ihrem Integralmittelwert*, Comment. Math. Helv. 10 (1938), 226-227.
13. B.G. Pachpatte, *On Ostrowski–Grüss–Čebyšev type inequalities for functions whose modulus of derivatives are convex*, JIPAM. J. Inequal. Pure Appl. Math. 6(4) (2005), Article 128.
14. B.G. Pachpatte, *New Ostrowski type inequalities involving the product of two functions*, JIPAM. J. Inequal. Pure Appl. Math. 7(3) (2006), Article 104.
15. B.G. Pachpatte, *On a new generalization of Ostrowski type inequality*, Tamkang J. Math. 38(4) (2007), 335-339.

Chapter 12

General Fractional Hermite–Hadamard Inequalities Using m-Convexity and (s, m)-Convexity

Here we present general fractional Hermite-Hadamard type inequalities using m-convexity and (s, m)-convexity. These inequalities are with respect to generalized Riemann-Liouville fractional integrals. Our work is motivated by and expands [8] to the greatest generality and all possible directions. It follows [1].

12.1 Background

We use a lot here the following generalized fractional integrals.

Definition 12.1. (see also [4, p. 99]) The left and right fractional integrals, respectively, of a function f with respect to given function g are defined as follows:

Let $a, b \in \mathbb{R}$, $a < b$, $\alpha > 0$. Here $g \in AC([a, b])$ (absolutely continuous functions) and is strictly increasing, $f \in L_\infty([a, b])$. We set

$$\left(I^\alpha_{a+;g} f\right)(x) = \frac{1}{\Gamma(\alpha)} \int_a^x (g(x) - g(t))^{\alpha-1} g'(t) f(t)\, dt, \quad x \geq a, \tag{12.1}$$

clearly $\left(I^\alpha_{a+;g} f\right)(a) = 0$,

and

$$\left(I^\alpha_{b-;g} f\right)(x) = \frac{1}{\Gamma(\alpha)} \int_x^b (g(t) - g(x))^{\alpha-1} g'(t) f(t)\, dt, \quad x \leq b, \tag{12.2}$$

clearly $\left(I^\alpha_{b-;g} f\right)(b) = 0$.

When g is the identity function id, we get that $I^\alpha_{a+;id} = I^\alpha_{a+}$ and $I^\alpha_{b-;id} = I^\alpha_{b-}$ the ordinary left and right Riemann-Liouville fractional integrals, where

$$\left(I^\alpha_{a+} f\right)(x) = \frac{1}{\Gamma(\alpha)} \int_a^x (x - t)^{\alpha-1} f(t)\, dt, \quad x \geq a, \tag{12.3}$$

$\left(I^\alpha_{a+} f\right)(a) = 0$, and

$$\left(I^\alpha_{b-} f\right)(x) = \frac{1}{\Gamma(\alpha)} \int_x^b (t - x)^{\alpha-1} f(t)\, dt, \quad x \leq b, \tag{12.4}$$

$\left(I^\alpha_{b-} f\right)(b) = 0$.

Remark 12.1. (see also [2]) We observe that

$$\left(I_{a+;g}^{\alpha}f\right)(x) = \frac{1}{\Gamma(\alpha)} \int_a^x \left(g(x) - g(t)\right)^{\alpha-1} \left(f \circ g^{-1}\right) \left(g(t)\right) g'(t) \, dt$$

(by change of variable for Lebesgue integrals)

$$= \frac{1}{\Gamma(\alpha)} \int_{g(a)}^{g(x)} \left(g(x) - z\right)^{\alpha-1} \left(f \circ g^{-1}\right)(z) \, dz = \left(I_{g(a)+}^{\alpha} \left(f \circ g^{-1}\right)\right) \left(g(x)\right), \quad x \geq a,$$

(12.5)

equivalently $g(x) \geq g(a)$.

That is in the terms and assumptions of Definition 12.1 we get

$$\left(I_{a+;g}^{\alpha}f\right)(x) = \left(I_{g(a)+}^{\alpha} \left(f \circ g^{-1}\right)\right)(g(x)), \quad \text{for } x \geq a. \tag{12.6}$$

Similarly we observe that

$$\left(I_{b-;g}^{\alpha}f\right)(x) = \frac{1}{\Gamma(\alpha)} \int_x^b \left(g(t) - g(x)\right)^{\alpha-1} \left(f \circ g^{-1}\right) \left(g(t)\right) g'(t) \, dt$$

$$= \frac{1}{\Gamma(\alpha)} \int_{g(x)}^{g(b)} \left(z - g(x)\right)^{\alpha-1} \left(f \circ g^{-1}\right)(z) \, dz = \left(I_{g(b)-}^{\alpha} \left(f \circ g^{-1}\right)\right) \left(g(x)\right), \quad (12.7)$$

for $x \leq b$.

That is

$$\left(I_{b-;g}^{\alpha}f\right)(x) = \left(I_{g(b)-}^{\alpha} \left(f \circ g^{-1}\right)\right)(g(x)), \quad \text{for } x \leq b. \tag{12.8}$$

So by (12.6) and (12.8) we have reduced the general fractional integrals to the ordinary left and right Riemann-Liouville fractional integrals.

When $g(x) = e^x$, $x \in [a, b]$ we have the application

Definition 12.2. The left and right fractional exponential integrals are defined as follows: Let $a, b \in \mathbb{R}$, $a < b$, $\alpha > 0$, $f \in L_\infty([a, b])$. We set

$$\left(I_{a+;e^x}^{\alpha}f\right)(x) = \frac{1}{\Gamma(\alpha)} \int_a^x \left(e^x - e^t\right)^{\alpha-1} e^t f(t) \, dt, \quad x \geq a, \tag{12.9}$$

and

$$\left(I_{b-;e^x}^{\alpha}f\right)(x) = \frac{1}{\Gamma(\alpha)} \int_x^b \left(e^t - e^x\right)^{\alpha-1} e^t f(t) \, dt, \quad x \leq b. \tag{12.10}$$

Note. We see that

$$\left(I_{a+;e^x}^{\alpha}f\right)(x) = \left(I_{e^a+}^{\alpha} (f \circ \ln)\right)(e^x), \quad x \geq a, \tag{12.11}$$

and

$$\left(I_{b-;e^x}^{\alpha}f\right)(x) = \left(I_{e^b-}^{\alpha} (f \circ \ln)\right)(e^x), \quad x \leq b. \tag{12.12}$$

Another example follows:

Definition 12.3. Let $a, b \in \mathbb{R}$, $a < b$, $\alpha > 0$, $f \in L_\infty([a, b])$, $A > 1$. We introduce the fractional integrals:

$$\left(I_{a+;A^x}^\alpha f\right)(x) = \frac{\ln A}{\Gamma(\alpha)} \int_a^x (A^x - A^t)^{\alpha-1} A^t f(t)\, dt, \quad x \ge a, \qquad (12.13)$$

and

$$\left(I_{b-;A^x}^\alpha f\right)(x) = \frac{\ln A}{\Gamma(\alpha)} \int_x^b (A^t - A^x)^{\alpha-1} A^t f(t)\, dt, \quad x \le b. \qquad (12.14)$$

We are motivated by

Theorem 12.1. *(1881, Hermite-Hadamard inequality, [5]) Let $f : I \subset \mathbb{R} \to \mathbb{R}$ be a convex function on the interval I of real numbers, and $a, b \in I$, with $a < b$. Then*

$$f\left(\frac{a+b}{2}\right) \le \frac{1}{b-a} \int_a^b f(t)\, dt \le \frac{f(a) + f(b)}{2}. \qquad (12.15)$$

Additionally to the classical convex functions, Toader [7], Hudzik and Maligranda [3] and Pinheiro [6] generalized the concepts of classical convex functions to the concepts of m-convex function and (s, m)-convex function.

Definition 12.4. The function $f : [0, b^*] \to \mathbb{R}$ is said to be m-convex, where $m \in [0, 1]$ and $b^* > 0$ if for every $x, y \in [0, b^*]$ and $t \in [0, 1]$, we have

$$f(tx + m(1 - t)y) \le tf(x) + m(1 - t)f(y). \qquad (12.16)$$

Definition 12.5. The function $f : [0, b^*] \to \mathbb{R}$ is said to be (s, m)-convex, where $(s, m) \in [0, 1]^2$ and $b^* > 0$, if for every $x, y \in [0, b^*]$ and $t \in [0, 1]$, we have

$$f(tx + m(1 - t)y) \le t^s f(x) + m(1 - t^s)f(y). \qquad (12.17)$$

We need the following list of Lemmas and Theorems from [8].

Lemma 12.1. *Let $\alpha > 0$, $f : [a, b] \to \mathbb{R}$ be a twice differentiable mapping on (a, b) with $a < b$. If $f'' \in L_1([a, b])$, then*

$$\frac{\Gamma(\alpha+1)}{2(b-a)^\alpha}\left[I_{a+}^\alpha f(b) + I_{b-}^\alpha f(a)\right] - f\left(\frac{a+b}{2}\right)$$

$$= \frac{(b-a)^2}{2} \int_0^1 m(t) f''(ta + (1-t)b)\, dt, \qquad (12.18)$$

where

$$m(t) = \begin{cases} t - \dfrac{1-(1-t)^{\alpha+1}-t^{\alpha+1}}{\alpha+1}, & t \in \left[0, \frac{1}{2}\right), \\[2mm] 1 - t - \dfrac{1-(1-t)^{\alpha+1}-t^{\alpha+1}}{\alpha+1}, & t \in \left[\frac{1}{2}, 1\right). \end{cases} \qquad (12.19)$$

Lemma 12.2. *Let* $\alpha > 0$, $f : [a, b] \to \mathbb{R}$ *be a twice differentiable mapping on* (a, b) *with* $a < b$. *If* $f'' \in L_1([a, b])$, $r > 0$, *then*

$$\frac{f(a) + f(b)}{r(r+1)} + \frac{2}{r+1} f\left(\frac{a+b}{2}\right) - \frac{\Gamma(\alpha+1)}{r(b-a)^\alpha}\left[I_{a+}^\alpha f(b) + I_{b-}^\alpha f(a)\right]$$

$$= (b-a)^2 \int_0^1 k(t) f''(ta + (1-t)b)\, dt, \qquad (12.20)$$

where

$$k(t) = \begin{cases} \frac{1-(1-t)^{\alpha+1}-t^{\alpha+1}}{r(\alpha+1)} - \frac{t}{r+1}, & t \in \left[0, \frac{1}{2}\right), \\ \frac{1-(1-t)^{\alpha+1}-t^{\alpha+1}}{r(\alpha+1)} - \frac{1-t}{r+1}, & t \in \left[\frac{1}{2}, 1\right). \end{cases} \qquad (12.21)$$

Lemma 12.3. *Let* $\alpha > 0$, $f : [a, b] \to \mathbb{R}$ *be a twice differentiable mapping on* (a, b) *with* $a < mb \le b$. *If* $f'' \in L_1([a, b])$, $r > 0$, *then*

$$\frac{f(a) + f(mb)}{r(r+1)} + \frac{2}{r+1} f\left(\frac{a+mb}{2}\right) - \frac{\Gamma(\alpha+1)}{r(mb-a)^\alpha}\left[I_{a+}^\alpha f(mb) + I_{mb-}^\alpha f(a)\right]$$

$$= (mb-a)^2 \int_0^1 k(t) f''(ta + m(1-t)b)\, dt, \qquad (12.22)$$

where $k(t)$ *is defined in (12.21).*

The following fractional m-convex Hermite-Hadamard type inequalities also come from [8].

Theorem 12.2. *Let* $f : [0, b^*] \to \mathbb{R}$ *be a twice differentiable mapping with* $b^* > 0$, $\alpha > 0$. *If* $|f''|^q$ *is measurable and* m-*convex on* $\left[a, \frac{b}{m}\right]$ *for some fixed* $q \ge 1$, $0 \le a < b$ *and* $m \in (0, 1]$ *with* $\frac{b}{m} \le b^*$, $r > 0$, *then*

$$H^m(f) := \left| \frac{f(a) + f(b)}{r(r+1)} + \frac{2}{r+1} f\left(\frac{a+b}{2}\right) - \frac{\Gamma(\alpha+1)}{r(b-a)^\alpha}\left[I_{a+}^\alpha f(b) + I_{b-}^\alpha f(a)\right] \right|$$

$$\le (b-a)^2 \left(\frac{\alpha}{r(\alpha+1)(\alpha+2)} + \frac{1}{4(r+1)} \right)$$

$$\cdot \left(\frac{|f''(a)|^q + m\left|f''\left(\frac{b}{m}\right)\right|^q}{2} \right)^{\frac{1}{q}} =: R_1^m(f). \qquad (12.23)$$

Theorem 12.3. *Let* $f : [0, b^*] \to \mathbb{R}$ *be a twice differentiable mapping with* $b^* > 0$, $\alpha > 0$. *If* $|f''|^q$ *is measurable and* m-*convex on* $\left[a, \frac{b}{m}\right]$ *for some fixed* $q > 1$, $0 \le a < b$ *and* $m \in (0, 1]$ *with* $\frac{b}{m} \le b^*$, $r > 0$, *then*

$$H^m(f) := \left| \frac{f(a) + f(b)}{r(r+1)} + \frac{2}{r+1} f\left(\frac{a+b}{2}\right) - \frac{\Gamma(\alpha+1)}{r(b-a)^\alpha}\left[I_{a+}^\alpha f(b) + I_{b-}^\alpha f(a)\right] \right|$$

$$\le \frac{(b-a)^2}{r(\alpha+1)} \left(1 - \frac{2}{p(\alpha+1)+1} \right)^{\frac{1}{p}} \left(\frac{|f''(a)|^q + m\left|f''\left(\frac{b}{m}\right)\right|^q}{2} \right)^{\frac{1}{q}} =: R_2^m(f), \qquad (12.24)$$

where $\frac{1}{p} + \frac{1}{q} = 1$.

Theorem 12.4. *Let* $f : [0, b^*] \to \mathbb{R}$ *be a twice differentiable mapping with* $b^* > 0$, $\alpha > 0$. *If* $|f''|^q$ *is measurable and* m*-convex on* $\left[a, \frac{b}{m}\right]$ *for some fixed* $q > 1$, $0 \le a < b$ *and* $m \in (0, 1]$ *with* $\frac{b}{m} \le b^*$, $r > 0$, *then*

$$H^m(f) := \left| \frac{f(a) + f(b)}{r(r+1)} + \frac{2}{r+1} f\left(\frac{a+b}{2}\right) - \frac{\Gamma(\alpha+1)}{r(b-a)^\alpha} \left[I^\alpha_{a+} f(b) + I^\alpha_{b-} f(a) \right] \right|$$

$$\le \frac{(b-a)^2}{r(\alpha+1)} \left(\frac{q(\alpha+1)-1}{q(\alpha+1)+1} \right)^{\frac{1}{q}} \left(\frac{|f''(a)|^q + m \left| f''\left(\frac{b}{m}\right) \right|^q}{2} \right)^{\frac{1}{q}} =: R_3^m(f). \quad (12.25)$$

Theorem 12.5. *Let* $f : [0, b^*] \to \mathbb{R}$ *be a twice differentiable mapping with* $b^* > 0$, $\alpha > 0$. *If* $|f''|^q$ *is measurable and* m*-convex on* $\left[a, \frac{b}{m}\right]$ *for some fixed* $q > 1$, $0 \le a < b$ *and* $m \in (0, 1]$ *with* $\frac{b}{m} \le b^*$, $r > 0$, *then*

$$H^m(f) := \left| \frac{f(a) + f(b)}{r(r+1)} + \frac{2}{r+1} f\left(\frac{a+b}{2}\right) - \frac{\Gamma(\alpha+1)}{r(b-a)^\alpha} \left[I^\alpha_{a+} f(b) + I^\alpha_{b-} f(a) \right] \right|$$

$$\le \left(\frac{2}{p+1} \right)^{\frac{1}{p}} \frac{(b-a)^2}{r+1} \left[\left(\frac{1}{2} + \frac{r+1}{r(\alpha+1)} \right)^{p+1} - \left(\frac{r+1}{r(\alpha+1)} \right)^{p+1} \right]^{\frac{1}{p}}$$

$$\cdot \left(\frac{|f''(a)|^q + m \left| f''\left(\frac{b}{m}\right) \right|^q}{2} \right)^{\frac{1}{q}} =: R_4^m(f), \quad (12.26)$$

where $\frac{1}{p} + \frac{1}{q} = 1$.

Theorem 12.6. *Let* $f : [0, b^*] \to \mathbb{R}$ *be a twice differentiable mapping with* $b^* > 0$, $\alpha > 0$. *If* $|f''|^q$ *is measurable and* m*-convex on* $\left[a, \frac{b}{m}\right]$ *for some fixed* $q > 1$, $0 \le a < b$ *and* $m \in (0, 1]$ *with* $\frac{b}{m} \le b^*$, $r > 0$, *then*

$$H^m(f) := \left| \frac{f(a) + f(b)}{r(r+1)} + \frac{2}{r+1} f\left(\frac{a+b}{2}\right) - \frac{\Gamma(\alpha+1)}{r(b-a)^\alpha} \left[I^\alpha_{a+} f(b) + I^\alpha_{b-} f(a) \right] \right|$$

$$\le \left(\frac{2}{q+1} \right)^{\frac{1}{q}} \frac{(b-a)^2}{r+1} \left[\left(\frac{1}{2} + \frac{r+1}{r(\alpha+1)} \right)^{q+1} - \left(\frac{r+1}{r(\alpha+1)} \right)^{q+1} \right]^{\frac{1}{q}}$$

$$\cdot \left(\frac{|f''(a)|^q + m \left| f''\left(\frac{b}{m}\right) \right|^q}{2} \right)^{\frac{1}{q}} =: R_5^m(f). \quad (12.27)$$

The following fractional (s, m)-convex Hermite-Hadamard type inequalities also come from [8].

Theorem 12.7. *Let* $f : [0, b] \to \mathbb{R}$ *be a twice differentiable mapping with* $0 \le a < mb \le b$, $\alpha > 0$. *If* $|f''|^q$ *is measurable and* (s, m)*-convex on* $[a, b]$ *for some fixed* $q \ge 1$ *and* $(s, m) \in (0, 1]^2$, $r > 0$, *then*

$$H_s^m(f) := \left| \frac{f(a) + f(mb)}{r(r+1)} + \frac{2}{r+1} f\left(\frac{a+mb}{2}\right) \right.$$

$$-\frac{\Gamma\left(\alpha+1\right)}{r\left(mb-a\right)^{\alpha}}\left[I_{a+}^{\alpha}f\left(mb\right)+I_{mb-}^{\alpha}f\left(a\right)\right]\Bigg|$$

$$\leq\left(mb-a\right)^{2}\left(\frac{\alpha}{r\left(\alpha+1\right)\left(\alpha+2\right)}+\frac{1}{4\left(r+1\right)}\right)^{1-\frac{1}{q}} \qquad (12.28)$$

$$\cdot\left[\left|f''\left(a\right)\right|^{q}I+m\left|f''\left(b\right)\right|^{q}\left(\frac{\alpha}{r\left(\alpha+1\right)\left(\alpha+2\right)}+\frac{1}{4\left(r+1\right)}-I\right)\right]^{\frac{1}{q}}=:R_{1s}^{m}\left(f\right),$$

where

$$I=\frac{1}{r\left(s+1\right)\left(s+\alpha+2\right)}-\frac{1}{r\left(\alpha+1\right)}B\left(s+1,\alpha+2\right)$$

$$+\frac{1}{\left(r+1\right)\left(s+1\right)\left(s+2\right)}\left(1-\left(\frac{1}{2}\right)^{s+1}\right).$$

Theorem 12.8. *Let* $f:[0,b]\rightarrow\mathbb{R}$ *be a twice differentiable mapping with* $0\leq a<mb\leq b$, $\alpha>0$. *If* $\left|f''\right|^{q}$ *is measurable and* (s,m)-*convex on* $[a,b]$ *for some fixed* $q>1$ *and* $(s,m)\in(0,1]^{2}$ $r>0$, *then*

$$H_{s}^{m}\left(f\right):=\left|\frac{f\left(a\right)+f\left(mb\right)}{r\left(r+1\right)}+\frac{2}{r+1}f\left(\frac{a+mb}{2}\right)\right.$$

$$\left.-\frac{\Gamma\left(\alpha+1\right)}{r\left(mb-a\right)^{\alpha}}\left[I_{a+}^{\alpha}f\left(mb\right)+I_{mb-}^{\alpha}f\left(a\right)\right]\right|$$

$$\leq\frac{\left(mb-a\right)^{2}}{r\left(\alpha+1\right)}\left(1-\frac{2}{p\left(\alpha+1\right)+1}\right)^{\frac{1}{p}}\left(\frac{1}{s+1}\left|f''\left(a\right)\right|^{q}+\frac{ms}{s+1}\left|f''\left(b\right)\right|^{q}\right)^{\frac{1}{q}} \qquad (12.29)$$

$$=:R_{2s}^{m}(f),$$

where $\frac{1}{p}+\frac{1}{q}=1$.

Theorem 12.9. *Let* $f:[0,b]\rightarrow\mathbb{R}$ *be a twice differentiable mapping with* $0\leq a<mb\leq b$, $\alpha>0$. *If* $\left|f''\right|^{q}$ *is measurable and* (s,m)-*convex on* $[a,b]$ *for some fixed* $q>1$ *and* $(s,m)\in(0,1]^{2}$, $r>0$, *then*

$$H_{s}^{m}\left(f\right):=\left|\frac{f\left(a\right)+f\left(mb\right)}{r\left(r+1\right)}+\frac{2}{r+1}f\left(\frac{a+mb}{2}\right)\right.$$

$$\left.-\frac{\Gamma\left(\alpha+1\right)}{r\left(mb-a\right)^{\alpha}}\left[I_{a+}^{\alpha}f\left(mb\right)+I_{mb-}^{\alpha}f\left(a\right)\right]\right|$$

$$\leq\frac{\left(mb-a\right)^{2}}{r\left(\alpha+1\right)}\left[\left|f''\left(a\right)\right|^{q}\left(\frac{1}{s+1}-\frac{1}{q\left(s+1\right)+s+1}-B\left(s+1,q\left(\alpha+1\right)+1\right)\right)\right.$$

$$(12.30)$$

$$+m\left|f''\left(b\right)\right|^{q}\left(\frac{s}{s+1}-\frac{2}{q\left(\alpha+1\right)+1}+\frac{1}{q\left(\alpha+1\right)+s+1}\right.$$

$$\left.\left.+B\left(s+1,q\left(\alpha+1\right)+1\right)\right)\right]=:R_{3s}^{m}\left(f\right).$$

Theorem 12.10. *Let* $f : [0, b] \to \mathbb{R}$ *be a twice differentiable mapping with* $0 \le a < mb \le b$, $\alpha > 0$. *If* $|f''|^q$ *is measurable and* (s, m)-*convex on* $[a, b]$ *for some fixed* $q > 1$ *and* $(s, m) \in (0, 1]^2$, $r > 0$, *then*

$$H_s^m (f) := \left| \frac{f(a) + f(mb)}{r(r+1)} + \frac{2}{r+1} f \left(\frac{a + mb}{2} \right) \right.$$

$$\left. - \frac{\Gamma(\alpha + 1)}{r(mb - a)^\alpha} \left[I_{a+}^\alpha f(mb) + I_{mb-}^\alpha f(a) \right] \right|$$

$$\le \frac{(mb - a)^2}{r+1} \left(\frac{2}{p+1} \right)^{\frac{1}{p}} \left[\left(\frac{1}{2} + \frac{r+1}{r(\alpha + 1)} \right)^{p+1} - \left(\frac{r+1}{r(\alpha + 1)} \right)^{p+1} \right]^{\frac{1}{p}}$$

$$\cdot \left(\frac{1}{s+1} |f''(a)|^q + \frac{ms}{s+1} |f''(b)|^q \right)^{\frac{1}{q}} =: R_{4s}^m (f), \tag{12.31}$$

where $\frac{1}{p} + \frac{1}{q} = 1$.

Theorem 12.11. *Let* $f : [0, b] \to \mathbb{R}$ *be a twice differentiable mapping with* $0 \le a < mb \le b$, $\alpha > 0$. *If* $|f''|^q$ *is measurable and* (s, m)-*convex on* $[a, b]$ *for some fixed* $q > 1$ *and* $(s, m) \in (0, 1]^2$, $r > 0$, *then*

$$H_s^m (f) := \left| \frac{f(a) + f(mb)}{r(r+1)} + \frac{2}{r+1} f \left(\frac{a + mb}{2} \right) \right.$$

$$\left. - \frac{\Gamma(\alpha + 1)}{r(mb - a)^\alpha} \left[I_{a+}^\alpha f(mb) + I_{mb-}^\alpha f(a) \right] \right|$$

$$\le \frac{(mb - a)^2}{r+1} \left[|f''(a)|^q H + m |f''(b)|^q \left(\frac{2}{q+1} \left(\frac{1}{2} + \frac{r+1}{r(\alpha + 1)} \right)^{q+1} \right. \right.$$

$$\left. \left. - \frac{2}{q+1} \left(\frac{r+1}{r(\alpha + 1)} \right)^{q+1} - H \right) \right] =: R_{5s}^m (f), \tag{12.32}$$

where

$$H = \int_0^{\frac{1}{2}} \left(\frac{r+1}{r(\alpha + 1)} + t \right)^q t^s dt + \int_{\frac{1}{2}}^1 \left(\frac{r+1}{r(\alpha + 1)} + 1 - t \right)^q t^s dt. \tag{12.33}$$

The aim of this chapter is to extend the results of [8] to generalized fractional integrals (12.1) and (12.2), in particular to fractional exponential integrals (12.9), (12.10) and to fractional trigonometric integrals (12.60), (12.61). That is to produce very general fractional m-convex and (s, m)-convex Hermite-Hadamard type inequalities.

12.2   Main Results

Combining Theorems 12.2-12.6 we get the following m-convex Hermite-Hadamard type inequality.

Theorem 12.12. *Let* $f : [0, b^*] \to \mathbb{R}$ *be a twice differentiable mapping with* $b^* > 0$, $\alpha > 0$. *If* $|f''|^q$ *is measurable and* m-convex on $\left[a, \frac{b}{m}\right]$ *for some fixed* $q > 1$, $0 \le a < b$ *and* $m \in (0, 1]$ *with* $\frac{b}{m} \le b^*$, $r > 0$, *then*

$$H^m (f) \le \min \left\{ R_1^m (f), R_2^m (f), R_3^m (f), R_4^m (f), R_5^m (f) \right\}. \tag{12.34}$$

Combining Theorems 12.7-12.11 we obtain the following (s, m)-convex Hermite-Hadamard type inequality.

Theorem 12.13. *Let* $f : [0, b] \to \mathbb{R}$ *be a twice differentiable mapping with* $0 \le a < mb \le b$, $\alpha > 0$. *If* $|f''|^q$ *is measurable and* (s, m)-convex on $[a, b]$ *for some fixed* $q > 1$ *and* $(s, m) \in (0, 1]^2$, $r > 0$, *then*

$$H_s^m (f) \le \min \left\{ R_{1s}^m (f), R_{2s}^m (f), R_{3s}^m (f), R_{4s}^m (f), R_{5s}^m (f) \right\}. \tag{12.35}$$

Next we generalize Lemmas 12.1-12.3.

Lemma 12.4. *Let* $\alpha > 0$, $a < b$, $f \in C ([a, b])$, $g \in C^1 ([a, b])$, g *strictly increasing on* $[a, b]$, $(f \circ g^{-1})$ *is twice differentiable function on* $(g (a), g (b))$ *with* $(f \circ g^{-1})'' \in L_1 ([g (a), g (b)])$. *Then*

$$\frac{\Gamma (\alpha + 1)}{2 (g (b) - g (a))^{\alpha}} \left[I_{a+;g}^{\alpha} f (b) + I_{b-;g}^{\alpha} f (a) \right] - (f \circ g^{-1}) \left(\frac{g (a) + g (b)}{2} \right)$$

$$= \frac{(g (b) - g (a))^2}{2} \int_0^1 m (t) (f \circ g^{-1})'' (t g (a) + (1 - t) g (b)) \, dt, \tag{12.36}$$

where $m (t)$ *as in (12.19).*

Lemma 12.5. *Let all as in Lemma 12.4,* $r > 0$. *Then*

$$\frac{f (a) + f (b)}{r (r + 1)} + \frac{2}{r + 1} (f \circ g^{-1}) \left(\frac{g (a) + g (b)}{2} \right)$$

$$- \frac{\Gamma (\alpha + 1)}{r (g (b) - g (a))^{\alpha}} \left[I_{a+;g}^{\alpha} f (b) + I_{b-;g}^{\alpha} f (a) \right]$$

$$= (g (b) - g (a))^2 \int_0^1 k (t) (f \circ g^{-1})'' (t g (a) + (1 - t) g (b)) \, dt, \tag{12.37}$$

where $k (t)$ *as in (12.21).*

Lemma 12.6. *Let all as Lemma 12.5, with* $g(a) < mg(b) \le g(b)$. *Then*

$$\frac{f(a) + (f \circ g^{-1})(mg(b))}{r(r+1)} + \frac{2}{r+1}(f \circ g^{-1})\left(\frac{g(a) + mg(b)}{2}\right)$$

$$-\frac{\Gamma(\alpha+1)}{r(mg(b) - g(a))^{\alpha}}\left[I_{g(a)+}^{\alpha}(f \circ g^{-1})(mg(b)) + I_{mg(b)-}^{\alpha}(f \circ g^{-1})(g(a))\right]$$

$$= (mg(b) - g(a))^2 \int_0^1 k(t)(f \circ g^{-1})''(tg(a) + m(1-t)g(b))\,dt, \qquad (12.38)$$

where $k(t)$ *as in (12.21).*

We apply Lemmas 12.4-12.6 to $g(x) = e^x$.

Lemma 12.7. *Let* $\alpha > 0$, $a < b$, $f \in C([a,b])$, $(f \circ \ln)$ *is twice differentiable function on* (e^a, e^b) *with* $(f \circ \ln)'' \in L_1([e^a, e^b])$. *Then*

$$\frac{\Gamma(\alpha+1)}{2(e^b - e^a)^{\alpha}}\left[I_{a+;e^x}^{\alpha}f(b) + I_{b-;e^x}^{\alpha}f(a)\right] - (f \circ \ln)\left(\frac{e^a + e^b}{2}\right)$$

$$= \frac{(e^b - e^a)^2}{2}\int_0^1 m(t)(f \circ \ln)''(te^a + (1-t)e^b)\,dt, \qquad (12.39)$$

where $m(t)$ *as in (12.19).*

Lemma 12.8. *Let all as in Lemma 12.7, $r > 0$. Then*

$$\frac{f(a) + f(b)}{r(r+1)} + \frac{2}{r+1}(f \circ \ln)\left(\frac{e^a + e^b}{2}\right) - \frac{\Gamma(\alpha+1)}{r(e^b - e^a)^{\alpha}}\left[I_{a+;e^x}^{\alpha}f(b) + I_{b-;e^x}^{\alpha}f(a)\right]$$

$$= (e^b - e^a)^2 \int_0^1 k(t)(f \circ \ln)''(te^a + (1-t)e^b)\,dt, \qquad (12.40)$$

where $k(t)$ *as in (12.21).*

Lemma 12.9. *Let all as in Lemma 12.8, with* $e^a < me^b \le e^b$. *Then*

$$\frac{f(a) + (f \circ \ln)(me^b)}{r(r+1)} + \frac{2}{r+1}(f \circ \ln)\left(\frac{e^a + me^b}{2}\right)$$

$$-\frac{\Gamma(\alpha+1)}{r(me^b - e^a)^{\alpha}}\left[I_{e^a+}^{\alpha}(f \circ \ln)(me^b) + I_{me^b-}^{\alpha}(f \circ \ln)(e^a)\right]$$

$$= (me^b - e^a)^2 \int_0^1 k(t)(f \circ \ln)''(te^a + (1-t)e^b)\,dt, \qquad (12.41)$$

where $k(t)$ *as in (12.21).*

We need

Notation 12.1. We denote by

$$H^m(f,g) := \left| \frac{f(a) + f(b)}{r(r+1)} + \frac{2}{r+1} (f \circ g^{-1}) \left(\frac{g(a) + g(b)}{2} \right) \right.$$

$$\left. - \frac{\Gamma(\alpha+1)}{r(g(b) - g(a))^\alpha} [I^\alpha_{a+;g} f(b) + I^\alpha_{b-;g} f(a)] \right|, \qquad (12.42)$$

$$R_1^m(f,g) := (g(b) - g(a))^2 \left(\frac{\alpha}{r(\alpha+1)(\alpha+2)} + \frac{1}{4(r+1)} \right)$$

$$\cdot \left(\frac{\left| (f \circ g^{-1})''(g(a)) \right|^q + m \left| (f \circ g^{-1})'' \left(\frac{g(b)}{m} \right) \right|^q}{2} \right)^{\frac{1}{q}}, \qquad (12.43)$$

$$R_2^m(f,g) := \frac{(g(b) - g(a))^2}{r(\alpha+1)} \left(1 - \frac{2}{p(\alpha+1)+1} \right)^{\frac{1}{p}}$$

$$\cdot \left(\frac{\left| (f \circ g^{-1})''(g(a)) \right|^q + m \left| (f \circ g^{-1})'' \left(\frac{g(b)}{m} \right) \right|^q}{2} \right)^{\frac{1}{q}}, \qquad (12.44)$$

where $\frac{1}{p} + \frac{1}{q} = 1$,

$$R_3^m(f,g) := \frac{(g(b) - g(a))^2}{r(\alpha+1)} \left(\frac{q(\alpha+1) - 1}{q(\alpha+1) + 1} \right)^{\frac{1}{q}}$$

$$\cdot \left(\frac{\left| (f \circ g^{-1})''(g(a)) \right|^q + m \left| (f \circ g^{-1})'' \left(\frac{g(b)}{m} \right) \right|^q}{2} \right)^{\frac{1}{q}}, \qquad (12.45)$$

$$R_4^m(f,g) := \left(\frac{2}{p+1} \right)^{\frac{1}{p}} \frac{(g(b) - g(a))^2}{r+1} \left[\left(\frac{1}{2} + \frac{r+1}{r(\alpha+1)} \right)^{p+1} \right.$$

$$\left. - \left(\frac{r+1}{r(\alpha+1)} \right)^{p+1} \right]^{\frac{1}{p}} \left(\frac{\left| (f \circ g^{-1})''(g(a)) \right|^q + m \left| (f \circ g^{-1})'' \left(\frac{g(b)}{m} \right) \right|^q}{2} \right)^{\frac{1}{q}}, \qquad (12.46)$$

where $\frac{1}{p} + \frac{1}{q} = 1$, and

$$R_5^m(f,g) := \left(\frac{2}{q+1} \right)^{\frac{1}{q}} \frac{(g(b) - g(a))^2}{r+1} \left[\left(\frac{1}{2} + \frac{r+1}{r(\alpha+1)} \right)^{q+1} \right.$$

$$\left. - \left(\frac{r+1}{r(\alpha+1)} \right)^{q+1} \right]^{\frac{1}{q}} \left(\frac{\left| (f \circ g^{-1})''(g(a)) \right|^q + m \left| (f \circ g^{-1})'' \left(\frac{g(b)}{m} \right) \right|^q}{2} \right)^{\frac{1}{q}}. \qquad (12.47)$$

We present the following fractional generalized m-convex Hermite-Hadamard type inequality.

Theorem 12.14. *Let all as in Notation 12.1. Here $\alpha > 0$, $b^* > 0$, $f \in C\left([0,b^*]\right)$, $g \in C^1\left([0,b^*]\right)$, g is strictly increasing on $[0,b^*]$ with $g\left(0\right) = 0$. Assume that $f \circ g^{-1}$: $[0, g\left(b^*\right)] \to \mathbb{R}$ is twice differentiable mapping. If $\left|\left(f \circ g^{-1}\right)''\right|^q$ is measurable and m-convex on $\left[g\left(a\right), \frac{g(b)}{m}\right]$ for some fixed $q > 1$, $0 \le a < b \le b^*$ and $m \in (0,1]$ with $\frac{g(b)}{m} \le g\left(b^*\right)$, $r > 0$, then*

$$H^m\left(f,g\right) \le \min\left\{R_1^m\left(f,g\right),\ R_2^m\left(f,g\right),\ R_3^m\left(f,g\right),\ R_4^m\left(f,g\right),\ R_5^m\left(f,g\right)\right\}.$$
$$(12.48)$$

Proof. By Theorem 12.12. ∎

We need

Notation 12.2. We denote by

$$H_s^m\left(f,g\right) := \left| \frac{f\left(a\right) + \left(f \circ g^{-1}\right)\left(mg\left(b\right)\right)}{r\left(r+1\right)} + \frac{2}{r+1}\left(f \circ g^{-1}\right)\left(\frac{g\left(a\right) + mg\left(b\right)}{2}\right) \right.$$

$$\left. - \frac{\Gamma\left(\alpha+1\right)}{r\left(mg\left(b\right) - g\left(a\right)\right)^\alpha} \left[I_{g(a)+}^\alpha\left(f \circ g^{-1}\right)\left(mg\left(b\right)\right) + I_{mg(b)-}^\alpha\left(f \circ g^{-1}\right)\left(g\left(a\right)\right)\right] \right|,$$
$$(12.49)$$

$$R_{1s}^m\left(f,g\right) := \left(mg\left(b\right) - g\left(a\right)\right)^2 \left(\frac{\alpha}{r\left(\alpha+1\right)\left(\alpha+2\right)} + \frac{1}{4\left(r+1\right)}\right)^{1-\frac{1}{q}}$$

$$\cdot \left[\left|\left(f \circ g^{-1}\right)''\left(g\left(a\right)\right)\right|^q I + m\left|\left(f \circ g^{-1}\right)''\left(g\left(b\right)\right)\right|^q \right. \qquad (12.50)$$

$$\left. \cdot \left(\frac{\alpha}{r\left(\alpha+1\right)\left(\alpha+2\right)} + \frac{1}{4\left(r+1\right)} - I\right)\right]^{\frac{1}{q}},$$

where

$$I := \frac{1}{r\left(s+1\right)\left(s+\alpha+2\right)} - \frac{1}{r\left(\alpha+1\right)}B\left(s+1,\alpha+2\right)$$

$$+ \frac{1}{\left(r+1\right)\left(s+1\right)\left(s+2\right)}\left(1 - \left(\frac{1}{2}\right)^{s+1}\right), \qquad (12.51)$$

$$R_{2s}^m\left(f,g\right) := \frac{\left(mg\left(b\right) - g\left(a\right)\right)^2}{r\left(\alpha+1\right)} \left(1 - \frac{2}{p\left(\alpha+1\right)+1}\right)^{\frac{1}{p}}$$

$$\cdot \left(\frac{1}{s+1}\left|\left(f \circ g^{-1}\right)''\left(g\left(a\right)\right)\right|^q + \frac{ms}{s+1}\left|\left(f \circ g^{-1}\right)''\left(g\left(b\right)\right)\right|^q\right)^{\frac{1}{q}}, \qquad (12.52)$$

where $\frac{1}{p} + \frac{1}{q} = 1$,

$$R_{3s}^m (f,g) := \frac{(mg(b) - g(a))^2}{r(\alpha+1)} \left[\left| (f \circ g^{-1})'' (g(a)) \right|^q \left(\frac{1}{s+1} - \frac{1}{q(\alpha+1)+s+1} \right. \right.$$

$$-B(s+1, q(\alpha+1)+1)) + m \left| (f \circ g^{-1})'' (g(b)) \right|^q \left(\frac{s}{s+1} - \frac{2}{q(\alpha+1)+1} \right. \quad (12.53)$$

$$\left. \left. + \frac{1}{q(\alpha+1)+s+1} + B(s+1, q(\alpha+1)+1) \right) \right],$$

$$R_{4s}^m (f,g) := \frac{(mg(b) - g(a))^2}{r+1} \left(\frac{2}{p+1} \right)^{\frac{1}{p}} \left[\left(\frac{1}{2} + \frac{r+1}{r(\alpha+1)} \right)^{p+1} \right.$$

$$\left. - \left(\frac{r+1}{r(\alpha+1)} \right)^{p+1} \right]^{\frac{1}{p}} \left(\frac{1}{s+1} \left| (f \circ g^{-1})'' (g(a)) \right|^q + \frac{ms}{s+1} \left| (f \circ g^{-1})'' (g(b)) \right|^q \right)^{\frac{1}{q}},$$

$$(12.54)$$

where $\frac{1}{p} + \frac{1}{q} = 1$, and

$$R_{5s}^m (f,g) := \frac{(mg(b) - g(a))^2}{r+1} \left[\left| (f \circ g^{-1})'' (g(a)) \right|^q H + m \left| (f \circ g^{-1})'' (g(b)) \right|^q \right.$$

$$\cdot \left. \left(\frac{2}{q+1} \left(\frac{1}{2} + \frac{r+1}{r(\alpha+1)} \right)^{q+1} - \frac{2}{q+1} \left(\frac{r+1}{r(\alpha+1)} \right)^{q+1} - H \right) \right], \quad (12.55)$$

where

$$H = \int_0^{\frac{1}{2}} \left(\frac{r+1}{r(\alpha+1)} + t \right)^q t^s dt + \int_{\frac{1}{2}}^1 \left(\frac{r+1}{r(\alpha+1)} + 1 - t \right)^q t^s dt. \quad (12.56)$$

Next we present a fractional generalized (s,m)-convex Hermite-Hadamard type inequality.

Theorem 12.15. *Here all as in Notation 12.2. Let $\alpha > 0$, $b > 0$, $f \in C([0,b])$, $g \in C^1([0,b])$, g is strictly increasing on $[0,b]$ with $g(0) = 0$. Assume that $f \circ g^{-1}$: $[0, g(b)] \to \mathbb{R}$ is twice differentiable mapping, with $0 \le g(a) < mg(b) \le g(b)$, $a \in [0,b]$. If $\left| (f \circ g^{-1})'' \right|^q$ is measurable and (s,m)-convex on $[g(a), g(b)]$ for some fixed $q > 1$ and $(s,m) \in (0,1]^2$, $r > 0$, then*

$$H_s^m (f,g) \le \min \{ R_{1s}^m (f,g), \ R_{2s}^m (f,g), \ R_{3s}^m (f,g), \ R_{4s}^m (f,g), \ R_{5s}^m (f,g) \}. \quad (12.57)$$

Proof.   By Theorem 12.13.   □

The case $q = 1$ is met separately.

Proposition 12.1. *Here* $H^m (f, g)$ *as in (12.42) of Notation 12.1. The rest of the assumptions as in Theorem 12.14 with* $q = 1$. *Then*

$$H^m (f, g) \leq (g (b) - g (a))^2 \left(\frac{\alpha}{r (\alpha + 1) (\alpha + 2)} + \frac{1}{4 (r + 1)} \right)$$

$$\cdot \left(\frac{\left| (f \circ g^{-1})'' (g (a)) \right| + m \left| (f \circ g^{-1})'' \left(\frac{g(b)}{m} \right) \right|}{2} \right). \tag{12.58}$$

Proof. By Theorem 12.2. $\qquad\qquad\qquad\qquad\qquad\qquad\qquad\qquad\qquad\qquad\qquad\qquad$ $\square$

Proposition 12.2. *Here* $H^m_s (f, g)$ *as in (12.49) of Notation 12.2. The rest of the assumptions as in Theorem 12.15 with* $q = 1$. *Then*

$$H^m_s (f, g) \leq (mg (b) - g (a))^2 \left[\left| (f \circ g^{-1})'' (g (a)) \right| I + m \left| (f \circ g^{-1})'' (g (b)) \right| \right.$$

$$\left. \cdot \left(\frac{\alpha}{r (\alpha + 1) (\alpha + 2)} + \frac{1}{4 (r + 1)} - I \right) \right], \tag{12.59}$$

where I *as in (12.51).*

Proof. By Theorem 12.7. $\qquad\qquad\qquad\qquad\qquad\qquad\qquad\qquad\qquad\qquad\qquad\qquad$ $\square$

We need

Definition 12.6. Let $a, b \in \left[0, \frac{\pi}{2} \right]$, $a < b$, $\alpha > 0$, $f \in L_\infty ([a, b])$. We consider the left and right fractional trigonometric integrals of f with respect to sine function denoted by sin :

$$\left(I^\alpha_{a+;\sin} f \right) (x) = \frac{1}{\Gamma (\alpha)} \int_a^x (\sin x - \sin t)^{\alpha - 1} \cos t f (t) \, dt, \quad x \geq a, \tag{12.60}$$

and

$$\left(I^\alpha_{b-;\sin} f \right) (x) = \frac{1}{\Gamma (\alpha)} \int_x^b (\sin t - \sin x)^{\alpha - 1} \cos t f (t) \, dt, \quad x \leq b. \tag{12.61}$$

We need

Notation 12.3. We denote by

$$H^m_* (f, \sin) := \left| \frac{f (a) + f (b)}{r (r + 1)} + \frac{2}{r + 1} (f \circ \sin^{-1}) \left(\frac{\sin (a) + \sin (b)}{2} \right) \right.$$

$$\left. - \frac{\Gamma (\alpha + 1)}{r (\sin (b) - \sin (a))^\alpha} \left[I^\alpha_{a+;\sin} f (b) + I^\alpha_{b-;\sin} f (a) \right] \right|, \tag{12.62}$$

$$R^m_{1*} (f, \sin) := (\sin (b) - \sin (a))^2 \left(\frac{\alpha}{r (\alpha + 1) (\alpha + 2)} + \frac{1}{4 (r + 1)} \right).$$

$$\cdot \left(\frac{\left| \left(f \circ \sin^{-1} \right)'' \left(\sin \left(a \right) \right) \right|^q + m \left| \left(f \circ \sin^{-1} \right)'' \left(\frac{\sin(b)}{m} \right) \right|^q}{2} \right)^{\frac{1}{q}}, \tag{12.63}$$

$$R_{2*}^m \left(f, \sin \right) := \frac{\left(\sin \left(b \right) - \sin \left(a \right) \right)^2}{r \left(\alpha + 1 \right)} \left(1 - \frac{2}{p \left(\alpha + 1 \right) + 1} \right)^{\frac{1}{p}}$$

$$\cdot \left(\frac{\left| \left(f \circ \sin^{-1} \right)'' \left(\sin \left(a \right) \right) \right|^q + m \left| \left(f \circ \sin^{-1} \right)'' \left(\frac{\sin(b)}{m} \right) \right|^q}{2} \right)^{\frac{1}{q}}, \tag{12.64}$$

where $\frac{1}{p} + \frac{1}{q} = 1$,

$$R_{3*}^m \left(f, \sin \right) := \frac{\left(\sin \left(b \right) - \sin \left(a \right) \right)^2}{r \left(\alpha + 1 \right)} \left(\frac{q \left(\alpha + 1 \right) - 1}{q \left(\alpha + 1 \right) + 1} \right)^{\frac{1}{q}}$$

$$\cdot \left(\frac{\left| \left(f \circ \sin^{-1} \right)'' \left(\sin \left(a \right) \right) \right|^q + m \left| \left(f \circ \sin^{-1} \right)'' \left(\frac{\sin(b)}{m} \right) \right|^q}{2} \right)^{\frac{1}{q}}, \tag{12.65}$$

$$R_{4*}^m \left(f, \sin \right) := \left(\frac{2}{p+1} \right)^{\frac{1}{p}} \frac{\left(\sin \left(b \right) - \sin \left(a \right) \right)^2}{r+1} \left[\left(\frac{1}{2} + \frac{r+1}{r \left(\alpha + 1 \right)} \right)^{p+1} \right.$$

$$\left. - \left(\frac{r+1}{r \left(\alpha + 1 \right)} \right)^{p+1} \right]^{\frac{1}{p}} \left(\frac{\left| \left(f \circ \sin^{-1} \right)'' \left(\sin \left(a \right) \right) \right|^q + m \left| \left(f \circ \sin^{-1} \right)'' \left(\frac{\sin(b)}{m} \right) \right|^q}{2} \right)^{\frac{1}{q}}, \tag{12.66}$$

where $\frac{1}{p} + \frac{1}{q} = 1$, and

$$R_{5*}^m \left(f, \sin \right) := \left(\frac{2}{q+1} \right)^{\frac{1}{q}} \frac{\left(\sin \left(b \right) - \sin \left(a \right) \right)^2}{r+1} \left[\left(\frac{1}{2} + \frac{r+1}{r \left(\alpha + 1 \right)} \right)^{q+1} \right.$$

$$\left. - \left(\frac{r+1}{r \left(\alpha + 1 \right)} \right)^{q+1} \right]^{\frac{1}{q}} \left(\frac{\left| \left(f \circ \sin^{-1} \right)'' \left(\sin \left(a \right) \right) \right|^q + m \left| \left(f \circ \sin^{-1} \right)'' \left(\frac{\sin(b)}{m} \right) \right|^q}{2} \right)^{\frac{1}{q}}. \tag{12.67}$$

We present the following fractional generalized m-convex Hermite-Hadamard type inequality for sin function. So here $g \left(x \right) = \sin \left(x \right)$, $x \in \left[0, \frac{\pi}{2} \right]$.

Theorem 12.16. *Let all as in Notation 12.3. Here $\alpha > 0$, $f \in C \left(\left[0, \frac{\pi}{2} \right] \right)$. Assume that $f \circ \sin^{-1} : \left[0, 1 \right] \to \mathbb{R}$ is twice differentiable mapping. If $\left| \left(f \circ \sin^{-1} \right)'' \right|^q$ is measurable and m-convex on $\left[\sin \left(a \right), \frac{\sin(b)}{m} \right]$ for some fixed $q > 1$, $0 \le a < b \le \frac{\pi}{2}$ and $m \in \left(0, 1 \right]$ with $\sin \left(b \right) \le m$, $r > 0$, then*

$$H_*^m \left(f, \sin \right)$$

$$\le \min \left\{ R_{1*}^m \left(f, \sin \right), \ R_{2*}^m \left(f, \sin \right), \ R_{3*}^m \left(f, \sin \right), \ R_{4*}^m \left(f, \sin \right), \ R_{5*}^m \left(f, \sin \right) \right\}. \tag{12.68}$$

Proof. By Theorem 12.14. □

We need

Notation 12.4. We denote by

$$H^m_{s*}(f,\sin) := \left| \frac{f(a) + (f \circ \sin^{-1})(m\sin(b))}{r(r+1)} \right.$$

$$+ \frac{2}{r+1}(f \circ \sin^{-1})\left(\frac{\sin(a) + m\sin(b)}{2}\right) - \frac{\Gamma(\alpha+1)}{r(m\sin(b) - \sin(a))^\alpha}$$

$$\cdot \left[I^\alpha_{\sin(a)+}(f \circ \sin^{-1})(m\sin(b)) + I^\alpha_{m\sin(b)-}(f \circ \sin^{-1})(\sin(a)) \right] \right|, \qquad (12.69)$$

$$R^m_{1s*}(f,\sin) := (m\sin(b) - \sin(a))^2 \left(\frac{\alpha}{r(\alpha+1)(\alpha+2)} + \frac{1}{4(r+1)} \right)^{1-\frac{1}{q}}$$

$$\cdot \left[\left| (f \circ \sin^{-1})''(\sin(a)) \right|^q I + m \left| (f \circ \sin^{-1})''(\sin(b)) \right|^q \right. \qquad (12.70)$$

$$\left. \cdot \left(\frac{\alpha}{r(\alpha+1)(\alpha+2)} + \frac{1}{4(r+1)} - I \right) \right]^{\frac{1}{q}},$$

where

$$I := \frac{1}{r(s+1)(s+\alpha+2)} - \frac{1}{r(\alpha+1)} B(s+1, \alpha+2)$$

$$+ \frac{1}{(r+1)(s+1)(s+2)}\left(1 - \left(\frac{1}{2}\right)^{s+1}\right), \qquad (12.71)$$

$$R^m_{2s*}(f,\sin) := \frac{(m\sin(b) - \sin(a))^2}{r(\alpha+1)}\left(1 - \frac{2}{p(\alpha+1)+1}\right)^{\frac{1}{p}}$$

$$\cdot \left(\frac{1}{s+1}\left| (f \circ \sin^{-1})''(\sin(a)) \right|^q + \frac{ms}{s+1}\left| (f \circ \sin^{-1})''(\sin(b)) \right|^q \right)^{\frac{1}{q}}, \qquad (12.72)$$

where $\frac{1}{p} + \frac{1}{q} = 1$,

$$R^m_{3s*}(f,\sin) := \frac{(m\sin(b) - \sin(a))^2}{r(\alpha+1)}$$

$$\cdot \left[\left| (f \circ \sin^{-1})''(\sin(a)) \right|^q \left(\frac{1}{s+1} - \frac{1}{q(\alpha+1)+s+1} \right) \right.$$

$$-B(s+1, q(\alpha+1)+1)) + m \left| (f \circ \sin^{-1})''(\sin(b)) \right|^q \left(\frac{s}{s+1} - \frac{2}{q(\alpha+1)+1} \right.$$

$$(12.73)$$

$$+\frac{1}{q\left(\alpha+1\right)+s+1}+B\left(s+1,q\left(\alpha+1\right)+1\right)\right],$$

$$R_{4s*}^{m}\left(f,\sin\right):=\frac{\left(m\sin\left(b\right)-\sin\left(a\right)\right)^{2}}{r+1}\left(\frac{2}{p+1}\right)^{\frac{1}{p}}\left[\left(\frac{1}{2}+\frac{r+1}{r\left(\alpha+1\right)}\right)^{p+1}\right.$$

$$\left.-\left(\frac{r+1}{r\left(\alpha+1\right)}\right)^{p+1}\right]^{\frac{1}{p}}\left(\frac{1}{s+1}\left|\left(f\circ\sin^{-1}\right)''\left(\sin\left(a\right)\right)\right|^{q}\right.$$

$$\left.+\frac{ms}{s+1}\left|\left(f\circ\sin^{-1}\right)''\left(\sin\left(b\right)\right)\right|^{q}\right)^{\frac{1}{q}},\tag{12.74}$$

where $\frac{1}{p}+\frac{1}{q}=1$, and

$$R_{5s*}^{m}\left(f,\sin\right):=\frac{\left(m\sin\left(b\right)-\sin\left(a\right)\right)^{2}}{r+1}$$

$$\cdot\left[\left|\left(f\circ\sin^{-1}\right)''\left(\sin\left(a\right)\right)\right|^{q}H+m\left|\left(f\circ\sin^{-1}\right)''\left(\sin\left(b\right)\right)\right|^{q}\right.$$

$$\cdot\left(\frac{2}{q+1}\left(\frac{1}{2}+\frac{r+1}{r\left(\alpha+1\right)}\right)^{q+1}-\frac{2}{q+1}\left(\frac{r+1}{r\left(\alpha+1\right)}\right)^{q+1}-H\right)\right],\tag{12.75}$$

where

$$H=\int_{0}^{\frac{1}{2}}\left(\frac{r+1}{r\left(\alpha+1\right)}+t\right)^{q}t^{s}dt+\int_{\frac{1}{2}}^{1}\left(\frac{r+1}{r\left(\alpha+1\right)}+1-t\right)^{q}t^{s}dt.\tag{12.76}$$

Next we present a fractional generalized (s,m)-convex Hermite-Hadamard type inequality involving $g\left(x\right)=\sin x$, $x\in\left[0,\frac{\pi}{2}\right]$.

Theorem 12.17. *Here all as in Notation 12.4. Let $\alpha>0$, $a,b\in\left[0,\frac{\pi}{2}\right]$, $a<b$, $f\in C\left(\left[0,b\right]\right)$. Assume that $f\circ\sin^{-1}:\left[0,\sin\left(b\right)\right]\rightarrow\mathbb{R}$ is twice differentiable mapping, with $0\leq\sin\left(a\right)<m\sin\left(b\right)\leq\sin\left(b\right)$. If $\left|\left(f\circ\sin^{-1}\right)''\right|^{q}$ is measurable and (s,m)-convex on $\left[\sin\left(a\right),\sin\left(b\right)\right]$ for some fixed $q>1$ and $(s,m)\in\left(0,1\right]^{2}$, $r>0$, then*

$$H_{s*}^{m}\left(f,\sin\right)$$

$$\leq\min\left\{R_{1s*}^{m}\left(f,\sin\right),\ R_{2s*}^{m}\left(f,\sin\right),\ R_{3s*}^{m}\left(f,\sin\right),\ R_{4s*}^{m}\left(f,\sin\right),\ R_{5s*}^{m}\left(f,\sin\right)\right\}.\tag{12.77}$$

Proof.   By Theorem 12.15.   □

Finally we treat the case of $q = 1$ when $g(x) = \sin x$, $x \in \left[0, \frac{\pi}{2}\right]$.

Proposition 12.3. *Here $H_*^m(f, \sin)$ as in (12.62) of Notation 12.3. The rest of the assumptions as in Theorem 12.16 with $q = 1$. Then*

$$H_*^m(f, \sin) \leq (\sin(b) - \sin(a))^2 \left(\frac{\alpha}{r(\alpha+1)(\alpha+2)} + \frac{1}{4(r+1)}\right)$$

$$\cdot \left(\frac{\left|(f \circ \sin^{-1})''(\sin(a))\right| + m\left|(f \circ \sin^{-1})''\left(\frac{\sin(b)}{m}\right)\right|}{2}\right). \tag{12.78}$$

Proof. By Proposition 12.1. ∎

Proposition 12.4. *Here $H_{s*}^m(f, \sin)$ as in (12.69) of Notation 12.4. The rest of the assumptions as in Theorem 12.17 with $q = 1$. Then*

$$H_{s*}^m(f, \sin) \leq (m\sin(b) - \sin(a))^2 \left[\left|(f \circ \sin^{-1})''(\sin(a))\right| I\right.$$

$$\left. + m\left|(f \circ \sin^{-1})''(\sin(b))\right| \left(\frac{\alpha}{r(\alpha+1)(\alpha+2)} + \frac{1}{4(r+1)} - I\right)\right], \tag{12.79}$$

where I as in (12.51).

Proof. By Proposition 12.2. ∎

Bibliography

1. G.A. Anastassiou, *Generalised fractional Hermite-Hadamard inequalities involving m-convexity and (s, m)-convexity*, Scientific J. Facta Universitatis Ser. Math. Inform. 28(2) (2013), 107-126.
2. G.A. Anastassiou, *The reduction method in fractional Calculus and fractional Ostrowski type inequalities*, Indian J. Math., accepted 2013.
3. H. Hudzik, L. Maligranda, *Some remarks on s-convex functions*, Aequationes Math. 48 (1994), 100-111.
4. A.A. Kilbas, H.M. Srivastava, J.J. Trujillo, *Theory and Applications of Fractional Differential Equations*, North-Holland Mathematics Studies, Vol. 204, Elsevier, New York, 2006.
5. D.S. Mitrinović, I.B. Lacković, *Hermite and convexity*, Aequationes Math. 28 (1985), 229-232.
6. M.R. Pinheiro, *Exploring the concept of s-convexity*, Aequationes Math. 74 (2007), 201-209.
7. G.H. Toader, *Some generalisations of the convexity*, Proc. Colloq. Approx. Optim., (1984), 329-338.
8. Y. Zhang, J.R. Wang, *On some new Hermite-Hadamard inequalities involving Riemann-Liouville fractional integrals*, J. Inequalities Appl. 2013 (2013), 220, doi: 10.1186/1029-242X-2013-220.

Chapter 13

About the Reduction Method in Fractional Calculus and Fractional Ostrowski Inequalities

Here we study generalized fractional integrals and fractional derivatives. We present the reduction method of Fractional Calculus and we reduce them to basic fractional integrals and fractional derivatives. We give a series of generalized Ostrowski type fractional inequalities involving s-convexity. We apply all of the above to Hadamard and Erdélyi-Kober fractional integrals and fractional derivatives. We produce also important generalized fractional Taylor formulae. It follows [4].

13.1   The Reduction Method in Fractional Calculus

We use a lot here the following generalized fractional integrals.

Definition 13.1. (see also [9, p. 99]) The left and right fractional integrals, respectively, of a function f with respect to given function g are defined as follows:

Let $a, b \in \mathbb{R}$, $a < b$, $\alpha > 0$. Here $g \in AC([a, b])$ (absolutely continuous functions) and is strictly increasing, $f \in L_\infty([a, b])$. We set

$$\left(I_{a+;g}^\alpha f\right)(x) = \frac{1}{\Gamma(\alpha)} \int_a^x (g(x) - g(t))^{\alpha-1} g'(t) f(t)\, dt, \quad x \geq a, \qquad (13.1)$$

clearly $\left(I_{a+;g}^\alpha f\right)(a) = 0$,
and

$$\left(I_{b-;g}^\alpha f\right)(x) = \frac{1}{\Gamma(\alpha)} \int_x^b (g(t) - g(x))^{\alpha-1} g'(t) f(t)\, dt, \quad x \leq b, \qquad (13.2)$$

clearly $\left(I_{b-;g}^\alpha f\right)(b) = 0$.

When g is the identity function id, we get that $I_{a+;id}^\alpha = I_{a+}^\alpha$ and $I_{b-;id}^\alpha = I_{b-}^\alpha$ the ordinary left and right Riemann-Liouville fractional integrals, where

$$\left(I_{a+}^\alpha f\right)(x) = \frac{1}{\Gamma(\alpha)} \int_a^x (x - t)^{\alpha-1} f(t)\, dt, \quad x \geq a, \qquad (13.3)$$

$\left(I_{a+}^\alpha f\right)(a) = 0$, and

$$\left(I_{b-}^\alpha f\right)(x) = \frac{1}{\Gamma(\alpha)} \int_x^b (t - x)^{\alpha-1} f(t)\, dt, \quad x \leq b, \qquad (13.4)$$

$\left(I_{b-}^\alpha f\right)(b) = 0$.

We need

Lemma 13.1. *Let $g \in AC\left([a,b]\right)$ which is strictly increasing and Borel measurable $f \in L_\infty\left([a,b]\right)$. Then*

$$\|f\|_{\infty,[a,b]} \geq \|f \circ g^{-1}\|_{\infty,[g(a),g(b)]}, \qquad (13.5)$$

i.e. $\left(f \circ g^{-1}\right) \in L_\infty\left([g\left(a\right),g\left(b\right)]\right).$
   If additionally $g^{-1} \in AC\left([g\left(a\right),g\left(b\right)]\right)$ *then*

$$\|f\|_{\infty,[a,b]} = \|f \circ g^{-1}\|_{\infty,[g(a),g(b)]}. \qquad (13.6)$$

Proof.   Here m stands for the Lebesgue measure. By definition we have

$$\|f\|_{\infty,[a,b]} = \operatorname{ess\,sup}|f\left(t\right)|$$

$$= \inf\left\{M : m\left\{t : \left|\left(f \circ g^{-1}\right)\left(g\left(t\right)\right)\right| > M\right\} = m\left\{t : |f\left(t\right)| > M\right\} = 0\right\}. \quad (13.7)$$

Furthermore we have

$$\|f \circ g^{-1}\|_{\infty,[g(a),g(b)]} = \inf\left\{L : m\left\{g\left(t\right) : \left|\left(f \circ g^{-1}\right)\left(g\left(t\right)\right)\right| > L\right\} = 0\right\}. \quad (13.8)$$

Because g is absolutely continuous and strictly increasing function on $[a,b]$, by [10, p. 108] exercise 14, we get that

$$m\left\{g\left(t\right) : \left|\left(f \circ g^{-1}\right)\left(g\left(t\right)\right)\right| > M\right\} = m\left(g\left(\left\{t : \left|\left(f \circ g^{-1}\right)\left(g\left(t\right)\right)\right| > M\right\}\right)\right) = 0, \quad (13.9)$$

given that $m\left\{t : \left|\left(f \circ g^{-1}\right)\left(g\left(t\right)\right)\right| > M\right\} = 0.$
Therefore each M of (13.7) fulfills $M \in \left\{L : m\{g(t) : \left|(f \circ g^{-1})(g(t))\right| > L\} = 0\right\}.$
   The last implies (13.5). Similarly arguing reverse we derive (13.6). $\square$

Under the spirit that a Lebesgue measurable function equals a Borel measurable function almost everywhere we use (13.5) in the next

Remark 13.1. We observe that

$$\left(I_{a+;g}^\alpha f\right)\left(x\right) = \frac{1}{\Gamma\left(\alpha\right)}\int_a^x \left(g\left(x\right) - g\left(t\right)\right)^{\alpha-1}\left(f \circ g^{-1}\right)\left(g\left(t\right)\right)g'\left(t\right)dt$$

(by change of variable for Lebesgue integrals)

$$= \frac{1}{\Gamma\left(\alpha\right)}\int_{g(a)}^{g(x)} \left(g\left(x\right) - z\right)^{\alpha-1}\left(f \circ g^{-1}\right)\left(z\right)dz = \left(I_{g(a)+}^\alpha \left(f \circ g^{-1}\right)\right)\left(g\left(x\right)\right), \quad x \geq a, \quad (13.10)$$

equivalently $g\left(x\right) \geq g\left(a\right).$
   That is in the terms and assumptions of Definition 13.1 we get

$$\left(I_{a+;g}^\alpha f\right)\left(x\right) = \left(I_{g(a)+}^\alpha \left(f \circ g^{-1}\right)\right)\left(g\left(x\right)\right), \quad \text{for } x \geq a. \quad (13.11)$$

Similarly we observe that

$$\left(I_{b-;g}^\alpha f\right)\left(x\right) = \frac{1}{\Gamma\left(\alpha\right)}\int_x^b \left(g\left(t\right) - g\left(x\right)\right)^{\alpha-1}\left(f \circ g^{-1}\right)\left(g\left(t\right)\right)g'\left(t\right)dt$$

$$= \frac{1}{\Gamma(\alpha)} \int_{g(x)}^{g(b)} (z - g(x))^{\alpha-1} (f \circ g^{-1})(z)\, dz = \left(I_{g(b)-}^{\alpha} (f \circ g^{-1})\right)(g(x)), \quad (13.12)$$

for $x \leq b$.

That is

$$\left(I_{b-;g}^{\alpha} f\right)(x) = \left(I_{g(b)-}^{\alpha} (f \circ g^{-1})\right)(g(x)), \quad \text{for } x \leq b. \qquad (13.13)$$

So by (13.11) and (13.13) we have reduced the general fractional integrals to the ordinary left and right Riemann-Liouville fractional integrals.

We need

Definition 13.2. ([8]) Let $0 < a < b < \infty$, $\alpha > 0$. The left and right Hadamard fractional integrals of order α are given by

$$\left(J_{a+}^{\alpha} f\right)(x) = \frac{1}{\Gamma(\alpha)} \int_{a}^{x} \left(\ln \frac{x}{y}\right)^{\alpha-1} \frac{f(y)}{y}\, dy, \quad x \geq a, \qquad (13.14)$$

and

$$\left(J_{b-}^{\alpha} f\right)(x) = \frac{1}{\Gamma(\alpha)} \int_{x}^{b} \left(\ln \frac{y}{x}\right)^{\alpha-1} \frac{f(y)}{y}\, dy, \quad x \leq b, \qquad (13.15)$$

respectively.

Here we take $f \in L_{\infty}([a,b])$.

Comparing to Definition 13.1 we have $g(x) = \ln x$ on $[a,b]$, $0 < a < b < \infty$. Comparing to (13.11) and (13.13) we get

$$\left(J_{a+}^{\alpha} f\right)(x) = \left(I_{(\ln a)+}^{\alpha} (f \circ \exp)\right)(\ln x), \quad \text{for } x \geq a, \qquad (13.16)$$

and

$$\left(J_{b-}^{\alpha} f\right)(x) = \left(I_{(\ln b)-}^{\alpha} (f \circ \exp)\right)(\ln x), \quad \text{for } x \leq b. \qquad (13.17)$$

We also consider

Definition 13.3. Let $0 < a < b < \infty$; $\alpha, \sigma > 0$ and $\eta > -1$. Let $f \in L_{\infty}([a,b])$. We mention here the left and right Erdélyi-Kober type fractional integrals, respectively: as in [11] we define

$$\left(I_{a+;\sigma,\eta}^{\alpha} f\right)(x) = \frac{\sigma x^{-\sigma(\alpha+\eta)}}{\Gamma(\alpha)} \int_{a}^{x} (x^{\sigma} - t^{\sigma})^{\alpha-1} t^{\sigma(\eta+1)-1} f(t)\, dt, \quad x \geq a, \qquad (13.18)$$

and similarly we also define

$$\left(I_{b-;\sigma,\eta}^{\alpha} f\right)(x) = \frac{\sigma x^{-\sigma(\alpha+\eta)}}{\Gamma(\alpha)} \int_{x}^{b} (t^{\sigma} - x^{\sigma})^{\alpha-1} t^{\sigma(\eta+1)-1} f(t)\, dt, \quad x \leq b. \qquad (13.19)$$

Remark 13.2. (following Definition 13.3) The above give rise to the following generalized weighted left and right fractional integrals.

We set

$$\left(K^{\alpha}_{a+;\sigma,\eta}f\right)(x) = x^{\sigma(\alpha+\eta)}\left(I^{\alpha}_{a+;\sigma,\eta}f\right)(x) \tag{13.20}$$

$$= \frac{\sigma}{\Gamma(\alpha)}\int_{a}^{x}(x^{\sigma}-t^{\sigma})^{\alpha-1}\,t^{\sigma(\eta+1)-1}f(t)\,dt$$

$$= \frac{1}{\Gamma(\alpha)}\int_{a}^{x}(x^{\sigma}-t^{\sigma})^{\alpha-1}\,(t^{\sigma\eta}f(t))\,\sigma t^{\sigma-1}dt$$

$$= \frac{1}{\Gamma(\alpha)}\int_{a}^{x}(x^{\sigma}-t^{\sigma})^{\alpha-1}\,(t^{\sigma\eta}f(t))\,dt^{\sigma}$$

(setting $z = t^{\sigma}$)

$$= \frac{1}{\Gamma(\alpha)}\int_{a^{\sigma}}^{x^{\sigma}}(x^{\sigma}-z)^{\alpha-1}\left(z^{\eta}f\left(z^{\frac{1}{\sigma}}\right)\right)dz = \left(I^{\alpha}_{a^{\sigma}+}\left(z^{\eta}f\left(z^{\frac{1}{\sigma}}\right)\right)\right)(x^{\sigma}),\ \ x \geq a,$$

(13.21)

that is

$$\left(K^{\alpha}_{a+;\sigma,\eta}f\right)(x) = \left(I^{\alpha}_{a^{\sigma}+}\left(z^{\eta}f\left(z^{\frac{1}{\sigma}}\right)\right)\right)(x^{\sigma}),\ \ \ x \geq a. \tag{13.22}$$

Similarly we put

$$\left(K^{\alpha}_{b-;\sigma,\eta}f\right)(x) = x^{\sigma(\alpha+\eta)}\left(I^{\alpha}_{b-;\sigma,\eta}f\right)(x)$$

$$= \frac{1}{\Gamma(\alpha)}\int_{x^{\sigma}}^{b^{\sigma}}(z-x^{\sigma})^{\alpha-1}\left(z^{\eta}f\left(z^{\frac{1}{\sigma}}\right)\right)dz = \left(I^{\alpha}_{b^{\sigma}-}\left(z^{\eta}f\left(z^{\frac{1}{\sigma}}\right)\right)\right)(x^{\sigma}),\ \ x \leq b,$$

(13.23)

that is

$$\left(K^{\alpha}_{b-;\sigma,\eta}f\right)(x) = \left(I^{\alpha}_{b^{\sigma}-}\left(z^{\eta}f\left(z^{\frac{1}{\sigma}}\right)\right)\right)(x^{\sigma}),\ \ \ x \leq b. \tag{13.24}$$

Comparing to Definition 13.1 here, we have that $g(x) = x^{\sigma} \in C^{1}([a,b])$, thus $x^{\sigma} \in AC([a,b])$ and it is strictly increasing. Clearly $g^{-1}(z) = z^{\frac{1}{\sigma}}$, $z \in [a^{\sigma}, b^{\sigma}]$. We set $F(t) = t^{\sigma\eta}f(t)$, $t \in [a,b]$. Clearly we have $F \in L_{\infty}([a,b])$. Notice that $F \circ g^{-1} = F \circ (id)^{\frac{1}{\sigma}}$, and $F(t) = (F \circ g^{-1})(g(t)) = (F \circ g^{-1})(z) = z^{\eta}f\left(z^{\frac{1}{\sigma}}\right)$. Thus a formal description of (13.22) and (13.24) follows.

We have

$$\left(K^{\alpha}_{a+;\sigma,\eta}f\right)(x) = \left(I^{\alpha}_{a^{\sigma}+}\left(F \circ (id)^{\frac{1}{\sigma}}\right)\right)(x^{\sigma}),\ \ \ x \geq a, \tag{13.25}$$

and

$$\left(K^{\alpha}_{b-;\sigma,\eta}f\right)(x) = \left(I^{\alpha}_{b^{\sigma}-}\left(F \circ (id)^{\frac{1}{\sigma}}\right)\right)(x^{\sigma}),\ \ \ x \leq b, \tag{13.26}$$

where $F(x) = x^{\sigma\eta}f(x)$, $x \in [a,b]$.

We introduce

Definition 13.4. Let $\alpha > 0$, $m = [\alpha]$, $\beta = \alpha - m$, $0 < \beta < 1$, $f \in C([a, b])$, $[a, b] \subset \mathbb{R}$; $g \in AC([a, b])$, g is strictly increasing. We define the subspace $C_{a+;g}^{\alpha}([a, b])$ of $C^m([a, b])$:

$$C_{a+;g}^{\alpha}([a, b]) = \left\{ f \in C^m([a, b]) : \left(I_{a+;g}^{1-\beta} f^{(m)} \right) \in C^1([a, b]) \right\}. \tag{13.27}$$

Denote $C_{a+}^{\alpha} = C_{a+;id}^{\alpha}$.

For $f \in C_{a+;g}^{\alpha}([a, b])$, we define the left g-generalized α-fractional derivative of f over $[a, b]$ as

$$D_{a+;g}^{\alpha}(f) = \left(I_{a+;g}^{1-\beta} f^{(m)} \right)'. \tag{13.28}$$

When $g = id$, we denote

$$D_{a+}^{\alpha} f = \left(I_{a+}^{1-\beta} f^{(m)} \right)', \tag{13.29}$$

called the left generalized α-fractional derivative of f over $[a, b]$, see [5], [2], p. 24.

We also introduce

Definition 13.5. Let $\alpha > 0$, $m = [\alpha]$, $\beta = \alpha - m$, $f \in C([a, b])$, $[a, b] \subset \mathbb{R}$; $g \in AC([a, b])$, g is strictly increasing. We define the subspace $C_{b-;g}^{\alpha}([a, b])$ of $C^m([a, b])$:

$$C_{b-;g}^{\alpha}([a, b]) = \left\{ f \in C^m([a, b]) : \left(I_{b-;g}^{1-\beta} f^{(m)} \right) \in C^1([a, b]) \right\}. \tag{13.30}$$

Denote $C_{b-}^{\alpha} = C_{b-;id}^{\alpha}$.

For $f \in C_{b-;g}^{\alpha}([a, b])$, we define the right g-generalized α-fractional derivative of f over $[a, b]$ as

$$D_{b-;g}^{\alpha}(f) = (-1)^{m-1} \left(I_{b-;g}^{1-\beta} f^{(m)} \right)'. \tag{13.31}$$

When $g = id$, we denote

$$D_{b-}^{\alpha} f = (-1)^{m-1} \left(I_{b-}^{1-\beta} f^{(m)} \right)', \tag{13.32}$$

called the right generalized α-fractional derivative of f over $[a, b]$, see [3].

Regarding fractional derivatives in this chapter from now on we consider only $0 < \alpha < 1$, i.e. $m = 0$ and $\beta = \alpha$.

So in this case we get

$$\left(D_{a+;g}^{\alpha} f \right)(x) = \frac{1}{\Gamma(1-\alpha)} \frac{d}{dx} \int_a^x (g(x) - g(t))^{-\alpha} g'(t) f(t)\, dt, \tag{13.33}$$

$$\left(D_{a+}^{\alpha} f \right)(x) = \frac{1}{\Gamma(1-\alpha)} \frac{d}{dx} \int_a^x (x - t)^{-\alpha} f(t)\, dt, \tag{13.34}$$

and

$$\left(D^\alpha_{b-;g}f\right)(x) = -\frac{1}{\Gamma(1-\alpha)}\frac{d}{dx}\int_x^b (g(t)-g(x))^{-\alpha}\,g'(t)\,f(t)\,dt, \qquad (13.35)$$

$$\left(D^\alpha_{b-}f\right)(x) = -\frac{1}{\Gamma(1-\alpha)}\frac{d}{dx}\int_x^b (t-x)^{-\alpha}\,f(t)\,dt, \qquad (13.36)$$

for any $x \in [a,b]$.

We mention the following fractional Taylor formulae.

Theorem 13.1.

1) *(see [2], pp. 8-10, [5]) Let $f \in C^\alpha_{a+}([a,b])$, $0 < \alpha < 1$. Then*

$$f(x) = \frac{1}{\Gamma(\alpha)}\int_a^x (x-t)^{\alpha-1}\left(D^\alpha_{a+}f\right)(t)\,dt = \left(I^\alpha_{a+}\left(D^\alpha_{a+}f\right)\right)(x), \quad x \in [a,b].$$
$$(13.37)$$

2) *(see [3]) Let $f \in C^\alpha_{b-}([a,b])$, $0 < \alpha < 1$. Then*

$$f(x) = \frac{1}{\Gamma(\alpha)}\int_x^b (t-x)^{\alpha-1}\left(D^\alpha_{b-}f\right)(t)\,dt = \left(I^\alpha_{b-}\left(D^\alpha_{b-}f\right)\right)(x), \quad x \in [a,b].$$
$$(13.38)$$

We make

Remark 13.3. Here $0 < \alpha < 1$ and $g \in C^1([a,b])$, g is strictly increasing. Furthermore we assume that $\left(D^\alpha_{g(a)+}\left(f\circ g^{-1}\right)\right)(g(x))$ exists. By (13.11) we have

$$\left(I^{1-\alpha}_{a+;g}f\right)(x) = \left(I^{1-\alpha}_{g(a)+}\left(f\circ g^{-1}\right)\right)(g(x)), \quad x \in [a,b]. \qquad (13.39)$$

Hence there exists

$$\left(D^\alpha_{a+;g}(f)\right)(x) = \left(I^{1-\alpha}_{a+;g}f\right)'(x) \stackrel{(13.39)}{=} \left(I^{1-\alpha}_{g(a)+}\left(f\circ g^{-1}\right)\right)'(g(x))\,g'(x)$$

$$= \left(D^\alpha_{g(a)+}\left(f\circ g^{-1}\right)\right)(g(x))\,g'(x), \quad x \in [a,b]. \qquad (13.40)$$

We have established that there exists

$$\left(D^\alpha_{a+;g}(f)\right)(x) = \left(D^\alpha_{g(a)+}\left(f\circ g^{-1}\right)\right)(g(x))\,g'(x), \quad x \in [a,b], \ f \in C([a,b]). \qquad (13.41)$$

Next we assume that there exists $\left(D^\alpha_{g(b)-}\left(f\circ g^{-1}\right)\right)(g(x))$. By (13.13) we get

$$\left(I^{1-\alpha}_{b-;g}f\right)(x) = \left(I^{1-\alpha}_{g(b)-}\left(f\circ g^{-1}\right)\right)(g(x)), \quad x \in [a,b]. \qquad (13.42)$$

Hence there exists

$$\left(D^\alpha_{b-;g}(f)\right)(x) = -\left(I^{1-\alpha}_{b-;g}f\right)'(x) \stackrel{(13.42)}{=} -\left(I^{1-\alpha}_{g(b)-}\left(f\circ g^{-1}\right)\right)'(g(x))\,g'(x)$$

$$= \left(D^\alpha_{g(b)-}\left(f\circ g^{-1}\right)\right)(g(x))\,g'(x), \quad x \in [a,b]. \qquad (13.43)$$

We have proved that there exists

$$\left(D^\alpha_{b-;g}(f)\right)(x) = \left(D^\alpha_{g(b)-}\left(f\circ g^{-1}\right)\right)(g(x))\,g'(x), \quad x \in [a,b], \ f \in C([a,b]). \qquad (13.44)$$

Next we apply (13.41) and (13.44).

We make

Remark 13.4. (all as in Definition 13.2) We introduce the following Hadamard type fractional derivatives, see (13.45), (13.46). Here $f \in C\left([a, b]\right)$. Let $0 < \alpha < 1$, and that $\left(D^\alpha_{(\ln a)+} \left(f \circ \exp\right)\right) (\ln x)$ exists for $x \in [a, b]$, $a > 0$.

Then by (13.41), we get

$$\left(D^\alpha_{a+;\ln} (f)\right) (x) = \frac{\left(D^\alpha_{(\ln a)+} \left(f \circ \exp\right)\right) (\ln x)}{x}, \quad x \in [a, b]. \tag{13.45}$$

Assume next that $\left(D^\alpha_{(\ln b)-} \left(f \circ \exp\right)\right) (\ln x)$ exists for $x \in [a, b]$.

Then by (13.44), we find

$$\left(D^\alpha_{b-;\ln} (f)\right) (x) = \frac{\left(D^\alpha_{(\ln b)-} \left(f \circ \exp\right)\right) (\ln x)}{x}, \quad x \in [a, b]. \tag{13.46}$$

We make

Remark 13.5. (refer to Definition 13.3, Remark 13.2) Let $0 < \alpha < 1$ By (13.25) we get

$$\left(K^{1-\alpha}_{a+;\sigma,\eta} f\right) (x) = \left(I^{1-\alpha}_{a^\sigma+} \left(F \circ (id)^{\frac{1}{\sigma}}\right)\right) (x^\sigma), \quad x \in [a, b]. \tag{13.47}$$

And by (13.26)

$$\left(K^{1-\alpha}_{b-;\sigma,\eta} f\right) (x) = \left(I^{1-\alpha}_{b^\sigma-} \left(F \circ (id)^{\frac{1}{\sigma}}\right)\right) (x^\sigma), \quad x \in [a, b]. \tag{13.48}$$

Above $F(x) = x^{\sigma\eta} f(x)$, $x \in [a, b]$.

Assume that

$$\frac{d \left(I^{1-\alpha}_{a^\sigma+} \left(F \circ (id)^{\frac{1}{\sigma}}\right)\right) (x^\sigma)}{dx^\sigma} \tag{13.49}$$

and

$$\frac{d \left(I^{1-\alpha}_{b^\sigma-} \left(F \circ (id)^{\frac{1}{\sigma}}\right)\right) (x^\sigma)}{dx^\sigma} \tag{13.50}$$

exist and are continuous in $x^\sigma \in [a^\sigma, b^\sigma]$, $f \in C\left([a, b]\right)$.

Then

$$\frac{d \left(K^{1-\alpha}_{a+;\sigma,\eta} f\right) (x)}{dx} = \frac{d \left(I^{1-\alpha}_{a^\sigma+} \left(F \circ (id)^{\frac{1}{\sigma}}\right)\right) (x^\sigma)}{dx^\sigma} \sigma x^{\sigma-1}, \tag{13.51}$$

and

$$\frac{d \left(K^{1-\alpha}_{b-;\sigma,\eta} f\right) (x)}{dx} = \frac{d \left(I^{1-\alpha}_{b^\sigma-} \left(F \circ (id)^{\frac{1}{\sigma}}\right)\right) (x^\sigma)}{dx^\sigma} \sigma x^{\sigma-1}, \tag{13.52}$$

exist and are continuous in $x \in [a, b]$.

So we introduce the modified Erdélyi-Kober type left and right fractional derivatives of $f \in C\left([a, b]\right)$, as follows:

$$\left(D_{a+:\sigma, \eta}^{\alpha} f\right)(x) = \frac{d\left(K_{a+;\sigma, \eta}^{1-\alpha} f\right)(x)}{dx}, \tag{13.53}$$

and

$$\left(D_{b-:\sigma, \eta}^{\alpha} f\right)(x) = -\frac{d\left(K_{b-;\sigma, \eta}^{1-\alpha} f\right)(x)}{dx}, \tag{13.54}$$

$x \in [a, b], \ 0 < \alpha < 1$.

That is, it holds

$$\left(D_{a+:\sigma, \eta}^{\alpha} f\right)(x) = \left(D_{a^{\sigma}+}^{\alpha}\left(F \circ (id)^{\frac{1}{\sigma}}\right)\right)(x^{\sigma}) \sigma x^{\sigma-1}, \tag{13.55}$$

and

$$\left(D_{b-:\sigma, \eta}^{\alpha} f\right)(x) = \left(D_{b^{\sigma}-}^{\alpha}\left(F \circ (id)^{\frac{1}{\sigma}}\right)\right)(x^{\sigma}) \sigma x^{\sigma-1}, \tag{13.56}$$

$x \in [a, b], \ 0 < \alpha < 1, \ a > 0$.

We make

Remark 13.6. (continuation of Remark 13.4) Hence $f \in C\left([a, b]\right)$. By (13.45) we get

$$\left(D_{(\ln a)+}^{\alpha}(f \circ \exp)\right)(\ln x) = x\left(D_{a+;\ln}^{\alpha}(f)\right)(x) = e^{\ln x}\left(D_{a+;\ln}^{\alpha}(f)\right)\left(e^{\ln x}\right), \tag{13.57}$$

$x \in [a, b]$.

Hence by (13.37) we obtain

$$f(x) = f\left(e^{\ln x}\right) = \left(I_{(\ln a)+}^{\alpha}\left(D_{(\ln a)+}^{\alpha}(f \circ \exp)\right)\right)(\ln x) \tag{13.58}$$

$$= \left(I_{(\ln a)+}^{\alpha}\left(e^{t}\left(D_{a+;\ln}^{\alpha}(f)\right)\left(e^{t}\right)\right)\right)(\ln x)$$

$$= \frac{1}{\Gamma(\alpha)} \int_{\ln a}^{\ln x} (\ln x - t)^{\alpha-1} e^{t}\left(D_{a+;\ln}^{\alpha}(f)\right)\left(e^{t}\right) dt, \tag{13.59}$$

$x \in [a, b]$.

By (13.46) we have

$$\left(D_{(\ln b)-}^{\alpha}(f \circ \exp)\right)(\ln x) = x\left(D_{b-;\ln}^{\alpha}(f)\right)(x) = e^{\ln x}\left(D_{b-;\ln}^{\alpha}(f)\right)\left(e^{\ln x}\right), \tag{13.60}$$

$x \in [a, b]$.

Hence by (13.38) we obtain

$$f(x) = f\left(e^{\ln x}\right) = \left(I_{(\ln b)-}^{\alpha}\left(D_{(\ln b)-}^{\alpha}(f \circ \exp)\right)\right)(\ln x)$$

$$= \left(I_{(\ln b)-}^{\alpha}\left(e^{t}\left(D_{b-;\ln}^{\alpha}(f)\right)\left(e^{t}\right)\right)\right)(\ln x)$$

$$= \frac{1}{\Gamma(\alpha)} \int_{\ln x}^{\ln b} (t - \ln x)^{\alpha-1} e^{t}\left(D_{b-;\ln}^{\alpha}(f)\right)\left(e^{t}\right) dt, \tag{13.61}$$

$x \in [a, b]$.

We have proved the following Taylor Hadamard type fractional formulae.

Theorem 13.2. *Let $0 < \alpha < 1$, and all as in Definition 13.2, $f \in C\left([a,b]\right)$, $a > 0$.*

1) Assume that $\left(D^\alpha_{(\ln a)+}\left(f \circ \exp\right)\right)(\ln x)$ exists and it is continuous, $x \in [a,b]$.
Then

$$f\left(x\right) = \left(I^\alpha_{(\ln a)+}\left(e^t\left(D^\alpha_{a+;\ln}\left(f\right)\right)\left(e^t\right)\right)\right)(\ln x)$$

$$= \frac{1}{\Gamma\left(\alpha\right)}\int_{\ln a}^{\ln x}\left(\ln x - t\right)^{\alpha-1}e^t\left(D^\alpha_{a+;\ln}\left(f\right)\right)\left(e^t\right)dt, \qquad (13.62)$$

$x \in [a,b]$.

2) Assume that $\left(D^\alpha_{(\ln b)-}\left(f \circ \exp\right)\right)(\ln x)$ exists and it is continuous, $x \in [a,b]$.
Then

$$f\left(x\right) = \left(I^\alpha_{(\ln b)-}\left(e^t\left(D^\alpha_{b-;\ln}\left(f\right)\right)\left(e^t\right)\right)\right)(\ln x)$$

$$= \frac{1}{\Gamma\left(\alpha\right)}\int_{\ln x}^{\ln b}\left(t - \ln x\right)^{\alpha-1}e^t\left(D^\alpha_{b-;\ln}\left(f\right)\right)\left(e^t\right)dt, \qquad (13.63)$$

$x \in [a,b]$.

We make

Remark 13.7. (continuation of Remark 13.5) By (13.55) and (13.56) we get

$$\left(D^\alpha_{a^\sigma+}\left(F \circ (id)^{\frac{1}{\sigma}}\right)\right)(x^\sigma) = \frac{x^{1-\sigma}}{\sigma}\left(D^\alpha_{a+;\sigma,\eta}f\right)(x)$$

$$= \frac{(x^\sigma)^{\left(\frac{1}{\sigma}-1\right)}}{\sigma}\left(D^\alpha_{a+;\sigma,\eta}f\right)\left((x^\sigma)^{\frac{1}{\sigma}}\right), \qquad (13.64)$$

and

$$\left(D^\alpha_{b^\sigma-}\left(F \circ (id)^{\frac{1}{\sigma}}\right)\right)(x^\sigma) = \frac{x^{1-\sigma}}{\sigma}\left(D^\alpha_{b-;\sigma,\eta}f\right)(x)$$

$$= \frac{(x^\sigma)^{\left(\frac{1}{\sigma}-1\right)}}{\sigma}\left(D^\alpha_{b-;\sigma,\eta}f\right)\left((x^\sigma)^{\frac{1}{\sigma}}\right), \qquad (13.65)$$

$x \in [a,b]$, $0 < \alpha < 1$, $f \in C\left([a,b]\right)$. Above assume $\left(D^\alpha_{a^\sigma+}\left(F \circ (id)^{\frac{1}{\sigma}}\right)\right)(x^\sigma)$, $\left(D^\alpha_{b^\sigma-}\left(F \circ (id)^{\frac{1}{\sigma}}\right)\right)(x^\sigma)$ exist and are continuous in $x^\sigma \in [a^\sigma, b^\sigma]$.

Hence, by (13.37) it holds

$$x^{\sigma\eta}f\left(x\right) = \left(F \circ (id)^{\frac{1}{\sigma}}\right)(x^\sigma) = \left(I^\alpha_{a^\sigma+}\left(D^\alpha_{a^\sigma+}\left(F \circ (id)^{\frac{1}{\sigma}}\right)\right)\right)(x^\sigma)$$

$$= \frac{1}{\sigma}\left(I^\alpha_{a^\sigma+}\left(t^{\left(\frac{1}{\sigma}-1\right)}\left(D^\alpha_{a+;\sigma,\eta}f\right)\left(t^{\frac{1}{\sigma}}\right)\right)\right)(x^\sigma) \qquad (13.66)$$

$$= \frac{1}{\sigma \Gamma(\alpha)} \int_{a^\sigma}^{x^\sigma} (x^\sigma - t)^{\alpha-1} t^{\left(\frac{1}{\sigma}-1\right)} \left(D_{a+;\sigma,\eta}^\alpha f\right) \left(t^{\frac{1}{\sigma}}\right) dt, \quad x \in [a, b]. \qquad (13.67)$$

Similarly, by (13.38) we derive

$$x^{\sigma\eta} f(x) = \left(F \circ (id)^{\frac{1}{\sigma}}\right)(x^\sigma) = \left(I_{b^\sigma-}^\alpha \left(D_{b^\sigma-}^\alpha \left(F \circ (id)^{\frac{1}{\sigma}}\right)\right)\right)(x^\sigma) \qquad (13.68)$$

$$= \frac{1}{\sigma} \left(I_{b^\sigma-}^\alpha \left(t^{\left(\frac{1}{\sigma}-1\right)} \left(D_{b-;\sigma,\eta}^\alpha f\right) \left(t^{\frac{1}{\sigma}}\right)\right)\right)(x^\sigma)$$

$$= \frac{1}{\sigma \Gamma(\alpha)} \int_{x^\sigma}^{b^\sigma} (t - x^\sigma)^{\alpha-1} t^{\left(\frac{1}{\sigma}-1\right)} \left(D_{b-;\sigma,\eta}^\alpha f\right) \left(t^{\frac{1}{\sigma}}\right) dt, \quad x \in [a, b]. \qquad (13.69)$$

We give the following Taylor Erdélyi-Kober type fractional formulae.

Theorem 13.3. *Let $0 < \alpha < 1$, all as in Definition 13.3, (13.20), (13.23), (13.53), (13.54), $f \in C([a, b])$, $a > 0$; $F(x) = x^{\sigma\eta} f(x)$, $x \in [a, b]$.*

1) Assume that $\left(D_{a^\sigma+}^\alpha \left(F \circ (id)^{\frac{1}{\sigma}}\right)\right)(x^\sigma)$ exists and it is continuous in $x^\sigma \in [a^\sigma, b^\sigma]$. Then

$$f(x) = \frac{x^{-\sigma\eta}}{\sigma} \left(I_{a^\sigma+}^\alpha \left(t^{\left(\frac{1}{\sigma}-1\right)} \left(D_{a+;\sigma,\eta}^\alpha f\right) \left(t^{\frac{1}{\sigma}}\right)\right)\right)(x^\sigma)$$

$$= \frac{x^{-\sigma\eta}}{\sigma \Gamma(\alpha)} \int_{a^\sigma}^{x^\sigma} (x^\sigma - t)^{\alpha-1} t^{\left(\frac{1}{\sigma}-1\right)} \left(D_{a+;\sigma,\eta}^\alpha f\right) \left(t^{\frac{1}{\sigma}}\right) dt, \quad x \in [a, b]. \qquad (13.70)$$

2) Assume that $\left(D_{b^\sigma-}^\alpha \left(F \circ (id)^{\frac{1}{\sigma}}\right)\right)(x^\sigma)$ exists and it is continuous in $x^\sigma \in [a^\sigma, b^\sigma]$. Then

$$f(x) = \frac{x^{-\sigma\eta}}{\sigma} \left(I_{b^\sigma-}^\alpha \left(t^{\left(\frac{1}{\sigma}-1\right)} \left(D_{b-;\sigma,\eta}^\alpha f\right) \left(t^{\frac{1}{\sigma}}\right)\right)\right)(x^\sigma)$$

$$= \frac{x^{-\sigma\eta}}{\sigma \Gamma(\alpha)} \int_{x^\sigma}^{b^\sigma} (t - x^\sigma)^{\alpha-1} t^{\left(\frac{1}{\sigma}-1\right)} \left(D_{b-;\sigma,\eta}^\alpha f\right) \left(t^{\frac{1}{\sigma}}\right) dt, \quad x \in [a, b]. \qquad (13.71)$$

13.2   Fractional Ostrowski Type Inequalities

We need the following

Lemma 13.2. *([12]) Let $f : [a, b] \to \mathbb{R}$ be a differentiable mapping on (a, b) with $a < b$. If $f' \in L_1([a, b])$, then for all $x \in [a, b]$ and $\alpha > 0$ we have:*

$$\left(\frac{(x-a)^\alpha + (b-x)^\alpha}{b-a}\right) f(x) - \frac{\Gamma(\alpha+1)}{(b-a)} \left[I_{x-}^\alpha f(a) + I_{x+}^\alpha f(b)\right]$$

$$= \frac{(x-a)^{\alpha+1}}{b-a} \int_0^1 t^\alpha f'(tx + (1-t)a) dt - \frac{(b-x)^{\alpha+1}}{b-a} \int_0^1 t^\alpha f'(tx + (1-t)b) dt. \qquad (13.72)$$

By (13.72), (13.11), (13.13), we obtain

Lemma 13.3. *Let* $f \in C([a,b])$, $g \in C^1([a,b])$, g *strictly increasing on* $[a,b]$, $f \circ g^{-1}$ *differentiable on* $(g(a), g(b))$ *with* $(f \circ g^{-1})' \in L_1([g(a), g(b)])$, $x \in [a,b]$, $a < b$, $a, b \in \mathbb{R}$, $\alpha > 0$. *Then*

$$\left(\frac{(g(x) - g(a))^\alpha + (g(b) - g(x))^\alpha}{g(b) - g(a)} \right) f(x)$$

$$- \frac{\Gamma(\alpha + 1)}{(g(b) - g(a))} \left[\left(I^\alpha_{x-;g} f \right)(a) + \left(I^\alpha_{x+;g} f \right)(b) \right]$$

$$= \frac{(g(x) - g(a))^{\alpha+1}}{(g(b) - g(a))} \int_0^1 t^\alpha \left(f \circ g^{-1} \right)' (tg(x) + (1-t) g(a)) \, dt \qquad (13.73)$$

$$- \frac{(g(b) - g(x))^{\alpha+1}}{(g(b) - g(a))} \int_0^1 t^\alpha \left(f \circ g^{-1} \right)' (tg(x) + (1-t) g(b)) \, dt.$$

We apply (13.73) for $g(x) = \ln x$, $x \in [a,b]$.

Lemma 13.4. *Let* $0 < a < b < \infty$, $\alpha > 0$. *Let* $f \in C([a,b])$, $(f \circ \exp)$ *is differentiable on* $(\ln a, \ln b)$ *with* $(f \circ \exp)' \in L_1([\ln a, \ln b])$, $x \in [a,b]$. *Then*

$$\left(\frac{\left(\ln \frac{x}{a} \right)^\alpha + \left(\ln \frac{b}{x} \right)^\alpha}{\ln \frac{b}{a}} \right) f(x) - \frac{\Gamma(\alpha+1)}{\ln \left(\frac{b}{a} \right)} \left[\left(J^\alpha_{x-} f \right)(a) + \left(J^\alpha_{x+} f \right)(b) \right]$$

$$= \frac{\left(\ln \frac{x}{a} \right)^{\alpha+1}}{\ln \frac{b}{a}} \int_0^1 t^\alpha (f \circ \exp)' (t \ln x + (1-t) \ln a) \, dt$$

$$- \frac{\left(\ln \frac{b}{x} \right)^{\alpha+1}}{\ln \frac{b}{a}} \int_0^1 t^\alpha (f \circ \exp)' (t \ln x + (1-t) \ln b) \, dt, \qquad (13.74)$$

where $J^\alpha_{x\pm} f$ *are the left and right Hadamard fractional integrals of order* α *anchored at* $x \in [a,b]$, *see* (13.14), (13.15).

We apply (13.73) for $g(x) = x^\sigma$, $\sigma > 0$, $x \in [a,b]$.

Lemma 13.5. *Let* $0 < a < b < \infty$, $\alpha > 0$, $f \in C([a,b])$. *Assume* $\left(F \circ (id)^{\frac{1}{\sigma}} \right)$ *is differentiable on* (a^σ, b^σ) *with* $\left(F \circ (id)^{\frac{1}{\sigma}} \right)' \in L_1([a^\sigma, b^\sigma])$, $x \in [a,b]$. *Here* $F(x) = x^{\sigma\eta} f(x)$, $x \in [a,b]$, $\eta > -1$. *Then*

$$\left(\frac{(x^\sigma - a^\sigma)^\alpha + (b^\sigma - x^\sigma)^\alpha}{b^\sigma - a^\sigma} \right) x^{\sigma\eta} f(x)$$

$$- \frac{\Gamma(\alpha+1)}{(b^\sigma - a^\sigma)} \left[\left(K^\alpha_{x-;\sigma,\eta} f \right)(a) + \left(K^\alpha_{x+;\sigma,\eta} f \right)(b) \right]$$

$$= \frac{(x^\sigma - a^\sigma)^{\alpha+1}}{(b^\sigma - a^\sigma)} \int_0^1 t^\alpha \left(F \circ (id)^{\frac{1}{\sigma}} \right)' (tx^\sigma + (1-t) a^\sigma) \, dt$$

$$- \frac{(b^\sigma - x^\sigma)^{\alpha+1}}{(b^\sigma - a^\sigma)} \int_0^1 t^\alpha \left(F \circ (id)^{\frac{1}{\sigma}} \right)' (tx^\sigma + (1-t) b^\sigma) \, dt, \qquad (13.75)$$

where $\left(K^\alpha_{x\pm;\sigma,\eta} f \right)$ *as in* (13.25), (13.26).

We need

Definition 13.6. ([7]) A function $f : [0, \infty) \to \mathbb{R}$ is said to be s-convex in the second sense if

$$f(\lambda x + (1 - \lambda) y) \le \lambda^s f(x) + (1 - \lambda)^s f(y), \qquad (13.76)$$

for all $x, y \in [0, \infty)$, $\lambda \in [0, 1]$ and for some fixed $s \in (0, 1]$.

This class of s-convex functions is denoted by K_s^2.

When $s = 1$, s-convexity reduces to ordinary convexity.

If "$\ge$" holds in (13.76), we talk about s-concavity in the second sense.

In our proofs it is used a lot and it is built in the following

Theorem 13.4. ([6]) *Suppose that $f : [0, \infty) \to [0, \infty)$ is an s-convex function in the second sense, where $s \in (0, 1]$, and let $a, b \in [0, \infty)$, $a < b$. If $f' \in L_1([a, b])$, then it holds*

$$2^{s-1} f\left(\frac{a + b}{2}\right) \le \frac{1}{b - a} \int_a^b f(x) \, dx \le \frac{f(a) + f(b)}{s + 1}, \qquad (13.77)$$

where the constant $\frac{1}{s+1}$ is the best possible in the second inequality.

We are also motivated by the following Ostrowski type inequality in

Theorem 13.5. ([1]) *Let $f : I \subset [0, \infty) \to \mathbb{R}$ be a differentiable mapping on I^0 such that $f' \in L_1([a, b])$, where $a, b \in I$, $a < b$. If $|f'|$ is s-convex in the second sense on $[a, b]$ for some fixed $s \in (0, 1]$ and $|f'(x)| \le M$, for all $x \in [a, b]$, then*

$$\left| f(x) - \frac{1}{b - a} \int_a^b f(t) \, dt \right| \le \frac{M}{b - a} \left[\frac{(x - a)^2 + (b - x)^2}{s + 1} \right], \qquad (13.78)$$

for each $x \in [a, b]$.

We need

Theorem 13.6. ([12]) *Let $f : [a, b] \subset [0, \infty) \to \mathbb{R}$, be a differentiable mapping on (a, b) with $a < b$, such that $f' \in L_1([a, b])$. If $|f'|$ is s-convex in the second sense on $[a, b]$ for some fixed $s \in (0, 1]$ and $|f'(x)| \le M$, $x \in [a, b]$, $\alpha > 0$, then*

$$\Delta_x(f) := \left| \left(\frac{(x - a)^\alpha + (b - x)^\alpha}{b - a} \right) f(x) - \frac{\Gamma(\alpha + 1)}{(b - a)} \left[I_{x-}^\alpha f(a) + I_{x+}^\alpha f(b) \right] \right|$$

$$\le \frac{M}{b - a} \left(1 + \frac{\Gamma(\alpha + 1)\Gamma(s + 1)}{\Gamma(\alpha + s + 1)} \right) \left[\frac{(x - a)^{\alpha+1} + (b - x)^{\alpha+1}}{\alpha + s + 1} \right]. \qquad (13.79)$$

We give the following general fractional Ostrowski type inequality. The proof comes by Lemma 13.3 and Theorem 13.6.

Theorem 13.7. *Let $f \in C([a, b])$, $g \in C^1([a, b])$, g strictly increasing on $[a, b]$, $f \circ g^{-1}$ differentiable on $(g(a), g(b))$ with $(f \circ g^{-1})' \in L_1([g(a), g(b)])$, $x \in [a, b]$,*

$a < b$, $a, b \in \mathbb{R}$, $\alpha > 0$. *Assume* $\left|\left(f \circ g^{-1}\right)'\right|$ *is s-convex in the second sense on* $[g(a), g(b)] \subset [0, \infty)$ *for some fixed* $s \in (0, 1]$ *and* $\left|\left(f \circ g^{-1}\right)'(g(x))\right| \leq M$, $x \in [a, b]$. *Then*

$$\Delta_{g(x)}(f) := \left|\left(\frac{(g(x) - g(a))^\alpha + (g(b) - g(x))^\alpha}{g(b) - g(a)}\right) f(x)\right.$$

$$\left. - \frac{\Gamma(\alpha + 1)}{(g(b) - g(a))} \left[\left(I^\alpha_{x-;g} f\right)(a) + \left(I^\alpha_{x+;g} f\right)(b)\right]\right|$$

$$\leq \frac{M}{(g(b) - g(a))} \left(1 + \frac{\Gamma(\alpha + 1)\Gamma(s + 1)}{\Gamma(\alpha + s + 1)}\right)$$

$$\cdot \left[\frac{(g(x) - g(a))^{\alpha+1} + (g(b) - g(x))^{\alpha+1}}{\alpha + s + 1}\right]. \tag{13.80}$$

We need

Theorem 13.8. *([12]) All as in Theorem 13.6, but here* $|f'|^q$ *is s-convex in the second sense on* $[a, b]$ *for some fixed* $s \in (0, 1]$, $p, q > 1 : \frac{1}{p} + \frac{1}{q} = 1$. *Then*

$$\Delta_x(f) \leq \frac{M}{(1 + p\alpha)^{\frac{1}{p}}} \left(\frac{2}{s + 1}\right)^{\frac{1}{q}} \left[\frac{(x - a)^{\alpha+1} + (b - x)^{\alpha+1}}{b - a}\right]. \tag{13.81}$$

We apply Theorem 13.8 and Lemma 13.3. We give the following fractional Ostrowski type inequality.

Theorem 13.9. *All as in Theorem 13.7, however here* $\left|\left(f \circ g^{-1}\right)'\right|^q$ *is s-convex in the second sense on* $[g(a), g(b)] \subset [0, \infty)$ *for some fixed* $s \in (0, 1]$, $p, q > 1 : \frac{1}{p} + \frac{1}{q} = 1$. *Then*

$$\Delta_{g(x)}(f) \leq \frac{M}{(1 + p\alpha)^{\frac{1}{p}}} \left(\frac{2}{s + 1}\right)^{\frac{1}{q}} \left[\frac{(g(x) - g(a))^{\alpha+1} + (g(b) - g(x))^{\alpha+1}}{g(b) - g(a)}\right].$$
$$\tag{13.82}$$

We need

Theorem 13.10. *([12]) All as in Theorem 13.6, but here* $|f'|^q$ *is s-convex in the second sense on* $[a, b]$ *for some fixed* $s \in (0, 1]$, *with* $q \geq 1$. *Then*

$$\Delta_x(f) \leq M \left(\frac{1}{1 + \alpha}\right)^{1 - \frac{1}{q}} \left(\frac{1}{\alpha + s + 1}\right)^{\frac{1}{q}}$$

$$\cdot \left(1 + \frac{\Gamma(\alpha + 1)\Gamma(s + 1)}{\Gamma(\alpha + s + 1)}\right)^{\frac{1}{q}} \left[\frac{(x - a)^{\alpha+1} + (b - x)^{\alpha+1}}{b - a}\right]. \tag{13.83}$$

We give with the use of (13.83) the following

Theorem 13.11. *Here all as in Theorem 13.9, however $q \geq 1$, p is not related. Then*

$$\Delta_{g(x)}(f) \leq M \left(\frac{1}{1+\alpha} \right)^{1-\frac{1}{q}} \left(\frac{1}{\alpha+s+1} \right)^{\frac{1}{q}}$$

$$\cdot \left(1 + \frac{\Gamma(\alpha+1)\Gamma(s+1)}{\Gamma(\alpha+s+1)} \right)^{\frac{1}{q}} \left[\frac{(g(x)-g(a))^{\alpha+1} + (g(b)-g(x))^{\alpha+1}}{g(b)-g(a)} \right]. \quad (13.84)$$

We need

Theorem 13.12. *([12]) All as in Theorem 13.6, but here $|f'|^q$ is s-concave in the second sense on $[a, b]$ for some fixed $s \in (0, 1]$, $p, q > 1 : \frac{1}{p} + \frac{1}{q} = 1$. Then*

$$\Delta_x(f) \leq \frac{2^{\frac{(s-1)}{q}}}{(1+p\alpha)^{\frac{1}{p}}(b-a)}$$

$$\cdot \left[(x-a)^{\alpha+1} \left| f'\left(\frac{x+a}{2} \right) \right| + (b-x)^{\alpha+1} \left| f'\left(\frac{b+x}{2} \right) \right| \right]. \quad (13.85)$$

Using (13.85) we get

Theorem 13.13. *All as in Theorem 13.7, but here $\left| (f \circ g^{-1})' \right|^q$ is s-concave in the second sense on $[g(a), g(b)] \subset [0, \infty)$ for some fixed $s \in (0, 1]$, $p, q > 1 : \frac{1}{p} + \frac{1}{q} = 1$. Then*

$$\Delta_{g(x)}(f) \leq \frac{2^{\frac{(s-1)}{q}}}{(1+p\alpha)^{\frac{1}{p}}(g(b)-g(a))}$$

$$\cdot \left[(g(x)-g(a))^{\alpha+1} \left| (f \circ g^{-1})'\left(\frac{g(x)+g(a)}{2} \right) \right| \right. \quad (13.86)$$

$$\left. + (g(b)-g(x))^{\alpha+1} \left| (f \circ g^{-1})'\left(\frac{g(b)+g(x)}{2} \right) \right| \right].$$

We make

Remark 13.8. Let $0 < a < b < \infty$, $\alpha > 0$. We have that

$$\Delta_{\ln x}(f) = \left| \left(\frac{(\ln \frac{x}{a})^{\alpha} + (\ln \frac{b}{x})^{\alpha}}{\ln \frac{b}{a}} \right) f(x) \right.$$

$$\left. - \frac{\Gamma(\alpha+1)}{\ln \frac{b}{a}} \left[(J^{\alpha}_{x-}f)(a) + (J^{\alpha}_{x+}f)(b) \right] \right|, \quad (13.87)$$

where $J^\alpha_{x\pm}f$ are the Hadamard fractional integrals, see (13.14), (13.15), and

$$\Delta_{x^\sigma}(f) = \left| \left(\frac{(x^\sigma - a^\sigma)^\alpha + (b^\sigma - x^\sigma)^\alpha}{b^\sigma - a^\sigma} \right) x^{\sigma\eta} f(x) \right.$$

$$\left. - \frac{\Gamma(\alpha+1)}{(b^\sigma - a^\sigma)} \left[(K^\alpha_{x-;\sigma,\eta}f)(a) + (K^\alpha_{x+;\sigma,\eta}f)(b) \right] \right|, \qquad (13.88)$$

where $K^\alpha_{x\pm;\sigma,\eta}(f)$ as in (13.25), (13.26), the modified Erdélyi-Kober type fractional integrals, see also (13.18), (13.19), (13.20), and (13.23), where $\sigma > 0$, $\eta > -1$.

Using Theorem 13.7 we get

Theorem 13.14. *Let* $0 < a < b < \infty$, $\alpha > 0$. *Let* $f \in C([a,b])$, $(f \circ \exp)$ *is differentiable on* $(\ln a, \ln b)$ *with* $(f \circ \exp)' \in L_1([\ln a, \ln b])$, $x \in [a,b]$. *Assume* $|(f \circ \exp)'|$ *is* s-*convex in the second sense on* $[\ln a, \ln b] \subset [0, \infty)$ *for some fixed* $s \in (0,1]$ *and* $|(f \circ \exp)'(\ln x)| \leq M$, $x \in [a,b]$. *Then*

$$\Delta_{\ln x}(f) \leq \frac{M}{\ln \frac{b}{a}} \left(1 + \frac{\Gamma(\alpha+1)\Gamma(s+1)}{\Gamma(\alpha+s+1)} \right)$$

$$\cdot \left[\frac{\left(\ln \frac{x}{a}\right)^{\alpha+1} + \left(\ln \frac{b}{x}\right)^{\alpha+1}}{\alpha+s+1} \right]. \qquad (13.89)$$

Using Theorem 13.9 we derive

Theorem 13.15. *All as in Theorem 13.14, but here* $|(f \circ \exp)'|^q$ *is* s-*convex in the second sense on* $[\ln a, \ln b] \subset [0, \infty)$ *for some fixed* $s \in (0,1]$, $p,q > 1 : \frac{1}{p} + \frac{1}{q} = 1$. *Then*

$$\Delta_{\ln x}(f) \leq \frac{M}{(1+p\alpha)^{\frac{1}{p}}} \left(\frac{2}{s+1} \right)^{\frac{1}{q}} \left[\frac{\left(\ln \frac{x}{a}\right)^{\alpha+1} + \left(\ln \frac{b}{x}\right)^{\alpha+1}}{\ln \frac{b}{a}} \right]. \qquad (13.90)$$

Using Theorem 13.11 we derive

Theorem 13.16. *All as in Theorem 13.15, however* $q \geq 1$, p *is not related. Then*

$$\Delta_{\ln x}(f) \leq M \left(\frac{1}{1+\alpha} \right)^{1-\frac{1}{q}} \left(\frac{1}{\alpha+s+1} \right)^{\frac{1}{q}}$$

$$\cdot \left(1 + \frac{\Gamma(\alpha+1)\Gamma(s+1)}{\Gamma(\alpha+s+1)} \right)^{\frac{1}{q}} \left[\frac{\left(\ln \frac{x}{a}\right)^{\alpha+1} + \left(\ln \frac{b}{x}\right)^{\alpha+1}}{\ln \frac{b}{a}} \right]. \qquad (13.91)$$

Based on Theorem 13.13 we produce

Theorem 13.17. *All as in Theorem 13.14, however here* $|(f \circ \exp)'|^q$ *is* s-*concave in the second sense on* $[\ln a, \ln b] \subset [0, \infty)$ *for some fixed* $s \in (0,1]$, $p,q > 1 : \frac{1}{p} + \frac{1}{q} = 1$. *Then*

$$\Delta_{\ln x}(f) \leq \frac{2^{\frac{(s-1)}{q}}}{(1+p\alpha)^{\frac{1}{p}} \left(\ln \frac{b}{a}\right)}$$

$$\cdot \left[\left(\ln \frac{x}{a}\right)^{\alpha+1} \left|(f \circ \exp)'\left(\frac{\ln(xa)}{2}\right)\right| + \left(\ln \frac{b}{x}\right)^{\alpha+1} \left|(f \circ \exp)'\left(\frac{\ln(bx)}{2}\right)\right| \right]. \qquad (13.92)$$

Based on Theorem 13.7 we give

Theorem 13.18. *Let $0 < a < b < \infty$, $f \in C([a,b])$, $\alpha, \sigma > 0$, $\eta > -1$. Assume $\left(F \circ (id)^{\frac{1}{\sigma}}\right)$ is differentiable on (a^{σ}, b^{σ}) with $\left(F \circ (id)^{\frac{1}{\sigma}}\right)' \in L_1([a^{\sigma}, b^{\sigma}])$, $x \in [a,b]$. Here $F(x) = x^{\sigma\eta} f(x)$, $x \in [a,b]$. Assume $\left|\left(F \circ (id)^{\frac{1}{\sigma}}\right)'\right|$ is s-convex in the second sense on $[a^{\sigma}, b^{\sigma}]$ for some fixed $s \in (0,1]$ and $\left|\left(F \circ (id)^{\frac{1}{\sigma}}\right)'(x^{\sigma})\right| \leq M$, $x \in [a,b]$. Then*

$$\Delta_{x^{\sigma}}(f) \leq \frac{M}{(b^{\sigma} - a^{\sigma})}\left(1 + \frac{\Gamma(\alpha + 1)\Gamma(s+1)}{\Gamma(\alpha + s + 1)}\right)$$
$$\cdot \left[\frac{(x^{\sigma} - a^{\sigma})^{\alpha+1} + (b^{\sigma} - x^{\sigma})^{\alpha+1}}{\alpha + s + 1}\right]. \tag{13.93}$$

By Theorem 13.9 we get

Theorem 13.19. *All as in Theorem 13.18, however here $\left|\left(F \circ (id)^{\frac{1}{\sigma}}\right)'\right|^q$ is s-convex in the second sense on $[a^{\sigma}, b^{\sigma}]$ for some fixed $s \in (0,1]$, $p, q > 1 : \frac{1}{p} + \frac{1}{q} = 1$. Then*

$$\Delta_{x^{\sigma}}(f) \leq \frac{M}{(1 + p\alpha)^{\frac{1}{p}}}\left(\frac{2}{s+1}\right)^{\frac{1}{q}}\left[\frac{(x^{\sigma} - a^{\sigma})^{\alpha+1} + (b^{\sigma} - x^{\sigma})^{\alpha+1}}{b^{\sigma} - a^{\sigma}}\right]. \tag{13.94}$$

Using Theorem 13.11 we get

Theorem 13.20. *Here all as in Theorem 13.19, however $q \geq 1$, p is not related. Then*

$$\Delta_{x^{\sigma}}(f) \leq M\left(\frac{1}{1+\alpha}\right)^{1-\frac{1}{q}}\left(\frac{1}{\alpha + s + 1}\right)^{\frac{1}{q}}$$
$$\cdot \left(1 + \frac{\Gamma(\alpha + 1)\Gamma(s+1)}{\Gamma(\alpha + s + 1)}\right)^{\frac{1}{q}}\left[\frac{(x^{\sigma} - a^{\sigma})^{\alpha+1} + (b^{\sigma} - x^{\sigma})^{\alpha+1}}{b^{\sigma} - a^{\sigma}}\right]. \tag{13.95}$$

Using Theorem 13.13 we obtain

Theorem 13.21. *All as in Theorem 13.18, however here $\left|\left(F \circ (id)^{\frac{1}{\sigma}}\right)'\right|^q$ is s-concave in the second sense on $[a^{\sigma}, b^{\sigma}]$ for some fixed $s \in (0,1]$, $p, q > 1 : \frac{1}{p} + \frac{1}{q} = 1$. Then*

$$\Delta_{x^{\sigma}}(f) \leq \frac{2^{\frac{(s-1)}{q}}}{(1 + p\alpha)^{\frac{1}{p}}(b^{\sigma} - a^{\sigma})}$$
$$\cdot \left[(x^{\sigma} - a^{\sigma})^{\alpha+1}\left|\left(F \circ (id)^{\frac{1}{\sigma}}\right)'\left(\frac{x^{\sigma} + a^{\sigma}}{2}\right)\right|\right.$$
$$\left. + (b^{\sigma} - x^{\sigma})^{\alpha+1}\left|\left(F \circ (id)^{\frac{1}{\sigma}}\right)'\left(\frac{b^{\sigma} + x^{\sigma}}{2}\right)\right|\right]. \tag{13.96}$$

13.3 Addendum

We make

Remark 13.9. Let $0 < \alpha < 1$, $f \in C\left([a,b]\right)$, $g \in C^1\left([a,b]\right)$, g strictly increasing; $\left(D^\alpha_{g(a)+}\left(f \circ g^{-1}\right)\right)(g(x))$, $\left(D^\alpha_{g(b)-}\left(f \circ g^{-1}\right)\right)(g(x))$ exist and are continuous on $[g(a),g(b)]$. Also assume $g'(x) \neq 0$, almost all $x \in [a,b]$.

Then by (13.41) we get

$$\left(D^\alpha_{g(a)+}\left(f \circ g^{-1}\right)\right)(g(x)) = \left(g'(x)\right)^{-1}\left(D^\alpha_{a+;g}(f)\right)(x), \tag{13.97}$$

almost all $x \in [a,b]$.

Also by (13.44) we get

$$\left(D^\alpha_{g(b)-}\left(f \circ g^{-1}\right)\right)(g(x)) = \left(g'(x)\right)^{-1}\left(D^\alpha_{b-;g}(f)\right)(x), \tag{13.98}$$

almost all $x \in [a,b]$.

Then by (13.37) and (13.38) we obtain

$$f(x) = \left(f \circ g^{-1}\right)(g(x)) = I^\alpha_{g(a)+}\left(D^\alpha_{g(a)+}\left(f \circ g^{-1}\right)\right)(g(x))$$

$$\overset{(13.97)}{=} \frac{1}{\Gamma(\alpha)} \int_{g(a)}^{g(x)} (g(x) - t)^{\alpha-1} \left(g'(t)\right)^{-1}\left(D^\alpha_{a+;g}(f)\right)(t)\,dt, \tag{13.99}$$

and

$$f(x) = \left(f \circ g^{-1}\right)(g(x)) = I^\alpha_{g(b)-}\left(D^\alpha_{g(b)-}\left(f \circ g^{-1}\right)\right)(g(x))$$

$$\overset{(13.98)}{=} \frac{1}{\Gamma(\alpha)} \int_{g(x)}^{g(b)} (t - g(x))^{\alpha-1} \left(g'(t)\right)^{-1}\left(D^\alpha_{b-;g}(f)\right)(t)\,dt, \tag{13.100}$$

for any $x \in [a,b]$.

We have proved the following generalized fractional Taylor formulae.

Theorem 13.22. Let $0 < \alpha < 1$, $f \in C\left([a,b]\right)$, $g \in C^1\left([a,b]\right)$, g strictly increasing; each of $\left(D^\alpha_{g(a)+}\left(f \circ g^{-1}\right)\right)(g(x))$, $\left(D^\alpha_{g(b)-}\left(f \circ g^{-1}\right)\right)(g(x))$ exists and it is continuous on $[g(a),g(b)]$. Assume that $g'(x) \neq 0$, for almost all $x \in [a,b]$. Then

1)

$$f(x) = I^\alpha_{g(a)+}\left(D^\alpha_{g(a)+}\left(f \circ g^{-1}\right)\right)(g(x))$$

$$= \frac{1}{\Gamma(\alpha)} \int_{g(a)}^{g(x)} (g(x) - t)^{\alpha-1} \left(g'(t)\right)^{-1}\left(D^\alpha_{a+;g}(f)\right)(t)\,dt, \tag{13.101}$$

and

2)

$$f(x) = I^\alpha_{g(b)-}\left(D^\alpha_{g(b)-}\left(f \circ g^{-1}\right)\right)(g(x))$$

$$= \frac{1}{\Gamma(\alpha)} \int_{g(x)}^{g(b)} (t - g(x))^{\alpha-1} \left(g'(t)\right)^{-1}\left(D^\alpha_{b-;g}(f)\right)(t)\,dt, \tag{13.102}$$

for any $x \in [a,b]$.

Bibliography

1. M. Alomari, M. Darus, S.S. Dragomir, P. Cerone, *Ostrowski type inequalities for functions whose derivatives are s-convex in the second sense*, Appl. Math. Lett. 23 (2010), 1071-1076.
2. G.A. Anastassiou, *Fractional Differentiation Inequalities*, Research Monograph, Springer, New York, 2009.
3. G.A. Anastassiou, *On right fractional calculus*, Chaos, Solitons Fractals 42 (2009), 365-376.
4. G.A. Anastassiou, *The reduction method in fractional calculus and fractional Ostrowski type inequalities*, Indian J. Math., accepted, 2013.
5. J.A. Canavati, *The Riemann-Liouville integral*, Nieuw Arch. Wiskd. 5 (1) (1987), 53-75.
6. S.S. Dragomir, S. Fitzpatrik, *The Hadamard's inequality for s-convex functions in the second sense*, Demonstratio Math. 32 (4) (1999), 687-696.
7. H. Hudzik, L. Maligranda, *Some remarks on s-convex functions*, Aequationes Math. 48 (1994), 100-111.
8. S. Iqbal, K. Krulic, J. Pecaric, *On an inequality of H.G. Hardy*, J. Inequalities Appl. 2010 (2010), Article ID 264347, 23 pages.
9. A.A. Kilbas, H.M. Srivastava, J.J. Trujillo, *Theory and Applications of Fractional Differential Equations*, vol. 204 of North-Holland Mathematics Studies, Elsevier, New York, NY, USA, 2006.
10. H.L. Royden, *Real Analysis*, Second Edition, Macmillan Publishing Co., Inc., New York, 1968.
11. S.G. Samko, A.A. Kilbas, O.I. Marichev, *Fractional Integral and Derivatives: Theory and Applications*, Gordon and Breach Science Publishers, Yverdon, Switzerland, 1993.
12. E. Set, *New inequalities of Ostrowski type for mappings whose derivatives are s-convex in the second sense via fractional integrals*, Comput. Math. Appl. 63 (2012), 1147-1154.

Index